ERROR ESTIMATION FOR PATTERN RECOGNITION

ERROR ESTIMATION FOR PATTERN RECOGNITION

ULISSES M. BRAGA-NETO
EDWARD R. DOUGHERTY

IEEE
IEEE PRESS

WILEY

For general information on our other products and services or for technical support, please contact our Customer Care Department within the United States at (800) 762-2974, outside the United States at (317) 572-3993 or fax (317) 572-4002.

Wiley also publishes its books in a variety of electronic formats. Some content that appears in print may not be available in electronic formats. For more information about Wiley products, visit our web site at www.wiley.com.

Library of Congress Cataloging-in-Publication Data is available.

ISBN: 978-1-118-99973-8

Printed in the United States of America

10 9 8 7 6 5 4 3 2 1

To our parents (in memoriam)
Jacinto and Consuêlo
and
Russell and Ann

And to our wives and children
Flávia, Maria Clara, and Ulisses
and
Terry, Russell, John, and Sean

CONTENTS

PREFACE

Error estimation determines the accuracy of the predictions that can be performed by an inferred classifier, and therefore plays a fundamental role in all scientific applications involving classification. And yet, we find that the subject is often neglected in most textbooks on pattern recognition and machine learning (with some admirable exceptions, such as the books by Devroye et al. (1996) and McLachlan (1992)). Most texts in the field describe a host of classification rules to train classifiers from data but treat the question of how to determine the accuracy of the inferred model only perfunctorily. Nevertheless, it is our belief, and the motivation behind this book, that the study of error estimation is at least as important as the study of classification rules themselves.

The error rate of interest could be the overall error rate on the population; however, it could also be the error rate on the individual classes. Were we to know the population distribution, we could in principle compute the error of a classifier exactly. Since we generally do not know the population distribution, we need to estimate the error from sample data. If the sample size is large enough relative to the number of features, an accurate error estimate is obtained by dividing the sample into training and testing sets, designing a classifier on the training set, and testing it on the test set. However, this is not possible if the sample size is small, in which case training and testing has to be performed on the same data. The focus of this book is on the latter case, which is common in situations where data are expensive, time-consuming, or difficult to obtain. In addition, in the case of "big data" applications, one is often dealing with small sample sizes, since the number of measurements far exceeds the number of sample points.

A classifier defines a relation between the features and the label (target random variable) relative to the joint feature–label distribution. There are two possibilities regarding the definition of a classifier: either it is derived from the feature–label distribution and therefore is intrinsic to it, as in the case of an optimal classifier, or it is not derived directly from the feature–label distribution, in which case it is extrinsic to the distribution. In both cases, classification error rates are intrinsic because they are derived from the distribution, given the classifier. Together, a classifier with a stated error rate constitute a *pattern recognition model*. If the classifier is intrinsic to the distribution, then there is no issue of error estimation because the feature–label distribution is known and the error rate can be derived. On the other hand, if the classifier is extrinsic to the distribution, then it is generally the case that the distribution is not known, the classifier is constructed from data via a *classification rule*, and the error rate of interest is estimated from the data via an *error estimation rule*. These rules together constitute a *pattern recognition rule*. (Feature selection, if present, is considered to be part of the classification rule, and thus of the pattern recognition rule.)

From a scientific perspective, that is, when a classifier is applied to phenomena, as with any scientific model, the validity of a pattern recognition model is characterized by the degree of concordance between predictions based on the model and corresponding observations.[1] The only prediction based on a pattern recognition model is the percentage of errors the classifier will make when it is applied. Hence, validity is based on concordance between the error rate in the model and the empirical error rate on the data when applied. Hence, when the distributions are not known and the model error must be estimated, model validity is completely dependent on the accuracy of the error estimation rule, so that the salient epistemological problem of classification is error estimation.[2]

Good classification rules produce classifiers with small average error rates. But is a classification rule good if its error rate cannot be estimated accurately? This is not just an abstract exercise in epistemology, but it has immediate practical consequences for applied science. For instance, the problem is so serious in genomic classification that the Director of the Center for Drug Evaluation and Research at the FDA has estimated that as much as 75% of published biomarker associations are not replicable. Absent an accurate error estimation rule, we lack knowledge of the error rate. Thus, the classifier is useless, no matter how sophisticated the classification rule that produced it. Therefore, one should speak about the goodness of a pattern recognition rule, as defined above, rather than that of a classification rule in isolation.

A common scenario in the literature these days proceeds in the following manner: propose an algorithm for classifier design; apply the algorithm to a few small-sample data sets, with no information pertaining to the distributions from which the data

[1]In the words of Richard Feynman, "It is whether or not the theory gives predictions that agree with experiment. It is not a question of whether a theory is philosophically delightful, or easy to understand, or perfectly reasonable from the point of view of common sense." (Feynman, 1985)

[2]From a scientific perspective, the salient issue is error estimation. One can imagine Harald Cramér leisurely sailing on the Baltic off the coast of Stockholm, taking in the sights and sounds of the sea, when suddenly a gene-expression classifier to detect prostate cancer pops into his head. No classification rule has been applied, nor is that necessary. All that matters is that Cramér's imagination has produced a classifier that operates on the population of interest with a sufficiently small error rate. Estimation of that rate requires an accurate error estimation rule.

sets have been sampled; and apply an error estimation scheme based on resampling the data, typically cross-validation. With regard to the third step, we are given no characterization of the accuracy of the error estimator and why it should provide a reasonably good estimate. Most strikingly, as we show in this book, we can expect it to be inaccurate in small-sample cases. Nevertheless, the claim is made that the proposed algorithm has been "validated." Very little is said about the accuracy of the error estimation step, except perhaps that cross-validation is close to being unbiased if not too many points are held out. But this kind of comment is misleading, given that a small bias may be of little consequence if the variance is large, which it usually is for small samples and large feature sets. In addition, the classical cross-validation unbiasedness theorem holds if sampling is random over the mixture of the populations. In situations where this is not the case, for example, the populations are sampled separately, bias is introduced, as it is shown in Chapter 5. These kinds of problems are especially detrimental in the current era of high-throughput measurement devices, for which it is now commonplace to be confronted with tens of thousands of features and very small sample sizes.

The subject of error estimation has in fact a long history and has produced a large body of literature; four main review papers summarize the major advances in the field up to 2000 (Hand, 1986; McLachlan, 1987; Schiavo and Hand, 2000; Toussaint, 1974); recent advances in error estimation since 2000 include work on model selection (Bartlett et al., 2002), bolstering (Braga-Neto and Dougherty, 2004a; Sima et al., 2005b), feature selection (Hanczar et al., 2007; Sima et al., 2005a; Xiao et al., 2007; Zhou and Mao, 2006), confidence intervals (Kaariainen, 2005; Kaariainen and Langford, 2005; Xu et al., 2006), model-based second-order properties (Zollanvari et al., 2011, 2012), and Bayesian error estimators (Dalton and Dougherty, 2011b,c). This book covers the classical studies as well as the recent developments. It discusses in detail nonparametric approaches, but gives special consideration, especially in the latter part of the book, to parametric, model-based approaches.

Pattern recognition plays a key role in many disciplines, including engineering, physics, statistics, computer science, social science, manufacturing, materials, medicine, biology, and more, so this book will be useful for researchers and practitioners in all these areas. This book can serve as a text at the graduate level, can be used as a supplement for general courses on pattern recognition and machine learning, or can serve as a reference for researchers across all technical disciplines where classification plays a major role, which may in fact be all technical disciplines.

The book is organized into eight chapters. Chapters 1 and 2 provide the foundation for the rest of the book and must be read first. Chapters 3, 4, and 8 stand on their own and can be studied separately. Chapter 5 provides the foundation for Chapters 6 and 7, so these chapters should be read in this sequence. For example, chapter sequences 1-2-3-4, 1-2-5-6-7, and 1-2-8 are all possible ways of reading the book. Naturally, the book is best read from beginning to end. Short descriptions of each chapter are provided next.

Chapter 1. Classification. To make the book self-contained, the first chapter covers basic topics in classification required for the remainder of the text: classifiers, population-based and sample-based discriminants, and classification rules. It defines a few basic classification rules: LDA, QDA, discrete classification, nearest-neighbors, SVMs, neural networks, and classification trees, closing with a section on feature selection.

Chapter 2. Error Estimation. This chapter covers the basics of error estimation: definitions, performance metrics for estimation rules, test-set error estimation, and training-set error estimation. It also includes a discussion of pattern recognition models. The test-set error estimator is straightforward and well-understood; there being efficient distribution-free bounds on performance. However, it assumes large sample sizes, which may be impractical. The thrust of the book is therefore on training-set error estimation, which is necessary for small-sample classifier design. We cover in this chapter the following error estimation rules: resubstitution, cross-validation, bootstrap, optimal convex estimation, smoothed error estimation, and bolstered error estimation.

Chapter 3. Performance Analysis. The main focus of the book is on performance characterization for error estimators, a topic whose coverage is inadequate in most pattern recognition texts. The fundamental entity in error analysis is the joint distribution between the true error and the error estimate. Of special importance is the regression between them. This chapter discusses the deviation distribution between the true and estimated error, with particular attention to the root-mean-square (RMS) error as the key metric of error estimation performance. The chapter covers various other issues: the impact of error estimation on feature selection, bias that can arise when considering multiple data sets or multiple rules, and measurement of performance reproducibility.

Chapter 4. Error Estimation for Discrete Classification. For discrete classification, the moments of resubstitution and the basic resampling error estimators, cross-validation and bootstrap, can be represented in finite-sample closed forms, as presented in the chapter. Using these representations, formulations of various performance measures are obtained, including the bias, variance, deviation variance, and the RMS and correlation between the true and estimated errors. Next, complete enumeration schemes to represent the joint and conditional distributions between the true and estimated errors are discussed. The chapter concludes by surveying large-sample performance bounds for various error estimators.

Chapter 5. Distribution Theory. This chapter provides general distributional results for discriminant-based classifiers. Here we also introduce the issue of separate vs. mixture sampling. For the true error, distributional results are in terms of conditional probability statements involving the discriminant. Passing on to resubstitution and the resampling-based estimators, the error estimators can be expressed in terms of error-counting expressions involving the discriminant. With these in hand, one can then go on to evaluate expected error rates and higher order moments of the error rates, all of which take combinatorial forms. Second-order moments are included, thus allowing computation of the RMS. The chapter concludes with analytical expressions for the sampling distribution for resubstitution and leave-one-out cross-validation.

Chapter 6. Gaussian Distribution Theory: Univariate Case. Historically, much effort was put into characterizing expected error rates for linear discrimination in the

Gaussian model for the true error, resubstitution, and leave-one-out. This chapter considers these, along with more recent work on the bootstrap. It then goes on to provide exact expressions for the first- and higher-order moments of various error rates, thereby providing an exact analytic expression for the RMS. The chapter provides exact expressions for the marginal sampling distributions of the resubstitution and leave-one-out error estimators. It closes with comments on the joint distribution of true and estimated error rates.

Chapter 7. Gaussian Distribution Theory: Multivariate Case. This chapter begins with statistical representations for multivariate discrimination, including Bowkers classical representation for Anderson's LDA discriminant, Moran's representations for John's LDA discriminant, and McFarland and Richards' QDA discriminant representation. A discussion follows on computational techniques for obtaining the error rate moments based on these representations. Large-sample methods are then treated in the context of double-asymptotic approximations where the sample size and feature dimension both approach infinity, in a comparable fashion. This leads to asymptotically-exact, finite-sample approximations to the first and second moments of various error rates, which lead to double-asymptotic approximations for the RMS.

Chapter 8. Bayesian MMSE Error Estimation. In small-sample situations it is virtually impossible to make any useful statements concerning error-estimation accuracy unless distributional assumptions are made. In previous chapters, these distributional assumptions were made from a frequentist point of view, whereas this chapter considers a Bayesian approach to the problem. Partial assumptions lead to an uncertainty class of feature–label distributions and, assuming a prior distribution over the uncertainty class, one can obtain a minimum-mean-square-error (MMSE) estimate of the error rate. In opposition to the classical case, a sample-conditioned MSE of the error estimate can be defined and computed. In addition to the general MMSE theory, this chapter provides specific representations of the MMSE error estimator for discrete classification and linear classification in the Gaussian model. It also discusses consistency of the estimate and calibration of non-Bayesian error estimators relative to a prior distribution governing model uncertainty.

We close by noting that for researchers in applied mathematics, statistics, engineering, and computer science, error estimation for pattern recognition offers a wide open field with fundamental and practically important problems around every turn. Despite this fact, error estimation has been a neglected subject over the past few decades, even though it spans a huge application domain and its understanding is critical for scientific epistemology. This text, which is believed to be the first book devoted entirely to the topic of error estimation for pattern recognition, is an attempt to correct the record in that regard.

ULISSES M. BRAGA-NETO
EDWARD R. DOUGHERTY

College Station, Texas
April 2015

ACKNOWLEDGMENTS

Much of the material in this book has been developed over a dozen years of our joint work on error estimation for pattern recognition. In this regard, we would like to acknowledge the students whose excellent work has contributed to the content: Chao Sima, Jianping Hua, Thang Vu, Lori Dalton, Mohammadmahdi Yousefi, Mohammad Esfahani, Amin Zollanvari, and Blaise Hanczar. We also would like to thank Don Geman for many hours of interesting conversation about error estimation. We appreciate the financial support we have received from the Translational Genomics Research Institute, the National Cancer Institute, and the National Science Foundation. Last, and certainly not least, we would like to acknowledge the encouragement and support provided by our families in this time-consuming endeavor.

U.M.B.N.
E.R.D.

LIST OF SYMBOLS

$\mathbf{X} = (X_1, \ldots, X_d)$	feature vector
Π_0, Π_1	populations or classes
Y	population or class label
$F_{\mathbf{X}Y}$	feature–label distribution
$c_0, c_1 = 1 - c_0$	population prior probabilities
$p_0(\mathbf{x}), p_1(\mathbf{x})$	class-conditional densities
$p(\mathbf{x}) = c_0 p_0(\mathbf{x}) + c_1 p_1(\mathbf{x})$	density across mixture of populations
$\eta_0(\mathbf{x}), \eta_1(\mathbf{x}) = 1 - \eta_0(\mathbf{x})$	population posterior probabilities
ψ, ε	classifier and classifier true error rate
$\psi_{\mathrm{bay}}, \varepsilon_{\mathrm{bay}}$	Bayes classifier and Bayes error
$\varepsilon^0, \varepsilon^1$	population-specific true error rates
D	population-based discriminant
$X \sim Z$	X has the same distribution as Z
$\boldsymbol{\mu}_0, \boldsymbol{\mu}_1$	population mean vectors
Σ_0, Σ_1	population covariance matrices
$\Sigma = \Sigma_0 = \Sigma_1$	population common covariance matrix
$\boldsymbol{\mu}_1, \boldsymbol{\mu}_0$	population mean vectors
δ	Mahalanobis distance
$\Phi(x)$	cdf of $N(0, 1)$ Gaussian random variable
n	sample size
$n_0, n_1 = n - n_0$	population-specific sample sizes
S	sample data
$\overline{S}$	sample data with class labels flipped
Ψ	classification rule
$S_n, \Psi_n, \psi_n, \varepsilon_n$	"n" indicates mixture sampling design

$S_{n_0,n_1}, \Psi_{n_0,n_1}, \psi_{n_0,n_1}, \varepsilon_{n_0,n_1}$	"n_0,n_1" indicates separate sampling design
$\mathcal{C}$	set of classification rules
$V_{\mathcal{C}}, \mathcal{S}(\mathcal{C},n)$	VC dimension and shatter coefficients
W	sample-based discriminant
$\bar{\mathbf{X}}_0, \bar{\mathbf{X}}_1$	population-specific sample mean vectors
$\hat{\Sigma}_0, \hat{\Sigma}_1$	population-specific sample covariance matrices
$\hat{\Sigma}$	pooled sample covariance matrix
$W_L, W_{\overline{L}}$	Anderson's and John's LDA discriminants
p_i, q_i, U_i, V_i	population-specific bin probabilities and counts
Υ	feature selection rule
Ξ	error estimation rule
$\hat{\varepsilon}$	error estimator
$\hat{\varepsilon}^0, \hat{\varepsilon}^1$	population-specific estimated error rates
$\mathrm{Bias}(\hat{\varepsilon})$	error estimator bias
$\mathrm{Var}_{\mathrm{dev}}(\hat{\varepsilon})$	error estimator deviation variance
$\mathrm{RMS}(\hat{\varepsilon})$	error estimator root mean square error
$V_{\mathrm{int}} = \mathrm{Var}(\hat{\varepsilon} \mid S)$	error estimator internal variance
$\rho(\varepsilon, \hat{\varepsilon})$	error estimator correlation coefficient
$\hat{\varepsilon}_n^r$	classical resubstitution estimator
$\hat{\varepsilon}^{r,01}$	resubstitution estimator with known c_0, c_1
$\hat{\varepsilon}_n^l$	classical leave-one-out estimator
$\hat{\varepsilon}^{l,01}$	leave-one-out estimator with known c_0, c_1
$\hat{\varepsilon}_n^{\mathrm{cv}(k)}$	classical k-fold cross-validation estimator
$\hat{\varepsilon}_{n_0,n_1}^{\mathrm{cv}(k_0,k_1),01}$	leave-(k_0,k_1)-out cross-validation estimator with known c_0, c_1
$\mathbf{C}, \mathbf{C}_0, \mathbf{C}_1$	multinomial (bootstrap) vectors
$S_n^*, S_n^{\mathbf{C}}, S_{n_0,n_1}^{\mathbf{C}_0,\mathbf{C}_1}$	bootstrap samples
$F_{\mathbf{X}Y}^*$	bootstrap empirical distribution
$\hat{\varepsilon}_n^{\mathrm{boot}}$	classical zero bootstrap estimator
$\hat{\varepsilon}_n^{\mathrm{zero}}$	classical complete zero bootstrap estimator
$\hat{\varepsilon}_n^{632}$	classical .632 bootstrap estimator
$\hat{\varepsilon}_{n_0,n_1}^{\mathrm{zero},01}$	complete zero bootstrap estimator with known c_0, c_1
$\hat{\varepsilon}_n^{\mathrm{br}}$	bolstered resubstitution estimator
$\hat{\varepsilon}_n^{\mathrm{bloo}}$	bolstered leave-one-out estimator
A_{est}	estimated comparative advantage
C_{bias}	comparative bias
$R_n(\rho, \tau)$	reproducibility index

CHAPTER 1

CLASSIFICATION

A classifier is designed and its error estimated from sample data taken from a population. Two fundamental issues arise. First, given a set of features (random variables), how does one design a classifier from the sample data that provides good classification over the general population? Second, how does one estimate the error of a classifier from the data? This chapter gives the essentials regarding the first question and the remainder of the book addresses the second.

1.1 CLASSIFIERS

Formally, classification involves a predictor vector $\mathbf{X} = (X_1, \dots, X_d) \in R^d$, also known as a *feature vector*, where each $X_i \in R$ is a *feature*, for $i = 1, \dots, d$. The feature vector $\mathbf{X}$ represents an individual from one of two populations Π_0 or Π_1.[1] The classification problem is to assign $\mathbf{X}$ correctly to its population of origin. Following the majority of the literature in pattern recognition, the populations are coded into a discrete *label* $Y \in \{0, 1\}$. Therefore, given a feature vector $\mathbf{X}$, classification attempts to predict the corresponding value of the label Y. The relationship between $\mathbf{X}$ and Y is assumed to be stochastic in nature; that is, we assume that there is a joint *feature–label distribution* $F_{\mathbf{X}Y}$ for the pair $(\mathbf{X}, Y)$.[2] The feature–label

[1]For simplicity and clarity, only the two-population case is discussed in the book; however, the extension to more than two populations is considered in the problem sections.
[2]A review of basic probability concepts used in the book is provided in Appendix A.

Error Estimation for Pattern Recognition, First Edition. Ulisses M. Braga-Neto and Edward R. Dougherty.

distribution completely characterizes the classification problem. It determines the probabilities $c_0 = P(\mathbf{X} \in \Pi_0) = P(Y = 0)$ and $c_1 = P(\mathbf{X} \in \Pi_1) = P(Y = 1)$, which are called the *prior probabilities*, as well as (if they exist) the probability densities $p_0(\mathbf{x}) = p(\mathbf{x} \mid Y = 0)$ and $p_1(\mathbf{x}) = p(\mathbf{x} \mid Y = 1)$, which are called the *class-conditional densities*. (In the discrete case, these are replaced by the corresponding probability mass functions.) To avoid trivialities, we shall always assume that $\min\{c_0, c_1\} \neq 0$. The *posterior probabilities* are defined by $\eta_0(\mathbf{x}) = P(Y = 0 \mid \mathbf{X} = \mathbf{x})$ and $\eta_1(\mathbf{x}) = P(Y = 1 \mid \mathbf{X} = \mathbf{x})$. Note that $\eta_0(\mathbf{x}) = 1 - \eta_1(\mathbf{x})$. If densities exist, then $\eta_0(\mathbf{x}) = c_0 p_0(\mathbf{x})/p(\mathbf{x})$ and $\eta_1(\mathbf{x}) = c_1 p_1(\mathbf{x})/p(\mathbf{x})$, where $p(\mathbf{x}) = c_0 p_0(\mathbf{x}) + c_1 p_1(\mathbf{x})$ is the density of $\mathbf{X}$ across the mixture of populations Π_0 and Π_1. Furthermore, in this binary-label setting, $\eta_1(\mathbf{x})$ is also the conditional expectation of Y given $\mathbf{X}$; that is, $\eta_1(\mathbf{x}) = E[Y|\mathbf{X} = \mathbf{x}]$.

A *classifier* is a measurable function $\psi : R^d \to \{0, 1\}$, which is to serve as a predictor of Y; that is, Y is to be predicted by $\psi(\mathbf{X})$. The error $\varepsilon[\psi]$ is the probability that the classification is erroneous, namely, $\varepsilon[\psi] = P(\psi(\mathbf{X}) \neq Y)$, which is a function both of ψ and the feature–label distribution $F_{\mathbf{X}Y}$. Owing to the binary nature of $\psi(\mathbf{X})$ and Y, we have that $\varepsilon[\psi] = E[|Y - \psi(\mathbf{X})|] = E[(Y - \psi(\mathbf{X}))^2]$, so that the classification error equals both the mean absolute deviation (MAD) and mean-square error (MSE) of predicting Y from $\mathbf{X}$ using ψ. If the class-conditional densities exist, then

$$\varepsilon[\psi] = \int_{\{\mathbf{x}|\psi(\mathbf{x})=0\}} \eta_1(\mathbf{x})p(\mathbf{x})\, d\mathbf{x} + \int_{\{\mathbf{x}|\psi(\mathbf{x})=1\}} \eta_0(\mathbf{x})p(\mathbf{x})\, d\mathbf{x}. \tag{1.1}$$

An optimal classifier ψ_{bay} is one having minimal classification error $\varepsilon_{\text{bay}} = \varepsilon[\psi_{\text{bay}}]$ — note that this makes ψ_{bay} also the minimum MAD and minimum MSE predictor of Y given $\mathbf{X}$. An optimal classifier ψ_{bay} is called a *Bayes classifier* and its error ε_{bay} is called the *Bayes error*. A Bayes classifier and the Bayes error depend on the feature–label distribution $F_{\mathbf{X}Y}$ — how well the labels are distributed among the variables being used to discriminate them, and how the variables are distributed in R^d. The Bayes error measures the inherent discriminability of the labels and is intrinsic to the feature–label distribution. Clearly, the right-hand side of (1.1) is minimized by the classifier

$$\psi_{\text{bay}}(\mathbf{x}) = \begin{cases} 1, & \eta_1(\mathbf{x}) \geq \eta_0(\mathbf{x}), \\ 0, & \text{otherwise.} \end{cases} \tag{1.2}$$

In other words, $\psi_{\text{bay}}(\mathbf{x})$ is defined to be 0 or 1 according to whether Y is more likely to be 0 or 1 given $\mathbf{X} = \mathbf{x}$. For this reason, a Bayes classifier is also known as a *maximum a posteriori* (MAP) classifier. The inequality can be replaced by strict inequality without affecting optimality; in fact, ties between the posterior probabilities can be broken arbitrarily, yielding a possibly infinite number of distinct Bayes classifiers. Furthermore, the derivation of a Bayes classifier does not require the existence of class-conditional densities, as has been assumed here; for an entirely general proof that ψ_{bay} in (1.2) is optimal, see Devroye et al. (1996, Theorem 2.1).

All Bayes classifiers naturally share the same Bayes error, which, from (1.1) and (1.2), is given by

$$\begin{aligned} \varepsilon_{\text{bay}} &= \int_{\{\mathbf{x}\mid\eta_1(\mathbf{x})<\eta_0(\mathbf{x})\}} \eta_1(\mathbf{x})p(\mathbf{x})\,d\mathbf{x} + \int_{\{\mathbf{x}\mid\eta_1(\mathbf{x})\geq\eta_0(\mathbf{x})\}} \eta_0(\mathbf{x})p(\mathbf{x})\,d\mathbf{x} \\ &= E\left[\min\{\eta_0(\mathbf{X}), \eta_1(\mathbf{X})\}\right]. \end{aligned} \tag{1.3}$$

By Jensen's inequality, it follows that $\varepsilon_{\text{bay}} \leq \min\{E[\eta_0(\mathbf{X})], E[\eta_1(\mathbf{X})]\}$. Therefore, if either posterior probability is uniformly small (for example, if one of the classes is much more likely *a priori* than the other), then the Bayes error is necessarily small.

1.2 POPULATION-BASED DISCRIMINANTS

Classifiers are often expressed in terms of a discriminant function, where the value of the classifier depends on whether or not the discriminant exceeds a threshold. Formally, a *discriminant* is a measurable function $D : R^d \to R$. A classifier ψ can be defined by

$$\psi(\mathbf{x}) = \begin{cases} 1, & D(\mathbf{x}) \leq k\,, \\ 0, & \text{otherwise}, \end{cases} \tag{1.4}$$

for a suitably chosen constant k. Such a classification procedure introduces three error rates: the population-specific error rates (assuming the existence of population densities)

$$\varepsilon^0 = P(D(\mathbf{X}) \leq k \mid Y = 0) = \int_{D(\mathbf{x})\leq k} p_0(\mathbf{x})\,d\mathbf{x}\,, \tag{1.5}$$

$$\varepsilon^1 = P(D(\mathbf{X}) > k \mid Y = 1) = \int_{D(\mathbf{x})> k} p_1(\mathbf{x})\,d\mathbf{x}\,, \tag{1.6}$$

and the overall error rate

$$\varepsilon = c_0\varepsilon^0 + c_1\varepsilon^1. \tag{1.7}$$

It is easy to show that this definition agrees with the previous definition of classification error.

It is well known (Anderson, 1984, Chapter 6) that a Bayes classifier is given by the likelihood ratio $p_0(\mathbf{x})/p_1(\mathbf{x})$ as the discriminant, with a threshold $k = c_1/c_0$. Since log is a monotone increasing function, an equivalent optimal procedure is to use the log-likelihood ratio as the discriminant,

$$D_{\text{bay}}(\mathbf{x}) = \log \frac{p_0(\mathbf{x})}{p_1(\mathbf{x})}, \tag{1.8}$$

and define a Bayes classifier as

$$\psi_{\text{bay}}(\mathbf{x}) = \begin{cases} 1, & D_{\text{bay}}(\mathbf{x}) \leq \log \frac{c_1}{c_0}, \\ 0, & \text{otherwise.} \end{cases} \tag{1.9}$$

In the absence of knowledge about c_0 and c_1, a Bayes procedure cannot be completely determined, nor is the overall error rate known; however, one can still consider the error rates ε^0 and ε^1 individually. In this context, a classifier ψ_1 is *at least as good as* classifier ψ_2 if the error rates ε^0 and ε^1 corresponding to ψ_1 are smaller or equal to those corresponding to ψ_2; and ψ_1 is *better* than ψ_2 if ψ_1 is at least as good as ψ_2 and at least one of these errors is strictly smaller. A classifier is *an admissible procedure* if no other classifier is better than it.

Now assume that one uses the log-likelihood ratio as the discriminant and adjusts the threshold k to obtain different error rates ε^0 and ε^1. Increasing k increases ε^0 but decreases ε^1, while decreasing k decreases ε^0 but increases ε^1. This corresponds to a *family of Bayes procedures*, as each procedure in the family corresponds to a Bayes classifier for a choice of c_0 and c_1 that gives $\log c_1/c_0 = k$. Under minimal conditions on the population densities, it can be shown that every admissible procedure is a Bayes procedure and vice versa, so that the class of admissible procedures and the class of Bayes procedures are the same (Anderson, 1984).

It is possible to define a best classifier in terms of a criterion based only on the population error rates ε^0 and ε^1. One well-known example of such a criterion is the *minimax criterion*, which seeks to find the threshold k_{mm} and corresponding classifier

$$\psi_{\text{mm}}(\mathbf{x}) = \begin{cases} 1, & \log \frac{p_0(\mathbf{x})}{p_1(\mathbf{x})} \leq k_{\text{mm}} \\ 0, & \text{otherwise} \end{cases} \tag{1.10}$$

that minimize the maximum of ε^0 and ε^1. For convenience, we denote below the dependence of the error rates ε^0 and ε^1 on k explicitly. The minimax threshold can thus be written as

$$k_{\text{mm}} = \arg\min_k \max\{\varepsilon^0(k), \varepsilon^1(k)\}. \tag{1.11}$$

It follows easily from the fact that the family of admissible procedures coincides with the family of Bayes procedures that the minimax procedure is a Bayes procedure. Furthermore, we have the following result, which appears, in a slightly different form, in Esfahani and Dougherty (2014a).

Theorem 1.1 *(Esfahani and Dougherty, 2014a) If the class-conditional densities are strictly positive and $\varepsilon^0(k)$, $\varepsilon^1(k)$ are continuous functions of k, then*

(a) There exists a unique point k_{mm} such that

$$\varepsilon^0(k_{\mathrm{mm}}) = \varepsilon^1(k_{\mathrm{mm}}), \tag{1.12}$$

and this point corresponds to the minimax solution defined in (1.11).

(b) Let c_{mm} be the value of the prior probability c_0 that maximizes the Bayes error:

$$\begin{aligned} c_{\mathrm{mm}} &= \arg\max_{c_0} \varepsilon_{\mathrm{bay}} \\ &= \arg\max_{c_0} \left\{ c_0 \varepsilon^0 \left(\log \frac{1-c_0}{c_0} \right) + (1-c_0)\varepsilon^1 \left(\log \frac{1-c_0}{c_0} \right) \right\}. \end{aligned} \tag{1.13}$$

Then

$$k_{\mathrm{mm}} = \log \left(\frac{1-c_{\mathrm{mm}}}{c_{\mathrm{mm}}} \right). \tag{1.14}$$

Proof: Part (a): Applying the monotone convergence theorem to (1.5) and (1.6) shows that $\varepsilon^0(k) \to 1$ and $\varepsilon^1(k) \to 0$ as $k \to \infty$ and $\varepsilon^0(k) \to 0$ and $\varepsilon^1(k) \to 1$ as $k \to -\infty$. Hence, since $\varepsilon^0(k)$ and $\varepsilon^1(k)$ are continuous functions of k, there is a point k_{mm} for which $\varepsilon^0(k_{\mathrm{mm}}) = \varepsilon^1(k_{\mathrm{mm}})$. Owing to $\varepsilon^0(k)$ and $\varepsilon^1(k)$ being strictly increasing and decreasing functions, respectively, this point is unique. We show that this point is the minimax solution via contradiction. Note that $\varepsilon^0(k_{\mathrm{mm}}) = \varepsilon^1(k_{\mathrm{mm}})$ implies that $\varepsilon_{\mathrm{mm}} = c_0\varepsilon^0(k_{\mathrm{mm}}) + c_1\varepsilon^1(k_{\mathrm{mm}}) = \varepsilon^0(k_{\mathrm{mm}}) = \varepsilon^1(k_{\mathrm{mm}})$, regardless of c_0, c_1. Suppose there exists $k' \neq k_{\mathrm{mm}}$ such that $\max\{\varepsilon^0(k'), \varepsilon^1(k')\} < \max\{\varepsilon^0(k_{\mathrm{mm}}), \varepsilon^1(k_{\mathrm{mm}})\} = \varepsilon_{\mathrm{mm}}$. First, consider $k' > k_{\mathrm{mm}}$. Since $\varepsilon^0(k)$ and $\varepsilon^1(k)$ are strictly increasing and decreasing functions, respectively, there are $\delta_0, \delta_1 > 0$ such that

$$\begin{cases} \varepsilon^0(k') = \varepsilon^0(k_{\mathrm{mm}}) + \delta_0 = \varepsilon_{\mathrm{mm}} + \delta_0, \\ \varepsilon^1(k') = \varepsilon^1(k_{\mathrm{mm}}) - \delta_1 = \varepsilon_{\mathrm{mm}} - \delta_1. \end{cases} \tag{1.15}$$

Therefore, $\max\{\varepsilon^0(k'), \varepsilon^1(k')\} = \varepsilon_{\mathrm{mm}} + \delta_0 > \varepsilon_{\mathrm{mm}}$, a contradiction. Similarly, if $k' < k_{\mathrm{mm}}$, there are $\delta_0, \delta_1 > 0$ such that $\varepsilon^0(k') = \varepsilon_{\mathrm{mm}} - \delta_0$ and $\varepsilon^1(k') = \varepsilon_{\mathrm{mm}} + \delta_1$, so that $\max\{\varepsilon^0(k'), \varepsilon^1(k')\} = \varepsilon_{\mathrm{mm}} + \delta_1 > \varepsilon_{\mathrm{mm}}$, again a contradiction.

Part (b): Define

$$\varepsilon(c, c_0) \stackrel{\Delta}{=} \left\{ c_0\varepsilon^0 \left(\log \frac{1-c}{c} \right) + (1-c_0)\varepsilon^1 \left(\log \frac{1-c}{c} \right) \right\}. \tag{1.16}$$

Let $k_{\mathrm{mm}} = \log(1-c_{\mathrm{mm}})/c_{\mathrm{mm}}$, where at this point c_{mm} is just the value that makes the equality true (given k_{mm}). We need to show that c_{mm} is given by the expression in part (b). We have that $\varepsilon_{\text{mm}} = \varepsilon(c_{\text{mm}}, c_0)$ is independent of c_0. Also, for a given value of c_0, the Bayes error is given by $\varepsilon_{\text{bay}} = \varepsilon(c_0, c_0)$. We need to show that $\varepsilon(c_{\text{mm}}, c_{\text{mm}}) = \max_{c_0} \varepsilon(c_0, c_0)$. The inequality $\varepsilon(c_{\text{mm}}, c_{\text{mm}}) \leq \max_{c_0} \varepsilon(c_0, c_0)$ is trivial. To establish the converse inequality, notice that $\varepsilon_{\text{mm}} = \varepsilon(c_{\text{mm}}, c_{\text{mm}}) \geq \varepsilon(c_0, c_0) = \varepsilon_{\text{bay}}$, for any value of c_0, by definition of the Bayes error, so that $\varepsilon(c_{\text{mm}}, c_{\text{mm}}) \geq \max_{c_0} \varepsilon(c_0, c_0)$. ■

The previous theorem has two immediate and important consequences: (1) The error $\varepsilon_{\text{mm}} = c_0\varepsilon^0(k_{\text{mm}}) + c_1\varepsilon^1(k_{\text{mm}})$ of the minimax classifier is such that $\varepsilon_{\text{mm}} = \varepsilon^0(k_{\text{mm}}) = \varepsilon^1(k_{\text{mm}})$, regardless of c_0, c_1. (2) The error of the minimax classifier is equal to the maximum Bayes error as c_0 varies: $\varepsilon_{\text{mm}} = \max_{c_0} \varepsilon_{\text{bay}}$.

Consider now the important special case of multivariate Gaussian populations $\Pi_0 \sim N_d(\boldsymbol{\mu}_0, \Sigma_0)$ and $\Pi_1 \sim N_d(\boldsymbol{\mu}_1, \Sigma_1)$ (see Section A.11 for basic facts about the multivariate Gaussian). The class-conditional densities take the form

$$p_i(\mathbf{x}) = \frac{1}{\sqrt{(2\pi)^n \det(\Sigma_i)}} \exp\left[-\frac{1}{2}(\mathbf{x} - \boldsymbol{\mu}_i)^T \Sigma_i^{-1}(\mathbf{x} - \boldsymbol{\mu}_i)\right], \quad i = 0, 1. \quad (1.17)$$

It follows from (1.8) that the optimal discriminant (using "D_Q" for quadratic discriminant) is given by

$$D_Q(\mathbf{x}) = -\frac{1}{2}(\mathbf{x} - \boldsymbol{\mu}_0)^T \Sigma_0^{-1}(\mathbf{x} - \boldsymbol{\mu}_0) + \frac{1}{2}(\mathbf{x} - \boldsymbol{\mu}_1)^T \Sigma_1^{-1}(\mathbf{x} - \boldsymbol{\mu}_1) + \log \frac{\det(\Sigma_1)}{\det(\Sigma_0)}. \quad (1.18)$$

This discriminant has the form $\mathbf{x}^T A \mathbf{x} - b^T \mathbf{x} + c = 0$, which describes a *hyperquadric* decision boundary in R^d (Duda and Hart, 2001). If $\Sigma_0 = \Sigma_1 = \Sigma$, then the optimal discriminant (using "D_L" for linear discriminant) reduces to

$$D_L(\mathbf{x}) = \left(\mathbf{x} - \frac{\boldsymbol{\mu}_0 + \boldsymbol{\mu}_1}{2}\right)^T \Sigma^{-1}\left(\boldsymbol{\mu}_0 - \boldsymbol{\mu}_1\right). \quad (1.19)$$

This discriminant has the form $A\mathbf{x}^T + b = 0$, which describes a hyperplane in R^d. The optimal classifiers based on discriminants $D_Q(\mathbf{x})$ and $D_L(\mathbf{x})$ define population-based *Quadratic Discriminant Analysis* (QDA) and population-based *Linear Discriminant Analysis* (LDA), respectively.

Using the properties of Gaussian random vectors (c.f. Section A.11), it follows from (1.19) that $D_L(\mathbf{X}) \mid Y = 0 \sim N\left(\frac{1}{2}\delta^2, \delta^2\right)$, whereas $D_L(\mathbf{X}) \mid Y = 1 \sim N\left(-\frac{1}{2}\delta^2, \delta^2\right)$, where

$$\delta = \sqrt{(\boldsymbol{\mu}_1 - \boldsymbol{\mu}_0)^T \Sigma^{-1}(\boldsymbol{\mu}_1 - \boldsymbol{\mu}_0)} \quad (1.20)$$

is the *Mahalanobis distance* between the populations. It follows that

$$\varepsilon_L^0(k) = P(D_L(\mathbf{X}) \le k \mid Y = 0) = \Phi\left(\frac{k - \frac{1}{2}\delta^2}{\delta}\right), \tag{1.21}$$

$$\varepsilon_L^1(k) = P(D_L(\mathbf{X}) > k \mid Y = 1) = \Phi\left(\frac{-k - \frac{1}{2}\delta^2}{\delta}\right), \tag{1.22}$$

where $\Phi(x)$ is the cumulative distribution function of a standard $N(0,1)$ Gaussian random variable. Regardless of the value of k, $\varepsilon_L^0(k)$ and $\varepsilon_L^1(k)$, and thus the overall error rate $\varepsilon_L(k) = c_0\varepsilon_L^0(k) + c_1\varepsilon_L^1(k)$, are decreasing functions of the Mahalanobis distance δ and $\varepsilon_L^0(k), \varepsilon_L^1(k), \varepsilon_L(k) \to 0$ as $\delta \to \infty$. The optimal classifier corresponds to the special case $k = k_{\text{bay}} = \log c_1/c_0$. Therefore, the Bayes error rates $\varepsilon_{L,\text{bay}}^0$, $\varepsilon_{L,\text{bay}}^1$, and $\varepsilon_{L,\text{bay}}$ are decreasing in δ, and converge to zero as $\delta \to \infty$. If the classes are equally likely, $c_0 = c_1 = 0.5$, then $k_{\text{bay}} = 0$, so that the Bayes error rates can be written simply as

$$\varepsilon_{L,\text{bay}}^0 = \varepsilon_{L,\text{bay}}^1 = \varepsilon_{L,\text{bay}} = \Phi\left(-\frac{\delta}{2}\right). \tag{1.23}$$

If the prior probabilities are not known, then, as seen in Theorem 1.1, the minimax rule is obtained by thresholding the discriminant $D_L(\mathbf{x})$ at a point k_{mm} that makes the two population-specific error rates ε_L^0 and ε_L^1 equal. From (1.21) and (1.22), we need to find the value k_{mm} for which

$$\Phi\left(\frac{k_{\text{mm}} - \frac{1}{2}\delta^2}{\delta}\right) = \Phi\left(\frac{-k_{\text{mm}} - \frac{1}{2}\delta^2}{\delta}\right). \tag{1.24}$$

Since Φ is monotone, the only solution to this equation is $k_{\text{mm}} = 0$. The minimax LDA classifier is thus given by

$$\psi_{L,\text{mm}}(\mathbf{x}) = \begin{cases} 1, & \left(\mathbf{x} - \frac{\boldsymbol{\mu}_0+\boldsymbol{\mu}_1}{2}\right)^T \Sigma^{-1}\left(\boldsymbol{\mu}_0 - \boldsymbol{\mu}_1\right) \le 0, \\ 0, & \text{otherwise.} \end{cases} \tag{1.25}$$

Clearly, the minimax LDA classifier is a Bayes LDA classifier when $c_0 = c_1$.

Under the general Gaussian model, the minimax threshold can differ substantially from 0. This is illustrated in Figure 1.1, which displays plots of the Bayes error as a function of the prior probability c_0 for different Gaussian models, including the case of distinct covariance matrices. Notice that, in accordance with Theorem 1.1, $k_{\text{mm}} = \log(1 - c_{\text{mm}})/c_{\text{mm}}$, where c_{mm} is the point where the plots reach the maximum. For the *homoskedastic* (equal-covariance) LDA model, $c_{\text{mm}} = 0.5$ and $k_{\text{mm}} = 0$, but k_{mm} becomes nonzero under the two *heteroskedastic* (unequal-covariance) models.

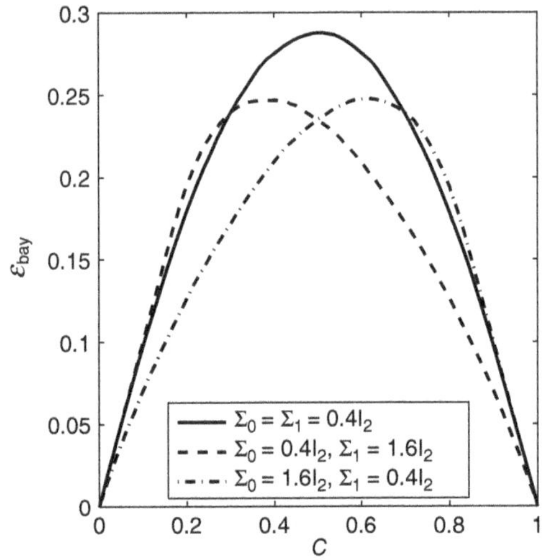

FIGURE 1.1 Bayes error as a function of c_0 under different Gaussian models, with $\boldsymbol{\mu}_0 = (0.3, 0.3)$ and $\boldsymbol{\mu}_1 = (0.8, 0.8)$ and indicated covariance matrices. Reproduced from Esfahani, S. and Dougherty (2014) with permission from Oxford University Press.

1.3 CLASSIFICATION RULES

In practice, the feature–label distribution is typically unknown and a classifier must be designed from sample data. An obvious approach would be to estimate the posterior probabilities from data, but often there is insufficient data to obtain good estimates. Fortunately, good classifiers can be obtained even when there is insufficient data for satisfactory density estimation.

We assume that we have a random sample $S_n = \{(\mathbf{X}_1, Y_1), \dots, (\mathbf{X}_n, Y_n)\}$ of feature vector–label pairs drawn from the feature–label distribution, meaning that the pairs $(\mathbf{X}_i, Y_i)$ are independent and identically distributed according to $F_{\mathbf{X}Y}$. The number of sample points from populations Π_0 and Π_1 are denoted by N_0 and N_1, respectively. Note that $N_0 = \sum_{i=1}^{n} I_{Y_i=0}$ and $N_1 = \sum_{i=1}^{n} I_{Y_i=1}$ are random variables such that $N_0 \sim \text{Binomial}(n, c_0)$ and $N_1 \sim \text{Binomial}(n, c_1)$, with $N_0 + N_1 = n$. A *classification rule* is a mapping $\Psi_n : S_n \mapsto \psi_n$, that is, given a sample S_n, a designed classifier $\psi_n = \Psi_n(S_n)$ is obtained according to the rule Ψ_n. Note that what we call a classification rule is really a sequence of classification rules depending on n. For simplicity, we shall not make the distinction formally.

We would like the error $\varepsilon_n = \varepsilon[\psi_n]$ of the designed classifier to be close to the the Bayes error. The *design error*,

$$\Delta_n = \varepsilon_n - \varepsilon_{\text{bay}}, \tag{1.26}$$

quantifies the degree to which a designed classifier fails to achieve this goal, where ε_n and Δ_n are sample-dependent random variables. The expected design error is $E[\Delta_n]$, the expectation being relative to all possible samples. The expected error of ψ_n is decomposed according to

$$E[\varepsilon_n] = \varepsilon_{\text{bay}} + E[\Delta_n]. \tag{1.27}$$

The quantity $E[\varepsilon_n]$, or alternatively $E[\Delta_n]$, measures the performance of the classification rule relative to the feature–label distribution, rather than the performance of an individual designed classifier. Nonetheless, a classification rule for which $E[\varepsilon_n]$ is small will also tend to produce designed classifiers possessing small errors.

Asymptotic properties of a classification rule concern behavior as sample size grows to infinity. A rule is said to be *consistent* for a feature–label distribution of $(\mathbf{X}, Y)$ if $\varepsilon_n \to \varepsilon_{\text{bay}}$ in probability as $n \to \infty$. The rule is said to be *strongly consistent* if convergence is with probability one. A classification rule is said to be *universally consistent* (resp. *universally strongly consistent*) if the previous convergence modes hold for any distribution of $(\mathbf{X}, Y)$. Note that the random sequence $\{\varepsilon_n; n = 1, 2, \dots\}$ is uniformly bounded in the interval [0, 1], by definition; hence, $\varepsilon_n \to \varepsilon_{\text{bay}}$ in probability implies that $E[\varepsilon_n] \to \varepsilon_{\text{bay}}$ as $n \to \infty$ (c.f. Corollary A.1). Furthermore, it is easy to see that the converse implication also holds in this case, because $\varepsilon_n - \varepsilon_{\text{bay}} = |\varepsilon_n - \varepsilon_{\text{bay}}|$, and thus $E[\varepsilon_n] \to \varepsilon_{\text{bay}}$ implies $E[|\varepsilon_n - \varepsilon_{\text{bay}}|] \to 0$, and hence L_1-convergence. This in turn is equivalent to convergence in probability (c.f. Theorem A.1). Therefore, a rule is consistent if and only if $E[\Delta_n] = E[\varepsilon_n] - \varepsilon_{\text{bay}} \to 0$ as $n \to \infty$, that is, the expected design error becomes arbitrarily small for a sufficiently large sample. If this condition holds for any distribution of $(\mathbf{X}, Y)$, then the rule is universally consistent, according to the previous definition.

Universal consistency is a useful property for large samples, but it is virtually of no consequence for small samples. In particular, the design error $E[\Delta_n]$ cannot be made universally small for fixed n. For any $\tau > 0$, fixed n, and classification rule Ψ_n, there exists a distribution of $(\mathbf{X}, Y)$ such that $\varepsilon_{\text{bay}} = 0$ and $E[\varepsilon_n] > \frac{1}{2} - \tau$ (Devroye et al., 1996, Theorem 7.1). Put another way, for fixed n, absent constraint on the feature–label distribution, a classification rule may have arbitrarily bad performance.

Furthermore, in addition to consistency of a classification rule, the rate at which $E[\Delta_n] \to 0$ is also critical. In other words, we would like a statement of the following form: for any $\tau > 0$, there exists $n(\tau)$ such that $E[\Delta_{n(\tau)}] < \tau$ for any distribution of $(\mathbf{X}, Y)$. Such a proposition is not possible; even if a classification rule is universally consistent, the design error converges to 0 arbitrarily slowly relative to all possible distributions. Indeed, it can be shown (Devroye et al., 1996, Theorem 7.2) that if $\{a_n; n = 1, 2, \dots\}$ is a decreasing sequence of positive numbers with $a_1 \leq \frac{1}{16}$, then for any classification rule Ψ_n, there exists a distribution of $(\mathbf{X}, Y)$ such that $\varepsilon_{\text{bay}} = 0$ and $E[\varepsilon_n] > a_n$. The situation changes if distributional assumptions are made — a scenario that will be considered extensively in this book.

A classification rule can yield a classifier that makes very few errors, or none, on the sample data from which it is designed, but performs poorly on the distribution as a whole, and therefore on new data to which it is applied. This situation, typically referred to as *overfitting*, is exacerbated by complex classifiers and small samples.

The basic point is that a classification rule should not cut up the space in a manner too complex for the amount of sample data available. This might improve the *apparent error rate* (i.e., the number of errors committed by the classifier using the training data as testing points), also called the *resubstitution error rate* (see the next chapter), but at the same time it will most likely worsen the true error of the classifier on independent future data (also called the *generalization error* in this context). The problem is not necessarily mitigated by applying an error-estimation rule — perhaps more sophisticated than the apparent error rate — to the designed classifier to see if it "actually" performs well, since when there is only a small amount of data available, error-estimation rules are usually inaccurate, and the inaccuracy tends to be worse for complex classification rules. Hence, a low error estimate is not sufficient to overcome our expectation of a large expected error when using a complex classifier with a small data set. These issues can be studied with the help of Vapnik–Chervonenkis theory, which is reviewed in Appendix B. In the remainder of this section, we apply several of the results in Appendix B to the problem of efficient classifier design.

Constraining classifier design means restricting the functions from which a classifier can be chosen to a class C. This leads to trying to find an optimal *constrained classifier* $\psi_C \in C$ having error ε_C. Constraining the classifier can reduce the expected design error, but at the cost of increasing the error of the best possible classifier. Since optimization in C is over a subclass of classifiers, the error ε_C of ψ_C will typically exceed the Bayes error, unless a Bayes classifier happens to be in C. This *cost of constraint* is

$$\Delta_C = \varepsilon_C - \varepsilon_{\text{bay}}. \tag{1.28}$$

A classification rule yields a classifier $\psi_{n,C} \in C$, with error $\varepsilon_{n,C}$, and $\varepsilon_{n,C} \geq \varepsilon_C \geq \varepsilon_{\text{bay}}$. The *design error* for constrained classification is

$$\Delta_{n,C} = \varepsilon_{n,C} - \varepsilon_C. \tag{1.29}$$

For small samples, this can be substantially less than Δ_n, depending on C and the classification rule. The error of a designed constrained classifier is decomposed as

$$\varepsilon_{n,C} = \varepsilon_{\text{bay}} + \Delta_C + \Delta_{n,C}. \tag{1.30}$$

Therefore, the expected error of the designed classifier from C can be decomposed as

$$E[\varepsilon_{n,C}] = \varepsilon_{\text{bay}} + \Delta_C + E[\Delta_{n,C}]. \tag{1.31}$$

The constraint is beneficial if and only if $E[\varepsilon_{n,C}] < E[\varepsilon_n]$, that is, if

$$\Delta_C < E[\Delta_n] - E[\Delta_{n,C}]. \tag{1.32}$$

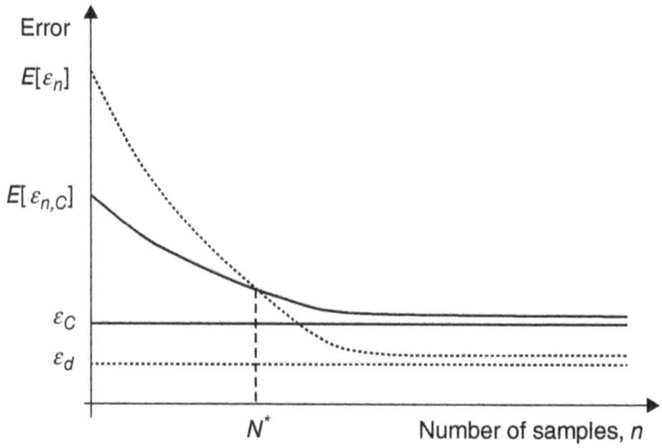

FIGURE 1.2 "Scissors Plot." Errors for constrained and unconstrained classification as a function of sample size.

If the cost of constraint is less than the decrease in expected design error, then the expected error of $\psi_{n,C}$ is less than that of ψ_n. The dilemma: strong constraint reduces $E[\Delta_{n,C}]$ at the cost of increasing ε_C, and vice versa. Figure 1.2 illustrates the design problem; this is known in pattern recognition as a "scissors plot" (Kanal and Chandrasekaran, 1971; Raudys, 1998). If n is sufficiently large, then $E[\varepsilon_n] < E[\varepsilon_{n,C}]$; however, if n is sufficiently small, then $E[\varepsilon_n] > E[\varepsilon_{n,C}]$. The point N^* at which the decreasing lines cross is the cut-off: for $n > N^*$, the constraint is detrimental; for $n < N^*$, it is beneficial. When $n < N^*$, the advantage of the constraint is the difference between the decreasing solid and dashed lines.

Clearly, how much constraint is applied to a classification rule is related to the size of C. The *Vapnik–Chervonenkis dimension* V_C yields a quantitative measure of the size of C. A larger VC dimension corresponds to a larger C and a greater potential for overfitting, and vice versa. The VC dimension is defined in terms of the *shatter coefficients* $\mathcal{S}(C, n)$, $n = 1, 2, \ldots$ The reader is referred to Appendix B for the formal definitions of VC dimension and shatter coefficients of a class C .

The Vapnik–Chervonenkis theorem (c.f. Theorem B.1) states that, regardless of the distribution of $(\mathbf{X}, Y)$, for all sample sizes and for all $\tau > 0$,

$$P\left(\sup_{\psi\in C} |\hat{\varepsilon}[\psi] - \varepsilon[\psi]| > \tau\right) \leq 8\mathcal{S}(C, n)e^{-n\tau^2/32}, \tag{1.33}$$

where $\varepsilon[\psi]$ and $\hat{\varepsilon}[\psi]$ are respectively the true and apparent error rates of a classifier $\psi \in C$. Since this holds for all $\psi \in C$, it holds in particular for the designed classifier $\psi = \psi_{n,C}$, in which case $\varepsilon[\psi] = \varepsilon_{n,C}$, the designed classifier error, and $\hat{\varepsilon}[\psi] = \hat{\varepsilon}_{n,C}$, the apparent error of the designed classifier. Thus, we may drop the sup to obtain

$$P\left(|\hat{\varepsilon}_{n,C} - \varepsilon_{n,C}| > \tau\right) \leq 8\mathcal{S}(C, n)e^{-n\tau^2/32}, \tag{1.34}$$

for all $\tau > 0$. Now, suppose that the classifier $\psi_{n,C}$ is designed by choosing the classifier in C that minimizes the apparent error. This is called the *empirical risk minimization* (ERM) principle (Vapnik, 1998). Many classification rules of interest satisfy this property, such as histogram rules, perceptrons, and certain neural network and classification tree rules. In this case, we can obtain the following inequality for the design error $\Delta_{n,C}$:

$$\begin{aligned}\Delta_{n,C} \;=\; \varepsilon_{n,C} - \varepsilon_C \;&=\; \varepsilon_{n,C} - \hat{\varepsilon}_{n,C} + \hat{\varepsilon}_{n,C} - \inf_{\psi\in C} \varepsilon[\psi] \\ &= \varepsilon[\psi_{n,C}] - \hat{\varepsilon}[\psi_{n,C}] + \inf_{\psi\in C} \hat{\varepsilon}[\psi] - \inf_{\psi\in C} \varepsilon[\psi] \\ &\le \varepsilon[\psi_{n,C}] - \hat{\varepsilon}[\psi_{n,C}] + \sup_{\psi\in C} |\hat{\varepsilon}[\psi] - \varepsilon[\psi]| \\ &\le 2 \sup_{\psi\in C} |\hat{\varepsilon}[\psi] - \varepsilon[\psi]| \,,\end{aligned} \tag{1.35}$$

where we used $\hat{\varepsilon}_{n,C} = \inf_{\psi\in C} \hat{\varepsilon}[\psi]$ (because of ERM) and the fact that $\inf f(x) - \inf g(x) \le \sup(f(x) - g(x)) \le \sup |f(x) - g(x)|$. Therefore,

$$P(\Delta_{n,C} > \tau) \;\le\; P\left(\sup_{\psi\in C} |\hat{\varepsilon}[\psi] - \varepsilon[\psi]| > \frac{\tau}{2} \right) . \tag{1.36}$$

We can now apply (1.34) to get

$$P(\Delta_{n,C} > \tau) \;\le\; 8S(C,n)e^{-n\tau^2/128}, \tag{1.37}$$

for all $\tau > 0$. We can also apply Lemma A.3 to (1.37) to obtain a bound on the expected design error:

$$E[\Delta_{n,C}] \;\le\; 16\sqrt{\frac{\log S(C,n) + 4}{2n}} . \tag{1.38}$$

If $V_C < \infty$, then we can use the inequality $S(C,n) \le (n+1)^{V_C}$ to replace $S(C,n)$ by $(n+1)^{V_C}$ in the previous equations. For example, from (1.34),

$$P(\Delta_{n,C} > \tau) \;\le\; 8(n+1)^{V_C} e^{-n\tau^2/128}, \tag{1.39}$$

for all $\tau > 0$. Note that in this case $\sum_{n=1}^{\infty} P(\Delta_{n,C} > \tau) < \infty$, for all $\tau > 0$, and all distributions, so that, by the First Borel–Cantelli lemma (c.f. Lemma A.1), $\Delta_{n,C} \to 0$ with probability 1, regardless of the distribution. In addition, (1.38) yields

$$E[\Delta_{n,C}] \;\le\; 16\sqrt{\frac{V_C \log(n+1) + 4}{2n}} . \tag{1.40}$$

Hence, $E[\Delta_{n,C}] = O(\sqrt{V_C \log n/n})$, and one must have $n \gg V_C$ for the bound to be small. In other words, sample size has to grow faster than the complexity of the classification rule in order to guarantee good performance. One may think that by not using the ERM principle in classifier design, one is not impacted by the VC dimension of the classification rule. In fact, independently of how to pick ψ_n from C, both empirical and analytical evidence suggest that this is not true. For example, it can be shown that, under quite general conditions, there is a distribution of the data that demands one to have $n \gg V_C$ for acceptable performance, regardless of the classification rule (Devroye et al., 1996, Theorem. 14.5)

A note of caution: all bounds in VC theory are worst-case, as there are no distributional assumptions, and thus they can be very loose for a particular feature–label distribution and in small-sample cases. Nonetheless, in the absence of distributional knowledge, the theory prescribes that increasing complexity is counterproductive unless there is a large sample available. Otherwise, one could easily end up with a very bad classifier whose error estimate is very small.

1.4 SAMPLE-BASED DISCRIMINANTS

A *sample-based discriminant* is a real function $W(S_n, \mathbf{x})$, where S_n is a given sample and $\mathbf{x} \in R^d$. According to this definition, in fact we have a sequence of discriminants indexed by n. For simplicity, we shall not make this distinction formally, nor shall we add a subscript "n" to the discriminant; what is intended will become clear by the context. Many useful classification rules are defined via sample-based discriminants. Formally, a sample-based discriminant W defines a designed classifier (and a classification rule, implicitly) via

$$\Psi_n(S_n)(\mathbf{x}) = \begin{cases} 1, & W(S_n, \mathbf{x}) \leq k, \\ 0, & \text{otherwise}, \end{cases} \tag{1.41}$$

where S_n is the sample and k is a threshold parameter, the choice of which can be made in different ways.

A *plug-in discriminant* is a discriminant obtained from the population discriminant of (1.8) by plugging in sample estimates. If the discriminant is a plug-in discriminant and the prior probabilities are known, then one should follow the optimal classification procedure (1.9) and set $k = \log c_1/c_0$. On the other hand, if the prior probabilities are not known, one can set $k = \log N_1/N_0$, which converges to $\log c_1/c_0$ as $n \to \infty$ with probability 1. However, with small sample sizes, this choice can be problematic, as it may introduce significant variance in classifier design. An alternative in these cases is to use a fixed parameter, perhaps the minimax choice based on a model assumption (see Section 1.2) — this choice is risky, as the model assumption may be incorrect (Esfahani and Dougherty, 2014a). In any case, the use of a fixed threshold parameter is fine provided that it is reasonably close to $\log c_1/c_0$.

1.4.1 Quadratic Discriminants

Recall the population-based QDA discriminant in (1.18). If part or all of the population parameters are not known, sample estimators can be plugged in to yield the sample-based QDA discriminant

$$\begin{aligned} W_Q(S_n, \mathbf{x}) &= -\frac{1}{2}(\mathbf{x} - \bar{\mathbf{X}}_0)^T \hat{\Sigma}_0^{-1} (\mathbf{x} - \bar{\mathbf{X}}_0) \\ &\quad + \frac{1}{2}(\mathbf{x} - \bar{\mathbf{X}}_1)^T \hat{\Sigma}_1^{-1} (\mathbf{x} - \bar{\mathbf{X}}_1) + \log \frac{|\hat{\Sigma}_1|}{|\hat{\Sigma}_0|}, \end{aligned} \tag{1.42}$$

where

$$\bar{\mathbf{X}}_0 = \frac{1}{N_0} \sum_{i=1}^n \mathbf{X}_i I_{Y_i=0} \quad \text{and} \quad \bar{\mathbf{X}}_1 = \frac{1}{N_1} \sum_{i=1}^n \mathbf{X}_i I_{Y_i=1} \tag{1.43}$$

are the sample means for each population, and

$$\hat{\Sigma}_0 = \frac{1}{N_0 - 1} \sum_{i=1}^n (\mathbf{X}_i - \bar{\mathbf{X}}_0)(\mathbf{X}_i - \bar{\mathbf{X}}_0)^T I_{Y_i=0}, \tag{1.44}$$

$$\hat{\Sigma}_1 = \frac{1}{N_1 - 1} \sum_{i=1}^n (\mathbf{X}_i - \bar{\mathbf{X}}_1)(\mathbf{X}_i - \bar{\mathbf{X}}_1)^T I_{Y_i=1}, \tag{1.45}$$

are the sample covariance matrices for each population.

The QDA discriminant is a special case of the general *quadratic discriminant*, which takes the form

$$W_{\text{quad}}(S_n, \mathbf{x}) = \mathbf{x}^T A(S_n) \mathbf{x} + \mathbf{b}(S_n)^T \mathbf{x} + c(S_n), \tag{1.46}$$

where the sample-dependent parameters are a $d \times d$ matrix $A(S_n)$, a d-dimensional vector $\mathbf{b}(S_n)$, and a scalar $c(S_n)$. The QDA discriminant results from the choices

$$\begin{aligned} A(S_n) &= -\frac{1}{2}\left(\hat{\Sigma}_1^{-1} - \hat{\Sigma}_0^{-1}\right), \\ \mathbf{b}(S_n) &= \hat{\Sigma}_1^{-1}\bar{\mathbf{X}}_1 - \hat{\Sigma}_0^{-1}\bar{\mathbf{X}}_0, \\ c(S_n) &= -\frac{1}{2}\left(\bar{\mathbf{X}}_1{}^T \hat{\Sigma}_1^{-1} \bar{\mathbf{X}}_1 - \bar{\mathbf{X}}_0{}^T \hat{\Sigma}_0^{-1} \bar{\mathbf{X}}_0\right) - \frac{1}{2} \log \frac{|\hat{\Sigma}_1|}{|\hat{\Sigma}_0|}. \end{aligned} \tag{1.47}$$

1.4.2 Linear Discriminants

Recall now the population-based LDA discriminant in (1.19). Replacing the population parameters by sample estimators yields the sample-based LDA discriminant

$$W_L(S_n,\mathbf{x}) = \left(\mathbf{x} - \frac{\bar{\mathbf{X}}_0 + \bar{\mathbf{X}}_1}{2}\right)^T \hat{\Sigma}^{-1}\left(\bar{\mathbf{X}}_0 - \bar{\mathbf{X}}_1\right), \tag{1.48}$$

where $\bar{\mathbf{X}}_0$ and $\bar{\mathbf{X}}_1$ are the sample means and

$$\hat{\Sigma} = \frac{1}{n-2}\sum_{i=1}^{n}\left[(\mathbf{X}_i - \bar{\mathbf{X}}_0)(\mathbf{X}_i - \bar{\mathbf{X}}_0)^T I_{Y_i=0} + (\mathbf{X}_i - \bar{\mathbf{X}}_1)(\mathbf{X}_i - \bar{\mathbf{X}}_1)^T I_{Y_1=1}\right] \tag{1.49}$$

is the *pooled sample covariance matrix*. This discriminant is also known as *Anderson's LDA discriminant* (Anderson, 1984).

If Σ is assumed to be known, it can be used in placed of $\hat{\Sigma}$ in (1.48) to obtain the discriminant

$$W_{\bar{L}}(S_n,\mathbf{x}) = \left(\mathbf{x} - \frac{\bar{\mathbf{X}}_0 + \bar{\mathbf{X}}_1}{2}\right)^T \Sigma^{-1}\left(\bar{\mathbf{X}}_0 - \bar{\mathbf{X}}_1\right). \tag{1.50}$$

This discriminant is also known as *John's LDA discriminant* (John, 1961; Moran, 1975). Anderson's and John's discriminants are studied in detail in Chapters 6 and 7.

If the common covariance matrix can be assumed to be diagonal, then only its diagonal elements, that is, the univariate feature variances, need to be estimated. This leads to the *diagonal LDA discriminant* (DLDA), an application of the "Naive Bayes" principle (Dudoit et al., 2002).

Finally, a simple form of LDA discriminant arises from assuming that $\Sigma = \sigma^2 I$, a scalar multiple of the identity matrix, in which case one obtains the *nearest-mean classifier* (NMC) discriminant (Duda and Hart, 2001).

All variants of LDA discriminants discussed above are special cases of the general *linear discriminant*, which takes the form

$$W_{\text{lin}}(S_n,\mathbf{x}) = \mathbf{a}(S_n)^T\mathbf{x} + b(S_n), \tag{1.51}$$

where the sample-dependent parameters are a d-dimensional vector $\mathbf{a}(S_n)$ and a scalar $b(S_n)$. The basic LDA discriminant results from the choices

$$\begin{aligned} \mathbf{a}(S_n) &= \hat{\Sigma}^{-1}(\bar{\mathbf{X}}_0 - \bar{\mathbf{X}}_1), \\ b(S_n) &= -\frac{1}{2}(\bar{\mathbf{X}}_0 + \bar{\mathbf{X}}_1)^T\hat{\Sigma}^{-1}(\bar{\mathbf{X}}_0 - \bar{\mathbf{X}}_1). \end{aligned} \tag{1.52}$$

Besides LDA, other examples of linear discriminants include perceptrons (Duda and Hart, 2001) and linear SVMs (c.f. Section 1.6.2). For any linear discriminant, the decision surface $W(\mathbf{x}) = k$ defines a hyperplane in the feature space R^d.

The QDA and LDA classification rules are derived from a Gaussian assumption; nevertheless, in practice they can perform well so long as the underlying populations are approximately unimodal, and one either knows the relevant covariance matrices or can obtain good estimates of them. In large-sample scenarios, QDA is preferable; however, LDA has been reported to be more robust relative to the underlying Gaussian assumption than QDA (Wald and Kronmal, 1977). Moreover, even when the covariance matrices are not identical, LDA can outperform QDA under small sample sizes owing to the difficulty of estimating the individual covariance matrices.

1.4.3 Kernel Discriminants

Kernel discriminants take the form

$$W_{\text{ker}}(S_n, \mathbf{x}) = \sum_{i=1}^{n} K\left(\frac{\mathbf{x} - \mathbf{X}_i}{h_n}\right)(I_{Y_i=0} - I_{Y_i=1}), \tag{1.53}$$

where the *kernel* $K(\mathbf{x})$ is a real function on R^d, which is usually (but not necessarily) nonnegative, and $h_n = H_n(S_n) > 0$ is the sample-based kernel bandwidth, defined by a rule H_n. From (1.53), we can see that the class label of a given point $\mathbf{x}$ is determined by the sum of the "contributions" $K((\mathbf{x} - \mathbf{X}_i)/h_n)$ of each training sample point $\mathbf{X}_i$, with positive sign if $\mathbf{X}_i$ is from class 0 and negative sign, otherwise. According to (1.41), with $k = 0$, $\psi_n(\mathbf{x})$ is equal to 0 or 1 according to whether the sum is positive or negative, respectively. The *Gaussian kernel* is defined by $K(\mathbf{x}) = e^{-||\mathbf{x}||^2}$, the *Epanechnikov kernel* is given by $K(\mathbf{x}) = (1 - ||\mathbf{x}||^2)I_{||\mathbf{x}|| \leq 1}$, and the *uniform kernel* is given by $I_{||\mathbf{x}|| \leq 1}$. Since the Gaussian kernel is never 0, all sample points get some weight. The Epanechnikov and uniform kernels possess bounded support, being 0 for sample points at a distance more than $h(S_n)$ from $\mathbf{x}$, so that only training points sufficiently close to $\mathbf{x}$ contribute to the definition of $\psi_n(\mathbf{x})$. The uniform kernel produces the *moving-window rule*, which takes the majority label among all sample points within distance $h(S_n)$ of $\mathbf{x}$. The decision surface $W_{\text{ker}}(S_n, \mathbf{x}) = k$ generally defines a complicated nonlinear manifold in R^d. The classification rules corresponding to the Gaussian, Epanechnikov, and uniform kernels are universally consistent (Devroye et al., 1996). Kernel discriminant rules also include as special cases Parzen-window classification rules (Duda and Hart, 2001) and nonlinear SVMs (c.f. Section 1.6.2).

1.5 HISTOGRAM RULE

Suppose that R^d is partitioned into cubes of equal side length r_n. For each point $\mathbf{x} \in R^d$, the *histogram rule* defines $\psi_n(\mathbf{x})$ to be 0 or 1 according to which is the majority among the labels for points in the cube containing $\mathbf{x}$. If the cubes are defined so that $r_n \to 0$ and $nr_n^d \to \infty$ as $n \to \infty$, then the rule is universally consistent (Gordon and Olshen, 1978).

An important special case of the histogram rule occurs when the feature space is finite. This discrete histogram rule is also referred to as *multinomial discrimination.* In this case, we can establish a bijection between all possible values taken on by $\mathbf{X}$ and a finite set of integers, so that there is no loss in generality in assuming a single feature X taking values in the set $\{1, \dots, b\}$. The value b provides a direct measure of the complexity of discrete classification — in fact, the VC dimension of the discrete histogram rule can be shown to be $V_C = b$ (c.f. Section B.3). The properties of the discrete classification problem are completely determined by the a priori class probabilities $c_0 = P(Y = 0)$, $c_1 = P(Y = 1)$, and the class-conditional probability mass functions $p_i = P(X = i|Y = 0)$, $q_i = P(X = i|Y = 1)$, for $i = 1, \dots, b$. Since $c_1 = 1 - c_0$, $p_b = 1 - \sum_{i=1}^{b-1} p_i$, and $q_b = 1 - \sum_{i=1}^{b-1} q_i$, the classification problem is determined by a $(2b - 1)$-dimensional vector $(c_0, p_1, \dots, p_{b-1}, q_1, \dots, q_{b-1}) \in R^{2b-1}$.

For a given a probability model, the error of a discrete classifier $\psi : \{1, \dots, b\} \to \{0, 1\}$ is given by

$$
\begin{aligned}
\varepsilon[\psi] &= \sum_{i=1}^{b} P(X = i,\, Y = 1 - \psi(i)) \\
&= \sum_{i=1}^{b} P(X = i|Y = 1 - \psi(i))P(Y = 1 - \psi(i)) \\
&= \sum_{i=1}^{b} \left[p_i c_0 I_{\psi(i)=1} + q_i c_1 I_{\psi(i)=0}\right] .
\end{aligned}
\tag{1.54}
$$

From this, it is clear that a Bayes classifier is given by

$$
\psi_{\text{bay}}(i) = \begin{cases} 1, & p_i c_0 < q_i c_1 , \\ 0, & \text{otherwise,} \end{cases}
\tag{1.55}
$$

for $i = 1, \dots, b$, with Bayes error

$$
\varepsilon_{\text{bay}} = \sum_{i=1}^{b} \min\{c_0 p_i, c_1 q_i\} .
\tag{1.56}
$$

Given a sample $S_n = \{(X_1, Y_1), \dots, (X_n, Y_n)\}$, the *bin counts* U_i and V_i are defined as the observed number of points with $X = i$ for class 0 and 1, respectively, for $i = 1, \dots, b$. The *discrete histogram classification rule* is given by majority voting between the bin counts over each bin:

$$
\Psi_n(S_n)(i) = I_{U_i < V_i} = \begin{cases} 1, & U_i < V_i \\ 0, & \text{otherwise,} \end{cases}
\tag{1.57}
$$

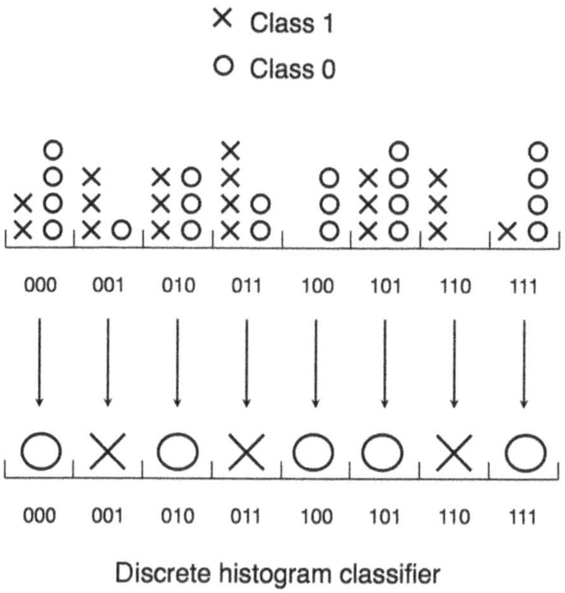

FIGURE 1.3 Discrete histogram rule: top shows distribution of the sample data in the bins; bottom shows the designed classifier.

for $i = 1, \ldots, b$. The rule is illustrated in Figure 1.3 where the top and bottom rows represent the sample data and the designed classifier, respectively. Note that in (1.57) the classifier is defined so that ties go to class 0, as happens in the figure.

The maximum-likelihood estimators of the distributional parameters are

$$\hat{c}_0 = \frac{N_0}{n}, \ \hat{c}_1 = \frac{N_1}{n}, \quad \text{and} \quad \hat{p}_i = \frac{U_i}{N_0}, \ \hat{q}_i = \frac{V_i}{N_1}, \ \text{for } i = 1, \ldots, b. \tag{1.58}$$

Plugging these into the expression of a Bayes classifier yields a histogram classifier, so that the histogram rule is the plug-in rule for discrete classification. For this reason, the fundamental rule is also called the *discrete-data plug-in rule.*

The classification error of the discrete histogram rule is given by

$$\varepsilon_n = \sum_{i=1}^{b} P(X = i, Y = 1 - \psi_n(i)) = \sum_{i=1}^{b} \left[c_0 p_i I_{V_i > U_i} + c_1 q_i I_{U_i \geq V_i} \right]. \tag{1.59}$$

Taking expectation leads to a remarkably simple expression for the expected classification error rate:

$$E[\varepsilon_n] = \sum_{i=1}^{b} \left[c_0 p_i P(V_i > U_i) + c_1 q_i P(U_i \geq V_i) \right]. \tag{1.60}$$

The probabilities in the previous equation can be easily computed by using the fact that the vector of bin counts $(U_1, \ldots, U_b, V_1, \ldots, V_b)$ is distributed multinomially with parameters $(n, c_0 p_1, \ldots, c_0 p_b, c_1 q_1, \ldots, c_1 q_b)$:

$$P(U_1 = u_1, \ldots, U_b = u_b, V_1 = v_1, \ldots, V_b = v_b) = \frac{n!}{u_1! \cdots u_b!\, v_1! \cdots v_b!} c_0^{\sum u_i} c_1^{\sum v_i} \prod_{i=1}^{b} p_i^{u_i} q_i^{v_i}. \tag{1.61}$$

The marginal distribution of the vector (U_i, V_i) is multinomial with parameters $(n, c_0 p_i, c_1 q_i, 1 - c_0 p_i - c_1 q_i)$, so that

$$\begin{aligned} P(V_i > U_i) &= \sum_{\substack{0 \le k+l \le n \\ k<l}} P(U_i = k, V_i = l) \\ &= \sum_{\substack{0 \le k+l \le n \\ k<l}} \frac{n!}{k!\, l!\, (n-k-l)!} (c_0 p_i)^k (c_1 q_i)^l (1 - c_0 p_i - c_1 q_i)^{n-k-l}, \end{aligned} \tag{1.62}$$

with $P(U_i \ge V_i) = 1 - P(V_i > U_i)$.

Using the Law of Large Numbers, it is easy to show that $\varepsilon_n \to \varepsilon_{\text{bay}}$ as $n \to \infty$ (for fixed b), with probability 1, for any distribution, that is, the discrete histogram rule is universally strongly consistent. As was mentioned in Section 1.3, this implies that $E[\varepsilon_n] \to \varepsilon_{\text{bay}}$.

As the number of bins b becomes comparable to the sample size n, the performance of the classifier degrades due to the increased likelihood of empty bins. Such bins have to be assigned an arbitrary label (here, the label zero), which clearly is suboptimal. This fact can be appreciated by examining (1.60), from which it follows that the expected error is bounded below in terms of the probabilities of empty bins:

$$E[\varepsilon_n] \ge \sum_{i=1}^{b} c_1 q_i P(U_i = 0, V_i = 0) = \sum_{i=1}^{b} c_1 q_i (1 - c_0 p_i - c_1 q_i)^n. \tag{1.63}$$

To appreciate the impact of the lower bound in (1.63), consider the following distribution, for even b: $c_0 = c_1 = \frac{1}{2}$, $q_i = \frac{2}{b}$, for $i = 1, \ldots, \frac{b}{2}$, $p_i = \frac{2}{b}$, for $i = \frac{b}{2} + 1, \ldots, b$, while q_i and p_i are zero, otherwise. In this case, it follows from (1.56) that $\varepsilon_{\text{bay}} = 0$. Despite an optimal error of zero, (1.63) yields the lower bound $E[\varepsilon_n] \ge \frac{1}{2}(1 - \frac{1}{b})^n$. With $b = n$ and $n \ge 4$, this bound gives $E[\varepsilon_n] > 0.15$, despite the classes being completely separated. This highlights the issue of small sample size n compared to the complexity b. Having n comparable to b results in a substantial increase in the expected error. But since $V_C = b$ for the discrete histogram rule, this simply reflects the need to have $n \gg V_C$ for acceptable performance, as argued in Section 1.3. While it is true that the previous discussion is based on a distribution-dependent bound, it is shown in Devroye et al. (1996) that, regardless of the distribution,

$$E[\varepsilon_n] \leq \varepsilon_{\text{bay}} + 1.075\sqrt{\frac{b}{n}}. \tag{1.64}$$

This confirms that n must greatly exceed b to guarantee good performance.

Another issue concerns the rate of convergence, that is, how fast the bounds converge to zero. This is not only a theoretical question, as the usefulness in practice of such results may depend on how large the sample size needs to be to guarantee that performance is close enough to optimality. The rate of convergence guaranteed by (1.64) is polynomial. It is shown next that exponential rates of convergence can be obtained, if one is willing to drop the distribution-free requirement. Provided that ties in bin counts are assigned a class randomly (with equal probability), the following exponential bound on the convergence of $E[\varepsilon_n]$ to ε_{bay} applies (Glick, 1973, Theorem A):

$$E[\varepsilon_n] - \varepsilon_{\text{bay}} \leq \left(\frac{1}{2} - \varepsilon_{\text{bay}}\right) e^{-cn}, \tag{1.65}$$

where the constant $c > 0$ is distribution-dependent:

$$c = \log \frac{1}{1 - \min_{\{i: c_0 p_i \neq c_1 q_i\}} |\sqrt{c_0 p_i} - \sqrt{c_1 q_i}|^2}. \tag{1.66}$$

Interestingly, the number of bins does not figure in this bound. The speed of convergence of the bound is determined by the minimum (nonzero) difference between the probabilities $c_0 p_i$ and $c_1 q_i$ over any one cell. The larger this difference is, the larger c is, and the faster convergence is. Conversely, the presence of a single cell where these probabilities are close slows down convergence of the bound.

1.6 OTHER CLASSIFICATION RULES

This section briefly describes other commonly employed classification rules that will be used in the text to illustrate the behavior of error estimators.

1.6.1 *k*-Nearest-Neighbor Rules

For the basic *nearest-neighbor* (NN) classification rule, ψ_n is defined for each $\mathbf{x} \in R^d$ by letting $\psi_n(\mathbf{x})$ take the label of the sample point closest to $\mathbf{x}$. For the NN classification rule, no matter the feature–label distribution of $(\mathbf{X}, Y)$, $\varepsilon_{\text{bay}} \leq \lim_{n\to\infty} E[\varepsilon_n] \leq 2\,\varepsilon_{\text{bay}}$ (Cover and Hart, 1967). It follows that $\lim_{n\to\infty} E[\Delta_n] \leq \varepsilon_{\text{bay}}$. Hence, the asymptotic expected design error is small if the Bayes error is small; however, this result does not give consistency. More generally, for the *k-nearest-neighbor* (*k*NN) classification rule, with k odd, the k points closest to $\mathbf{x}$ are selected and $\psi_n(\mathbf{x})$ is defined to be 0 or 1 according to which the majority is among the labels of these points. The 1NN rule is the NN rule defined previously. The Cover-Hart Theorem can be readily extended to the *k*NN case, which results in tighter bounds as k increases; for example, it can be shown that, for the 3NN rule, $\varepsilon_{\text{bay}} \leq \lim_{n\to\infty} E[\varepsilon_n] \leq 1.316\,\varepsilon_{\text{bay}}$, regardless of the distribution of $(\mathbf{X}, Y)$ (Devroye et al., 1996). The *k*NN rule is universally consistent if $k \to \infty$ and $k/n \to 0$ as $n \to \infty$ (Stone, 1977).

1.6.2 Support Vector Machines

The *support vector machine* (*SVM*) is based on the idea of a *maximal-margin hyperplane* (*MMH*) (Vapnik, 1998). Figure 1.4 shows a linearly separable data set and three hyperplanes (lines). The outer lines, which pass through points in the sample data, are the *margin hyperplanes*, whereas the third is the MMH, which is equidistant between the margin hyperplanes. The sample points closest to the MMH are on the margin hyperplanes and are called *support vectors* (the circled sample points in Figure 1.4). The distance from the MMH to any support vector is called the *margin*. The MMH defines a *linear SVM* classifier.

Let the equation for the MMH be $\mathbf{a}^T\mathbf{x} + b = 0$, where $\mathbf{a} \in R^d$ and $b \in R$ are the parameters of the linear SVM. It can be shown that the margin is given by $1/||\mathbf{a}||$ (see

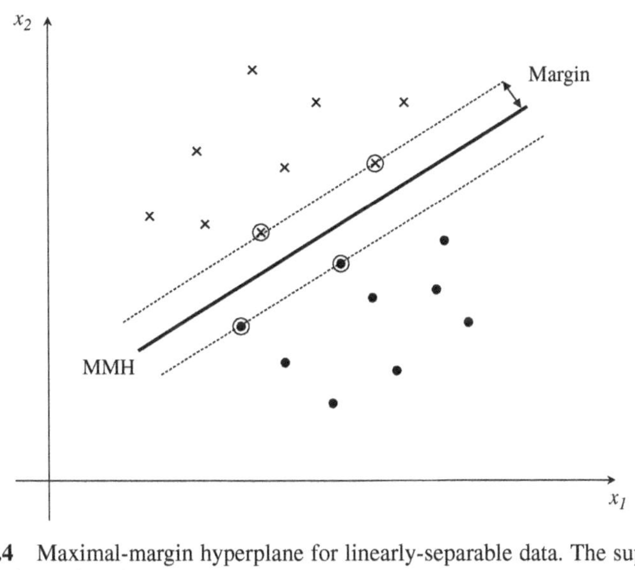

FIGURE 1.4 Maximal-margin hyperplane for linearly-separable data. The support vectors are the circled sample points.

Exercise 1.11). Since the goal is to maximize the margin, one finds the parameters of the MMH by solving the following quadratic optimization problem:

$$\min \frac{1}{2}||a||^2, \text{ subject to}$$
$$\begin{cases} a^T\mathbf{X}_i + b \geq 1\,, & \text{if } Y_i = 1\,, \\ a^T\mathbf{X}_i + b \leq -1\,, & \text{if } Y_i = 0\,, \end{cases} \quad \text{for } i = 1, \ldots, n\,. \tag{1.67}$$

If the sample is not linearly separable, then one has two choices: find a reasonable linear classifier or find a nonlinear classifier. In the first case, the preceding method can be modified by making appropriate changes to the optimization problem (1.67); in the second case, one can map the sample points into a higher dimensional space where they are linearly separable, find a hyperplane in that space, and then map back into the original space; see (Vapnik, 1998; Webb, 2002) for the details.

1.6.3 Neural Networks

A (feed-forward) two-layer *neural network* has the form

$$\psi(\mathbf{x}) = \mathrm{T}\left[c_0 + \sum_{i=1}^{k} c_i \sigma[\phi_i(\mathbf{x})]\right], \tag{1.68}$$

where T thresholds at 0, σ is a *sigmoid function* (that is, a nondecreasing function with limits -1 and $+1$ at $-\infty$ and ∞ , respectively), and

$$\phi_i(\mathbf{x}) = a_{i0} + \sum_{j=1}^{d} a_{ij} x_j. \tag{1.69}$$

Each operator in the sum of (1.68) is called a *neuron*. These form the hidden layer. We consider neural networks with the threshold sigmoid: $\sigma(t) = -1$ if $t \leq 0$ and $\sigma(t) = 1$ if $t > 0$. If $k \to \infty$ such that $(k \log n)/n \to 0$ as $n \to \infty$, then, as a class, neural networks are universally consistent (Farago and Lugosi, 1993), but one should beware of the increasing number of neurons required.

Any absolutely integrable function can be approximated arbitrarily closely by a sufficiently complex neural network. While this is theoretically important, there are limitations to its practical usefulness. Not only does one not know the function, in this case a Bayes classifier, whose approximation is desired, but even were we to know the function and how to find the necessary coefficients, a close approximation can require an extremely large number of model parameters. Given the neural-network structure, the task is to estimate the optimal weights. As the number of model parameters grows, use of the model for classifier design becomes increasingly intractable owing to the increasing amount of data required for estimation of the model parameters. Since the number of hidden units must be kept relatively small, thereby requiring significant constraint, when data are limited, there is no assurance that the optimal neural network of the prescribed form closely approximates the Bayes classifier. Model estimation is typically done by some iterative procedure, with advantages and disadvantages being associated with different methods (Bishop, 1995).

1.6.4 Classification Trees

The histogram rule partitions the space without reference to the actual data. One can instead partition the space based on the data, either with or without reference to the labels. Tree classifiers are a common way of performing data-dependent partitioning. Since any tree can be transformed into a binary tree, we need only consider binary classification trees. A tree is constructed recursively. If S represents the set of all data, then it is partitioned according to some rule into $S = S_1 \cup S_2$. There are four possibilities: (i) S_1 is split into $S_1 = S_{11} \cup S_{12}$ and S_2 is split into $S_2 = S_{21} \cup S_{22}$; (ii) S_1 is split into $S_1 = S_{11} \cup S_{12}$ and splitting of S_2 is terminated; (iii) S_2 is split into $S_2 = S_{21} \cup S_{22}$ and splitting of S_1 is terminated; and (iv) splitting of both S_1 and S_2 is terminated. In the last case, splitting is complete; in any of the others, it proceeds recursively until all branches end in termination, at which point the leaves on the tree represent the partition of the space. On each cell (subset) in the final partition, the designed classifier is defined to be 0 or 1, according to which the majority is among the labels of the points in the cell.

For *Classification and Regression Trees* (*CART*), splitting is based on an *impurity function*. For any rectangle R, let $N_0(R)$ and $N_1(R)$ be the numbers of 0-labeled and 1-labeled points, respectively, in R, and let $N(R) = N_0(R) + N_1(R)$ be the total number of points in R. The *impurity* of R is defined by

$$\kappa(R) = \zeta(p_R, 1 - p_R), \tag{1.70}$$

where $p_R = N_0(R)/N(R)$ is the proportion of 0 labels in R, and where $\zeta(p, 1-p)$ is a nonnegative function satisfying the following conditions: (1) $\zeta(0.5, 0.5) \geq \zeta(p, 1-p)$ for any $p \in [0, 1]$; (2) $\zeta(0, 1) = \zeta(1, 0) = 0$; and (3) as a function of p, $\zeta(p, 1-p)$ increases for $p \in [0, 0.5]$ and decreases for $p \in [0.5, 1]$. Several observations follow from the definition of ζ : (1) $\kappa(R)$ is maximum when the proportions of 0-labeled and 1-labeled points in R are equal (corresponding to maximum impurity); (2) $\kappa(R) = 0$ if R is pure; and (3) $\kappa(R)$ increases for greater impurity.

We mention three possible choices for ζ:

1. $\zeta_e(p, 1-p) = -p \log p - (1-p) \log(1-p)$ (*entropy impurity*)
2. $\zeta_g(p, 1-p) = p(1-p)$ (*Gini impurity*)
3. $\zeta_m(p, 1-p) = \min(p, 1-p)$ (*misclassification impurity*)

Regarding these three impurities: $\zeta_e(p, 1-p)$ provides an entropy estimate, $\zeta_g(p, 1-p)$ provides a variance estimate for a binomial distribution, and $\zeta_m(p, 1-p)$ provides a misclassification-rate estimate.

A splitting regimen is determined by the manner in which a split will cause an overall decrease in impurity. Let i be a coordinate, α be a real number, R be a rectangle to be split along the ith coordinate, $R^i_{\alpha,-}$ be the sub-rectangle resulting from the ith coordinate being less than or equal to α, and $R^i_{\alpha,+}$ be the sub-rectangle resulting from the ith coordinate being greater than α. Define the *impurity decrement* by

$$\Delta_i(R,\alpha) \overset{\Delta}{=} \kappa(R) - \frac{N\left(R^i_{\alpha,-}\right)}{N(R)}\kappa\left(R^i_{\alpha,-}\right) - \frac{N\left(R^i_{\alpha,+}\right)}{N(R)}\kappa\left(R^i_{\alpha,+}\right). \tag{1.71}$$

A good split will result in impurity reductions in the sub-rectangles. In computing $\Delta_i(R, \alpha)$, the new impurities are weighted by the proportions of points going into the sub-rectangles. CART proceeds iteratively by splitting a rectangle at $\hat{\alpha}$ on the ith coordinate if $\Delta_i(R, \alpha)$ is maximized for $\alpha = \hat{\alpha}$. There exist two splitting strategies: (i) the coordinate i is given and $\Delta_i(R, \alpha)$ is maximized over all α; (ii) the coordinate is not given and $\Delta_i(R, \alpha)$ is maximized over all i, α. Notice that only a finite number of candidate α need to be considered; say, the midpoints between adjacent training points. The search can therefore be exhaustive. Various stopping strategies are possible — for instance, stopping when maximization of $\Delta_i(R, \alpha)$ yields a value below a preset threshold, or when there are fewer than a specified number of sample points assigned to the node. One may also continue to grow the tree until all leaves are pure and then prune.

1.6.5 Rank-Based Rules

Rank-based classifiers utilize only the relative order between feature values, that is, their ranks. This produces classification rules that are very simple and thus are likely to avoid overfitting. The simplest example, the *Top-Scoring Pair* (TSP) classification rule, was introduced in Geman et al. (2004). Given two feature indices $1 \le i,j \le d$, with $i \neq j$, the TSP classifier is given by

$$\psi(\mathbf{x}) = \begin{cases} 1, & x_i < x_j, \\ 0, & \text{otherwise.} \end{cases} \tag{1.72}$$

Therefore, the TSP classifier depends only on the feature ranks, and not on their magnitudes. In addition, given (i,j), the TSP classifier is independent of the sample data; that is, there is minimum adjustment of the decision boundary. However, the pair (i,j) is selected by using the sample data $S_n = \{(X_{11}, \dots, X_{1d}, Y_1), \dots, (X_{n1}, \dots, X_{nd}, Y_n)\}$, as follows:

$$\begin{aligned} (i,j) &= \arg\max_{1\le i,j\le d} \left[\hat{P}(X_i < X_j \mid Y = 1) - \hat{P}(X_i < X_j \mid Y = 0)\right] \\ &= \arg\max_{1\le i,j\le d} \left[\frac{1}{N_0}\sum_{k=1}^{n} I_{X_{ki}<X_{kj}} I_{Y_k=1} - \frac{1}{N_1}\sum_{k=1}^{n} I_{X_{ki}<X_{kj}} I_{Y_k=0}\right]. \end{aligned} \tag{1.73}$$

This approach is easily extended to k pairs of features, $k = 1, 2, \ldots$, yielding a *kTSP* classification rule, and additional operations on ranks, such as the median, yielding a *Top Scoring Median* (TSM) classification rule (Afsari et al., 2014). If d is large, then the search in (1.73) is conducted using an efficient greedy algorithm; see Afsari et al. (2014) for the details.

1.7 FEATURE SELECTION

The motivation for increasing classifier complexity is to achieve finer discrimination, but as we have noted, increasing the complexity of classification rules can be detrimental because it can result in overfitting during the design process. One way of achieving greater discrimination is to utilize more features, but too many features when designing from sample data can result in increased expected classification error. This can be seen in the VC dimension of a linear classification rule, which has VC dimension $d + 1$ in R^d (c.f. Appendix B). With the proliferation of sensing technologies capable of measuring thousands or even millions of features (also known as "big data") at typically small sample sizes, this has become an increasingly serious issue. In addition to improving classification performance, other reasons to apply feature selection include the reduction of computational load, both in terms of execution time and data storage, and scientific interpretability of the designed classifiers.

A key issue is monotonicity of the expected classification error relative to the dimensionality of the feature vector. Let $\mathbf{X}' = \upsilon(\mathbf{X})$, where $\upsilon : R^D \to R^d$, with $D > d$, is a *feature selection transformation*, that is, it is an orthogonal projection from the larger dimensional space to the smaller one, which simply discards components from vector $\mathbf{X}$ to produce $\mathbf{X}'$. It can be shown that the Bayes error is monotone: $\varepsilon_{\text{bay}}(\mathbf{X}, Y) \leq \varepsilon_{\text{bay}}(\mathbf{X}', Y)$; for a proof, see Devroye et al (1996, Theorem 3.3). In other words, the optimal error rate based on the larger feature vector is smaller than or equal to the one based on the smaller one (this property in fact also applies to more general dimensionality reduction transformations). However, for the expected error rate of classifiers designed on sample data, this is not true: $E[\varepsilon_n(\mathbf{X}, Y)]$ could be smaller, equal, or larger than $E[\varepsilon_n(\mathbf{X}', Y)]$. Indeed, it is common for the expected error to decrease at first and then increase for increasingly large feature sets. This behavior is known as the *peaking phenomenon* or the *curse of dimensionality* (Hughes, 1968; Jain and Chandrasekaran, 1982). Peaking is illustrated in Figure 1.5, where the expected classification error rate is plotted against sample size and number of features, for different classification rules and Gaussian populations with blocked covariance matrices. The black line indicates where the error rate peaks (bottoms out) for each fixed sample size. For this experiment, the features are listed in a particular order, and they are added in that order to increase the size of the feature set — see Hua et al. (2005a) for the details. In fact, the peaking phenomenon can be quite complex. In Figure 1.5a, peaking occurs with very few features for sample sizes below 30, but exceeds 30 features for sample sizes above 90. In Figure 1.5b, even with a sample size of 200, the optimal number of features is only eight. The concave behavior and increasing number of optimal features in parts (a) and (b) of the figure,

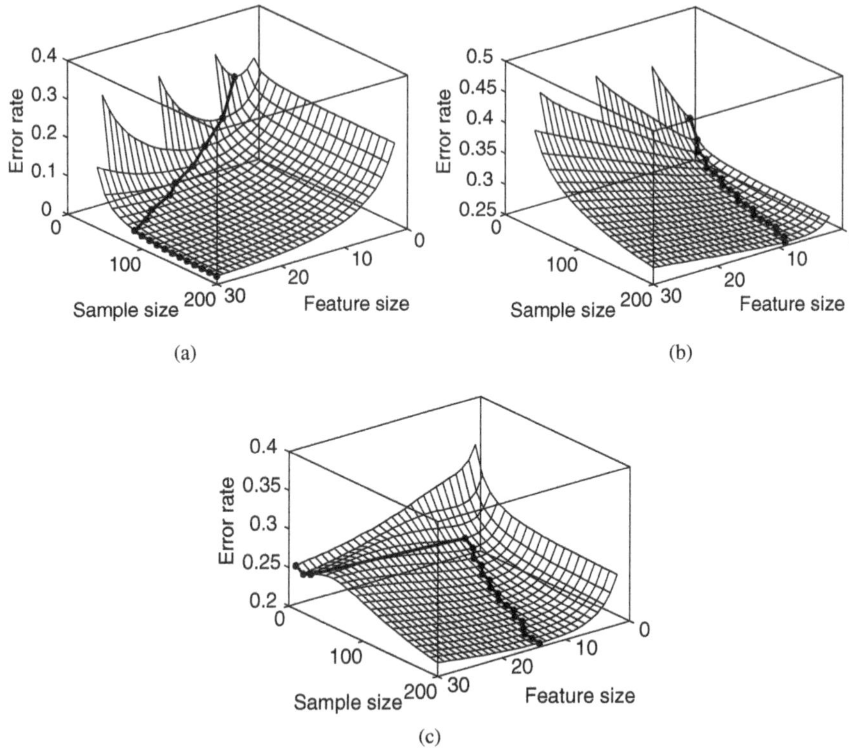

FIGURE 1.5 Expected classification error as a function of sample size and number of features, illustrating the peaking phenomenon. The black line indicates where the error rate peaks (bottoms out) for each fixed sample size. (a) LDA in model with common covariance matrix and slightly correlated features; (b) LDA in model with common covariance matrix and highly correlated features; (c) linear support vector machine in model with unequal covariance matrices. Reproduced from Hua et al. (2005) with permission from Oxford University Press.

which employs the LDA classification rule, corresponds to the usual understanding of peaking. But in Figure 1.5c, which employs a linear SVM, not only does the optimal number of features not increase as a function of sample size, but for fixed sample size the error curve is not concave. For some sizes the error decreases, increases, and then decreases again as the feature-set size grows, thereby forming a ridge across the expected-error surface. This behavior is not rare and applies to other classification rules. We can also see in Figure 1.5c what appears to be a peculiarity of the linear SVM classification rule: for small sample sizes, the error rate does not seem to peak at all. This odd phenomenon is supported by experimental evidence from other studies on linear SVMs (Afra and Braga-Neto, 2011).

Let S_n^D be the sample data including all D original features, and let Ψ_n^d be a classification rule defined on the low-dimensional sample space. A *feature selection rule* is a mapping $\Upsilon_n : (\Psi_n^d, S_n^D) \mapsto \upsilon_n$, that is, given a classification rule Ψ_n^d and sample

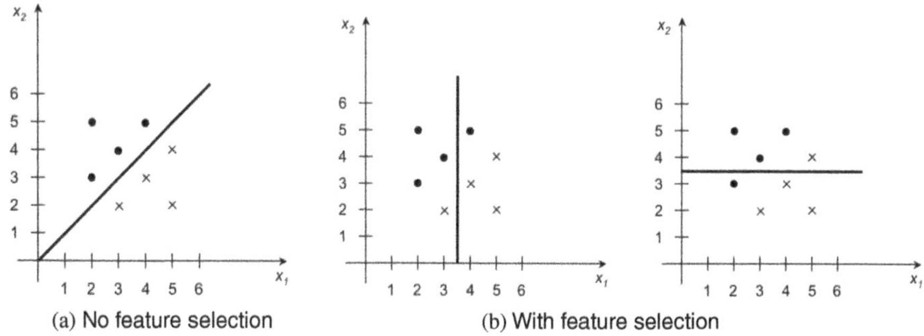

FIGURE 1.6 Constraint introduced by feature selection on a linear classification rule.

S_n^D, a feature selection transformation $\upsilon_n = \Upsilon_n(\Psi_n^d, S_n^D)$ is obtained according to the rule Υ_n. The reduced sample in the smaller space is $S_n^d = \{(\mathbf{X}_1', Y_1), \ldots, (\mathbf{X}_n', Y_n)\}$, where $\mathbf{X}_i' = \upsilon_n(\mathbf{X}_i)$, for $i = 1, \ldots, n$. Typically, feature selection is followed by classifier design using the selected feature set, according to the classification rule Ψ_n^d. This composition of feature selection and classifier design defines a new classification rule Ψ_n^D on the original high-dimensional feature space. If ψ_n^d denotes the classifier designed by Ψ_n^d on the smaller space, then the classifier designed by Ψ_n^D on the larger space is given by $\psi_n^D(\mathbf{X}) = \psi_n^d(\upsilon_n(\mathbf{X}))$. This means that feature selection becomes part of the classification rule in higher-dimensional space. This will be very important in Chapter 2 when we discuss error estimation rules in the presence of feature selection.

Since υ_n is an orthogonal projection, the decision boundary produced by ψ_n^D is an extension at right angles of the one produced by ψ_n^d. If the decision boundary in the lower-dimensional space is a hyperplane, then it is still a hyperplane in the higher-dimensional space, but a hyperplane that is not allowed to be oriented along any direction, but must be orthogonal to the lower-dimensional space. See Figure 1.6 for an illustration. Therefore, feature selection corresponds to a constraint on the classification rule in the sense of Section 1.3, and not just a reduction in dimensionality. As discussed in Section 1.3, there is a cost of constraint, which the designer is willing to incur if it is offset by a sufficient decrease in the expected design error.

Feature selection rules can be categorized into two broad classes: if the feature selection rule does not depend on Ψ_n^d, then it is called a *filter* rule, in which case $\upsilon_n = \Upsilon_n(S_n^D)$; otherwise, it is called a *wrapper* method. One could yet have a hybrid between the two, in case two rounds of feature selection are applied consecutively. In addition, feature selection rules can be *exhaustive*, which evaluate all $\binom{D}{d}$ possible projections from the higher-dimensional space to the lower-dimensional space, or *nonexhaustive*, otherwise. A typical exhaustive wrapper feature selection rule applies the classification rule Ψ_n^d on each of the $\binom{D}{d}$ possible feature sets of size d, obtains an estimate of the error of the resulting classifier using one of the techniques discussed in the next chapter, and then selects the feature set with the minimum estimated error. This creates a complex interaction between feature selection and error estimation (Sima et al., 2005a).

If D, and especially d, are large enough, then exhaustive feature selection is not possible due to computational reasons, and "greedy" nonexhaustive methods that only explore a subset of the search space must be used. Filter methods are typically of this kind. A simple and very popular example is based on finding individual features for which the class means are far apart relative to their pooled variance. The *t-test* statistic is typically employed for that purpose. An obvious problem with choosing features individually is that, while the selected features may perform individually the best, the approach is subject to choosing a feature set with a large number of redundant features and, in addition, features that perform poorly individually may do well in combination with other features; these will not be picked up. A classical counter example that illustrates this is *Toussaint's Counterexample*, where there are three original features and the best pair of individual features selected individually is the worst pair when considered together, according to the Bayes error; see Devroye et al. (1996) for the details.

As for nonexhaustive wrapper methods, the classical approaches are *branch-and-bound* algorithms, *sequential forward and backward selection*, and their variants (Webb, 2002). For example, sequential forward selection begins with a small set of features, perhaps one, and iteratively builds the feature set by adding more features as long as the estimated classification error is reduced. If features are allowed to be removed from the current feature set, in order to avoid bad features to be frozen in place, then one has a *floating* method. The standard greedy wrapper feature selection methods are *sequential forward selection* (SFS) and *sequential floating forward selection* (SFFS) (Pudil et al., 1994). The efficacy of greedy methods, that is, how fast and how close they get to the solution found by the corresponding exhaustive method, depends on the feature–label distribution, the classification rule and error estimation rule used, as well as the sample size and the original and target dimensionalities. A fundamental "no-free-lunch" result in Pattern Recognition specifies that, without distributional knowledge, no greedy method can always avoid the full complexity of the exhaustive search; that is, there will always be a feature–label distribution for which the algorithm must take as long as exhaustive search to find the best feature set. This result is known as the Cover–Campenhout Theorem (Cover and van Campenhout, 1977). Evaluation of methods is generally comparative and often based on simulations (Hua et al., 2009; Jain and Zongker, 1997; Kudo and Sklansky, 2000). Moreover, feature selection affects peaking. The classical way of examining peaking is to list the features in some order and then adjoin one at a time. This is how Figure 1.5 has been constructed. Peaking can be much more complicated when feature selection is involved (Sima and Dougherty, 2008).

EXERCISES

1.1 Suppose the class-conditional densities $p_0(\mathbf{x})$ and $p_1(\mathbf{x})$ are uniform distributions over a unit-radius disk centered at the origin and a unit-area square centered at $(\frac{1}{2}, \frac{1}{2})$, respectively. Find the Bayes classifier and the Bayes error.

1.2 Prove that $\varepsilon_{\text{bay}} = \frac{1}{2}(1 - E[|2\eta_1(\mathbf{X}) - 1|])$.

1.3 Consider Cauchy class-conditional densities:

$$p_i(x) = \frac{1}{\pi b} \frac{1}{1 + \left(\frac{x - a_i}{b}\right)^2}, \qquad i = 0, 1,$$

where $-\infty < a_0 < a_1 < \infty$ are location parameters (the Cauchy distribution does not have a mean), and $b > 0$ is a dispersion parameter. Assume that the classes are equally likely: $c_0 = c_1 = \frac{1}{2}$.

(a) Determine the Bayes classifier.

(b) Write the Bayes error as a function of the parameters a_0, a_1, and b.

(c) Plot the Bayes error as a function of $(a_1 - a_0)/b$ and explain what you see. In particular, what are the maximum and minimum (infimum) values of the curve and what do they correspond to?

1.4 This problem concerns the extension to the multiple-class case of concepts developed in Section 1.1. Let $Y \in \{0, 1, \ldots, c-1\}$, where c is the number of classes, and let

$$\eta_i(\mathbf{x}) = P(Y = i \mid \mathbf{X} = \mathbf{x}), \quad i = 0, 1, \ldots, c-1,$$

for each $\mathbf{x} \in R^d$. We need to remember that these probabilities are not independent, but satisfy $\eta_0(\mathbf{x}) + \eta_1(\mathbf{x}) + \cdots + \eta_{c-1}(\mathbf{x}) = 1$, for each $\mathbf{x} \in R^d$, so that one of the functions is redundant. Hint: you should answer the following items in sequence, using the previous answers in the solution of the following ones.

(a) Given a classifier $\psi : R^d \to \{0, 1, \ldots, c-1\}$, show that its *conditional error* at a point $\mathbf{x} \in R^d$ is given by

$$\begin{aligned} \varepsilon[\psi \mid \mathbf{X} = \mathbf{x}] &= P(\psi(\mathbf{X}) \neq Y \mid \mathbf{X} = \mathbf{x}) \\ &= 1 - \sum_{i=0}^{c-1} I_{\psi(\mathbf{x})=i}\, \eta_i(\mathbf{x}) = 1 - \eta_{\psi(\mathbf{x})}(\mathbf{x}). \end{aligned}$$

(b) Show that the classification error is given by

$$\varepsilon[\psi] = 1 - \sum_{i=0}^{c-1} \int_{\{\mathbf{x} \mid \psi(\mathbf{x}) = i\}} \eta_i(\mathbf{x}) p(\mathbf{x})\, d\mathbf{x}.$$

Hint: $\varepsilon[\psi] = E[\varepsilon[\psi \mid \mathbf{X}]]$.

(c) Prove that the Bayes classifier is given by

$$\psi_{\text{bay}}(\mathbf{x}) = \arg\max_{i=0,1,\ldots,c-1} \eta_i(\mathbf{x}), \quad \mathbf{x} \in R^d.$$

Hint: Start by considering the difference between conditional errors $P(\psi(\mathbf{X}) \neq Y \mid \mathbf{X} = \mathbf{x}) - P(\psi_{\text{bay}}(\mathbf{X}) \neq Y \mid \mathbf{X} = \mathbf{x})$.

(d) Show that the Bayes error is given by

$$\varepsilon_{\text{bay}} = 1 - E\left[\max_{i=0,1,\ldots,c-1} \eta_i(\mathbf{X})\right].$$

(e) Show that the results in parts (b), (c), and (d) reduce to (1.1), (1.2), and (1.3), respectively, in the two-class case $c = 2$.

1.5 Supose $\hat{\eta}_{0,n}$ and $\hat{\eta}_{1,n}$ are estimates of η_0 and η_1, respectively, for a sample of size n. The *plug-in rule* is defined by using $\hat{\eta}_{0,n}$ and $\hat{\eta}_{1,n}$ in place of η_0 and η_1, respectively, in (1.2). Derive the design error for the plug-in rule and show that it satisfies the bounds $\Delta_n \leq 2E[|\eta_1(\mathbf{X}) - \hat{\eta}_{1,n}(\mathbf{X})|]$ and $\Delta_n \leq 2E[|\eta_1(\mathbf{X}) - \hat{\eta}_{1,n}(\mathbf{X})|^2]^{1/2}$, thereby providing two criteria for consistency.

1.6 This problem concerns the Gaussian case, (1.17).

(a) Given a general linear discriminant $W_{\text{lin}}(\mathbf{x}) = \mathbf{a}^T\mathbf{x} + b$, where $\mathbf{a} \in R^d$ and $b \in R$ are arbitrary parameters, show that the error of the associated classifier

$$\psi(\mathbf{x}) = \begin{cases} 1, & W_{\text{lin}}(\mathbf{x}) = \mathbf{a}^T\mathbf{x} + b < 0 \\ 0, & \text{otherwise} \end{cases}$$

is given by

$$\varepsilon[\psi] = c_0\,\Phi\left(\frac{\mathbf{a}^T\boldsymbol{\mu}_0 + b}{\sqrt{\mathbf{a}^T\Sigma_0\mathbf{a}}}\right) + c_1\,\Phi\left(-\frac{\mathbf{a}^T\boldsymbol{\mu}_1 + b}{\sqrt{\mathbf{a}^T\Sigma_1\mathbf{a}}}\right), \qquad (1.74)$$

where Φ is the cumulative distribution of a standard normal random variable.

(b) Using the result from part (a) and the expression for the optimal linear discriminant in (1.19), show that if $\Sigma_0 = \Sigma_1 = \Sigma$ and $c_0 = c_1 = \frac{1}{2}$, then the Bayes error for the problem is given by (1.24):

$$\varepsilon_{\text{bay}} = \Phi\left(-\frac{\delta}{2}\right),$$

where $\delta = \sqrt{(\boldsymbol{\mu}_1 - \boldsymbol{\mu}_0)^T\Sigma^{-1}(\boldsymbol{\mu}_1 - \boldsymbol{\mu}_0)}$ is the Mahalanobis distance between the populations. Hence, there is a tight relationship (in fact, one-to-one) between the Mahalanobis distance and the Bayes error. What is the maximum and minimum (infimum) Bayes errors and when do they happen?

1.7 Determine the shatter coefficients and the VC dimension for

(a) The 1NN classification rule.

(b) A binary tree classification rule with data-independent splits and a fixed depth of k levels of splitting nodes.

(c) A classification rule that partitions the feature space into a finite number of b cells.

1.8 In this exercise we define two models to be used to generate synthetic data. In both cases, the dimension is d, the means are at the points $\boldsymbol{\mu}_0 = (0, \dots, 0)$ and $\boldsymbol{\mu}_1 = (u, \dots, u)$, and $\Sigma_0 = I$. For model M1,

$$\Sigma_1 = \sigma^2 \begin{pmatrix} 1 & \rho & \rho & \cdots & \rho \\ \rho & 1 & \rho & \cdots & \rho \\ \rho & \rho & 1 & \cdots & \rho \\ \vdots & \vdots & \vdots & \ddots & \rho \\ \rho & \rho & \rho & \cdots & 1 \end{pmatrix},$$

where Σ_1 is $d \times d$, and for model M2,

$$\Sigma_1 = \sigma^2 \begin{pmatrix} 1 & \rho & \rho & \cdots & 0 & 0 & 0 \\ \rho & 1 & \rho & \cdots & 0 & 0 & 0 \\ \rho & \rho & 1 & \cdots & 0 & 0 & 0 \\ \vdots & \vdots & \vdots & \ddots & \vdots & \vdots & \vdots \\ 0 & 0 & 0 & \cdots & 1 & \rho & \rho \\ 0 & 0 & 0 & \cdots & \rho & 1 & \rho \\ 0 & 0 & 0 & \cdots & \rho & \rho & 1 \end{pmatrix},$$

where Σ_1 is a $d \times d$ blocked matrix, consisting of 3×3 blocks along the diagonal. The prior class probabilities are $c_0 = c_1 = \frac{1}{2}$. For each model, the Bayes error is a function of d, u, ρ, and σ. For $d = 6$, $\sigma = 1, 2$, and $\rho = 0, 0.2, 0.8$, find for each model the value of u so that the Bayes error is $0.05, 0.15, 0.25, 0.35$.

1.9 For the models M1 and M2 of Exercise 1.8 with $d = 6$, $\sigma = 1, 2$, $\rho = 0, 0.2, 0.8$, and $n = 25, 50, 200$, plot $E[\varepsilon_n]$ against the Bayes error values $0.05, 0.15, 0.25, 0.35$, for the LDA, QDA, 3NN, and linear SVM classification rules. Hint: you will plot $E[\varepsilon_n]$ against the values of u corresponding to the Bayes error values $0.05, 0.15, 0.25, 0.35$.

1.10 The sampling mechanism considered so far is called "mixture sampling." It is equivalent to selecting Y_i from a Bernoulli random variable with mean $c_1 = P(Y = 1)$ and then sampling $\mathbf{X}_i$ from population Π_j if $Y_i = j$, for $i = 1, \dots, n$, $j = 0, 1$. The numbers of sample points N_0 and N_1 from each population are therefore random. Consider the case where the sample sizes n_0 and n_1 are not random, but fixed before sampling occurs. This mechanism is called "separate sampling," and will be considered in detail in Chapters 5–7. For the model M1 of Exercise 1.8, the LDA and 3NN classification rules, $d = 6$, $c_0 = c_1 = \frac{1}{2}$, $\sigma = 1$, $\rho = 0$, and Bayes error 0.25, find $E[\varepsilon_{n_0,n_1}]$ using separate sampling with $n_0 = 50$ and $n_1 = 50$, and compare the result to mixture sampling with $n = 100$. Repeat for $n_0 = 25$, $n_1 = 75$ and again compare to mixture sampling.

1.11 Consider a general linear discriminant $W_{\text{lin}}(\mathbf{x}) = \mathbf{a}^T\mathbf{x} + b$, where $\mathbf{a} \in R^d$ and $b \in R$ are arbitrary parameters.

(a) Show that the Euclidean distance of a point $\mathbf{x}_0$ to the hyperplane $W_{\text{lin}}(\mathbf{x}) = 0$ is given by $|W_{\text{lin}}(\mathbf{x}_0)|/||\mathbf{a}||$. Hint: Minimize the distance $||\mathbf{x} - \mathbf{x}_0||$ of a point $\mathbf{x}$ on the hyperplane to the point $\mathbf{x}_0$.

(b) Use part(a) to show that the margin for a linear SVM is $1/||\mathbf{a}||$.

1.12 This problem concerns a parallel between discrete classification and Gaussian classification. Let $\mathbf{X} = (X_1, \ldots, X_d) \in \{0, 1\}^d$ be a discrete feature vector, that is, all individual features are binary. Assume furthermore that the features are conditionally independent given $Y = 0$ and given $Y = 1$ — compare this to spherical class-conditional Gaussian densities, where the features are also conditionally independent given $Y = 0$ and given $Y = 1$.

(a) As in the Gaussian case with equal covariance matrices, prove that the Bayes classifier is linear, that is, show that $\psi_{\text{bay}}(\mathbf{x}) = I_{D_L(\mathbf{x})>0}$, for $\mathbf{x} \in \{0, 1\}^d$, where $D_L(\mathbf{x}) = \mathbf{a}^T\mathbf{x} + b$. Give the values of $\mathbf{a}$ and b in terms of the class-conditional distribution parameters $p_i = P(\mathbf{X}_i = 1|Y = 0)$ and $q_i = P(\mathbf{X}_i = 1|Y = 1)$, for $i = 1, \ldots, d$ (notice that these are different than the parameters p_i and q_i defined in Section 1.5), and the prior probabilities $c_0 = P(Y = 0)$ and $c_1 = P(Y = 1)$.

(b) Suppose that sample data $S_n = \{(\mathbf{X}_1, Y_1), \ldots, (\mathbf{X}_n, Y_n)\}$ is available, where $\mathbf{X}_i = (X_{i1}, \ldots, X_{id}) \in \{0, 1\}^d$, for $i = 1, \ldots, d$. Just as is done for LDA in the Gaussian case, obtain a sample discriminant $W_{\text{lin}}(\mathbf{x})$ from $D_L(\mathbf{x})$ in the previous item, by plugging in maximum-likelihood (ML) estimators $\hat{p}_i$, $\hat{q}_i$, $\hat{c}_0$, and $\hat{c}_1$ for the unknown parameters p_i, q_i, c_0 and c_1. The maximum-likelihood estimators in this case are the empirical frequencies (you may use this fact without showing it). Show that $W_{\text{lin}}(S_n, \mathbf{x}) = \mathbf{a}(S_n)^T\mathbf{x} + b(S_n)$, where $\mathbf{a}(S_n) \in R^d$ and $b(S_n) \in R$. Hence, this is a sample-based linear discriminant, just as in the case of sample-based LDA. Give the values of $\mathbf{a}(S_n)$ and $b(S_n)$ in terms of $S_n = \{(\mathbf{X}_1, Y_1), \ldots, (\mathbf{X}_n, Y_n)\}$.

1.13 Consider the simple CART classifier in R^2 depicted below.

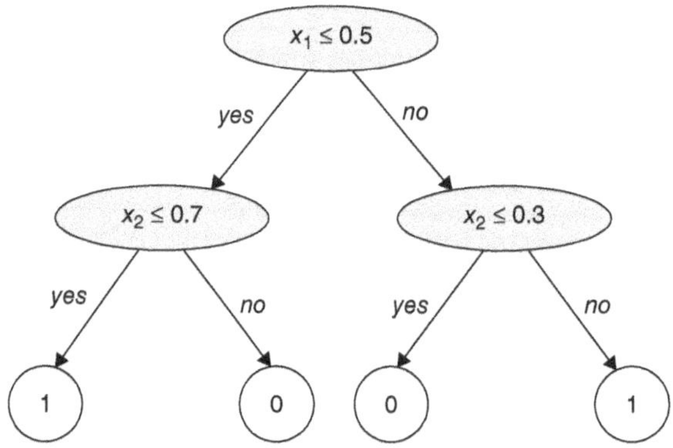

Design an equivalent two-hidden-layer neural network with threshold sigmoids, with three neurons in the first hidden layer and four neurons in the second hidden layer. Hint: Note the correspondence between the number of neurons and the number of splitting and leaf nodes.

1.14 Let $\mathbf{X} \in R^d$ be a feature set of size d, and let $X_0 \in R$ be an additional feature. Define the augmented feature set $\mathbf{X}' = (\mathbf{X}, X_0) \in R^{d+1}$. It was mentioned in Section 1.7 that the Bayes error satisfies the monotonicity property $\varepsilon_{\text{bay}}(\mathbf{X}, Y) \leq \varepsilon_{\text{bay}}(\mathbf{X}', Y)$. Show that a sufficient condition for the undesirable case where there is no improvement, that is, $\varepsilon_{\text{bay}}(\mathbf{X}, Y) = \varepsilon_{\text{bay}}(\mathbf{X}', Y)$, is that X_0 be independent of $(\mathbf{X}, Y)$ (in which case, the feature X_0 is redundant or "noise").

CHAPTER 2

ERROR ESTIMATION

As discussed in the Preface, error estimation is the salient epistemological issue in classification. It is a complex topic, because the accuracy of an error estimator depends on the classification rule, feature–label distribution, dimensionality, and sample size.

2.1 ERROR ESTIMATION RULES

A classifier together with a stated error rate constitute a *pattern recognition (PR) model*. Notice that a classifier by itself does not constitute a PR model. The validity of the PR model depends on the concordance between the stated error rate and the true error rate. When the feature–label distribution is unknown, the true error rate is unknown and the error rate stated in the model is estimated from data. Since the true error is unknown, *ipso facto,* it cannot be compared to the stated error. Hence, the validity of a given PR model cannot be ascertained and we take a step back and characterize validity in terms of the classification and error estimation procedures, which together constitute a *pattern recognition (PR) rule*.

Before proceeding, let us examine the epistemological problem more closely. A scientific model must provide predictions that can be checked against observations, so that comparison between the error estimate and the true error may appear to be a chimera. However, if one were to obtain a very large amount of test data, then the true error can be very accurately estimated; indeed, as we will see in Section 2.3, the proportion of errors made on test data converges with probability one to the true error as the amount of test data goes to infinity. Thus, in principle, the stated error in

Error Estimation for Pattern Recognition, First Edition. Ulisses M. Braga-Neto and Edward R. Dougherty.

a PR model can be compared to the true error with any desired degree of accuracy. Of course, if we could actually do this, then the test proportion could be used as the stated error in the model and the validity issue would be settled. But, in fact, the practical problem is that one does not have unlimited data, and often has very limited data.

Formally, a pattern recognition rule (Ψ_n, Ξ_n) consists of a classification rule Ψ_n and an *error estimation rule* $\Xi_n : (\Psi_n, S_n, \xi) \mapsto \hat{\varepsilon}_n$, where $0 \le \hat{\varepsilon}_n \le 1$ is an *error estimator*. Here, $S_n = \{(\mathbf{X}_1, Y_1), \ldots, (\mathbf{X}_n, Y_n)\}$ is a random sample of size n from the joint feature–label distribution $F_{\mathbf{X}Y}$ and ξ denotes internal random factors of Ξ_n that do not have to do with the sample data. Note that the definition implies a sequence of PR rules, error estimation rules, and error estimators depending on n; for simplicity, we shall not make the distinction formally. Provided with a realization of the data S_n and of the internal factors ξ, the PR rule produces a fixed PR model $(\psi_n, \hat{\varepsilon}_n)$, where $\psi_n = \Psi_n(S_n)$ is a fixed classifier and $\hat{\varepsilon}_n = \Xi_n(\Psi_n, S_n, \xi)$ is a fixed error estimate (for simplicity, we use here, as elsewhere in the book, the same notation for an error estimator, in which case the sample and internal factors are not specified, and an error estimate, in which case they are; the intended meaning will be either clear from the context or explicitly stated).

The goodness of a given PR rule relative to a specific feature–label distribution involves both how well the true error ε_n of the designed classifier $\psi_n = \Psi_n(S_n)$ approximates the optimal error ε_{bay} and how well the error estimate $\hat{\varepsilon}_n = \Xi_n(\Psi_n, S_n, \xi)$ approximates the true error ε_n, the latter approximation determining the validity of the PR rule. The validity of the PR rule depends therefore on the joint distribution between the random variables ε_n and $\hat{\varepsilon}_n$.

According to the nature of ξ, we distinguish two cases: (i) *randomized error estimation rules* possess random internal factors ξ — for fixed sample data S_n, the error estimation procedure Ξ_n produces random outcomes; (ii) *nonrandomized error estimation rules* produce a fixed value for fixed sample data S_n (in which case, there are no internal random factors). In these definitions, as everywhere else in the book, we are assuming that classification rules are nonrandomized so that the only sources of randomness for an error estimation rule are the sample data and the internal random factors, if any, which are not dependent on the classification rule.

For a concrete example, consider the error estimation rule given by

$$\Xi_n^r(\Psi_n, S_n) = \frac{1}{n} \sum_{i=1}^{n} |Y_i - \Psi_n(S_n)(\mathbf{X}_i)|. \tag{2.1}$$

This is known as the *resubstitution estimation rule* (Smith, 1947), to be studied in detail later. Basically, this error estimation rule produces the fraction of errors committed on S_n by the classifier designed by Ψ_n on S_n itself. For example, in Figure 1.6 of the last chapter, the resubstitution error estimate is zero in part (a) and 0.25 in part (b). Note that ξ is omitted in (2.1) because this is a nonrandomized error estimation rule. Suppose that Ψ_n is the Linear Discriminant Analysis (LDA) classification rule, discussed in the previous chapter. Then (Ψ_n, Ξ_n^r) is the *LDA-resubstitution* pattern recognition rule, and $\hat{\varepsilon}_n^r = \Xi_n^r(\Psi_n, S_n)$ is the *resubstitution error estimator* for the LDA classification rule. Notice that there is only one resubstitution error estimation rule, but multiple resubstitution error estimators, one for each given classification rule. This is an important distinction, because the properties of a particular error estimator are fundamentally dependent on the classification rule. Resubstitution is the prototypical nonrandomized error estimation rule. On the other hand, the prototypical randomized error estimation rule is *cross-validation* (Cover, 1969; Lachenbruch and Mickey, 1968; Stone, 1974; Toussaint and Donaldson, 1970); this error estimation rule will be studied in detail later. Basically, the sample is partitioned randomly into *k folds*; each fold is left out and the classification rule is applied on the remaining sample, with the left-out data used as a test set; the process is repeated k times (once for each fold), with the estimate being the average proportion of errors committed on all folds. This is generally a randomized error estimation rule, with internal random factors corresponding to the random selection of the folds.

Both the true error ε_n and the error estimator $\hat{\varepsilon}_n$ are random variables relative to the sampling distribution for S_n. If, in addition, the error estimation rule is randomized, then the error estimator $\hat{\varepsilon}_n$ is still random even after S_n is specified. This adds variability to the error estimator. We define the *internal variance* of $\hat{\varepsilon}_n$ by

$$V_{\text{int}} = \text{Var}(\hat{\varepsilon}_n | S_n). \tag{2.2}$$

The internal variance measures the variability due only to the internal random factors. The *full variance* $\text{Var}(\hat{\varepsilon}_n)$ of $\hat{\varepsilon}_n$ measures the variability due to both the sample S_n (which depends on the data distribution) and the internal random factors ξ. Letting $X = \hat{\varepsilon}_n$ and $Y = S_n$ in the Conditional Variance Formula, c.f. (A.58), one obtains

$$\text{Var}(\hat{\varepsilon}_n) = E[V_{\text{int}}] + \text{Var}(E[\hat{\varepsilon}_n | S_n]). \tag{2.3}$$

This variance decomposition equation illuminates the issue. The first term on the right-hand side contains the contribution of the internal variance to the total variance. For nonrandomized $\hat{\varepsilon}_n$, $V_{\text{int}} = 0$; for randomized $\hat{\varepsilon}_n$, $E[V_{\text{int}}] > 0$. Randomized error estimation rules typically attempt to make the internal variance small by averaging and intensive computation.

In the presence of feature selection from the original feature space R^D to a smaller space R^d, the previous discussion in this section holds without change, since, as discussed in Section 1.7, feature selection becomes part of the classification rule Ψ_n^D, defined on the original sample S_n^D, producing a classifier ψ_n^D in the original feature space R^D (albeit this is a constrained classifier, in the sense discussed in Section 1.7). The error estimation rule $\Xi_n^D : (\Psi_n^D, S_n^D, \xi) \mapsto \hat{\varepsilon}_n^D$ is also defined on the original sample S_n^D, with the classification rule Ψ_n^D, and the PR rule is (Ψ_n^D, Ξ_n^D). It is key to distinguish this from a corresponding rule (Ψ_n^d, Ξ_n^d), defined on a reduced sample S_n^d, producing a classifier ψ_n^d and error estimator $\hat{\varepsilon}_n^d$ in the reduced feature space R^d. This can become especially confusing because a classification rule Ψ_n^d is applied to the reduced sample S_n^d to obtain a classifier in the reduced feature space R^d, and in general is also used with a feature selection rule Υ_n to obtain the reduced sample — see Section 1.7 for a detailed discussion. In this case, a PR rule could equivalently be represented by a triple $(\Psi_n^d, \Upsilon_n, \Xi_n^D)$ consisting of a classification rule in the reduced space, a feature selection rule between the original and reduced spaces, and an estimation rule defined on the original space. We say that an error estimation rule is *reducible* if $\hat{\varepsilon}_n^d = \hat{\varepsilon}_n^D$. The prototypical example of reducible error estimation rule is resubstitution; considering Figure 1.6b, we can see that the resubstitution error estimate is 0.25 whether one computes it in the original 2D feature space, or projects the data down to the selected variable and then computes it. On the other hand, the prototypical non-reducible error estimation rule is cross-validation, as we will discuss in more detail in Section 2.5. Applying cross-validation in the reduced feature space is a serious mistake.

In this chapter we will examine several error estimation rules. In addition to the resubstitution and cross-validation error estimation rules mentioned previously, we will examine *bootstrap* (Efron, 1979, 1983), *convex* (Sima and Dougherty, 2006a), *smoothed* (Glick, 1978; Snapinn and Knoke, 1985), *posterior-probability* (Lugosi and Pawlak, 1994), and *bolstered* (Braga-Neto and Dougherty, 2004a) error estimation rules.

The aforementioned error estimation rules share the property that they do not involve prior knowledge regarding the feature–label distribution in their computation (although they may, in some cases, have been optimized relative to some collection of feature–label distributions prior to their use). In Chapter 8, we shall introduce another kind of error estimation rule that is optimal relative to the minimum mean-square error (MMSE) based on the sample data and a prior distribution governing an uncertainty class of feature–label distributions, which is assumed to contain the actual feature–label distribution of the problem (Dalton and Dougherty, 2011b,c). Since the obtained error estimators depend on a prior distribution on the unknown feature–label distribution, they are referred to as *Bayesian error estimators*. Relative to performance, a key difference between non-Bayesian and Bayesian error estimation rules is that the latter allow one to measure performance given the sample (Dalton and Dougherty, 2012a,b).

2.2 PERFORMANCE METRICS

Throughout this section, we assume that a pattern recognition rule (Ψ_n, Ξ_n) has been specified, and we would like to evaluate the performance of the corresponding error estimator $\hat{\varepsilon}_n = \Xi_n(\Psi_n, S_n, \xi)$ for the classification rule Ψ_n. In a non-Bayesian setting, it is not possible to know the accuracy of an error estimator given a fixed sample, and therefore estimation quality is judged via its "average" performance over all possible samples. Any such questions can be answered via the joint sampling distribution between the true error ε_n and the error estimator $\hat{\varepsilon}_n$, for the given classification rule. This section examines performance criteria for error estimators, including the

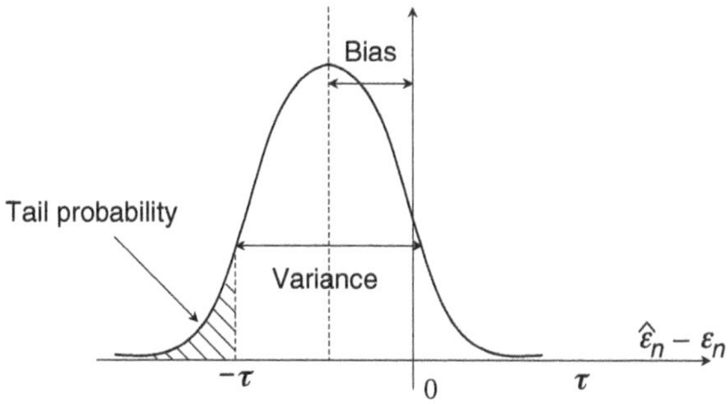

FIGURE 2.1 Deviation distribution and a few performance metrics of error estimation derived from it.

deviation with respect to the true error, consistency, correlation with the true error, regression on the true error, and confidence bounds. It is important to emphasize that the performance of an error estimator depends contextually on the specific feature–label distribution, classification rule, and sample size.

2.2.1 Deviation Distribution

Since the purpose of an error estimator is to approximate the true error, the distribution of the *deviation* $\hat{\varepsilon}_n - \varepsilon_n$ is central to accuracy characterization (see Figure 2.1). The deviation $\hat{\varepsilon}_n - \varepsilon_n$ is a function of the vector $(\varepsilon_n, \hat{\varepsilon}_n)$, and therefore the deviation distribution is determined by joint distribution between ε_n and $\hat{\varepsilon}_n$. It has the advantage of being a univariate distribution; it contains a useful subset of the information in the joint distribution.

Certain moments and probabilities associated with the deviation distribution play key roles as performance metrics for error estimation:

1. The *bias*,

$$\text{Bias}(\hat{\varepsilon}_n) = E[\hat{\varepsilon}_n - \varepsilon_n] = E[\hat{\varepsilon}_n] - E[\varepsilon_n]. \tag{2.4}$$

2. The *deviation variance*,

$$\text{Var}_{\text{dev}}(\hat{\varepsilon}_n) = \text{Var}(\hat{\varepsilon}_n - \varepsilon_n) = \text{Var}(\hat{\varepsilon}_n) + \text{Var}(\varepsilon_n) - 2\text{Cov}(\varepsilon_n, \hat{\varepsilon}_n). \tag{2.5}$$

3. The *root-mean-square error*,

$$\text{RMS}(\hat{\varepsilon}_n) = \sqrt{E\left[(\hat{\varepsilon}_n - \varepsilon_n)^2\right]} = \sqrt{\text{Bias}(\hat{\varepsilon}_n)^2 + \text{Var}_{\text{dev}}(\hat{\varepsilon}_n)}. \tag{2.6}$$

4. The *tail probabilities*,

$$P(\hat{\varepsilon}_n - \varepsilon_n \geq \tau) \text{ and } P(\hat{\varepsilon}_n - \varepsilon_n \leq -\tau), \quad \text{for } \tau > 0. \tag{2.7}$$

The estimator $\hat{\varepsilon}_n$ is said to be *optimistically biased* if $\text{Bias}(\hat{\varepsilon}_n) < 0$ and *pessimistically biased* if $\text{Bias}(\hat{\varepsilon}_n) > 0$. The optimal situation relative to bias corresponds to $\text{Bias}(\hat{\varepsilon}_n) = 0$, when the error estimator is said to be *unbiased*. The deviation variance, RMS, and tail probabilities are always nonnegative; in each case, a smaller value indicates better error estimation performance.

In some settings, it is convenient to speak not of the RMS but of its square, namely, the *mean-square error*,

$$\text{MSE}(\hat{\varepsilon}_n) = E\left[(\hat{\varepsilon}_n - \varepsilon_n)^2\right] = \text{Bias}(\hat{\varepsilon}_n)^2 + \text{Var}_{\text{dev}}(\hat{\varepsilon}_n). \tag{2.8}$$

The *correlation coefficient* between true and estimated error is given by

$$\rho(\varepsilon_n, \hat{\varepsilon}_n) = \frac{\text{Cov}(\varepsilon_n, \hat{\varepsilon}_n)}{\text{Std}(\varepsilon_n)\text{Std}(\hat{\varepsilon}_n)}, \tag{2.9}$$

where Std denotes standard deviation. Since $\hat{\varepsilon}_n$ is used to estimate ε_n, we would like $\hat{\varepsilon}_n$ and ε_n to be strongly correlated, that is, to have $\rho(\varepsilon_n, \hat{\varepsilon}_n) \approx 1$. Note that the deviation variance can be written as

$$\text{Var}_{\text{dev}}(\hat{\varepsilon}_n) = \text{Var}(\hat{\varepsilon}_n) + \text{Var}(\varepsilon_n) - 2\rho(\varepsilon_n, \hat{\varepsilon}_n)\text{Std}(\varepsilon_n)\text{Std}(\hat{\varepsilon}_n). \tag{2.10}$$

Therefore, strong correlation reduces the deviation variance, and therefore also the RMS. If n/d is large, then the individual variances tend to be small, so that the deviation variance is small; however, when n/d is small, the individual variances tend to be large, so that strong correlation is needed to offset these variances. Thus, the correlation between the true and estimated errors plays a key role in assessing the goodness of the error estimator (Hanczar et al., 2007). The deviation variance ranges from a minimum value of $(\text{Std}(\varepsilon_n) - \text{Std}(\hat{\varepsilon}_n))^2$ when $\rho(\varepsilon_n, \hat{\varepsilon}_n) = 1$ (perfect correlation), to a maximum value of $(\text{Std}(\varepsilon_n) + \text{Std}(\hat{\varepsilon}_n))^2$ when $\rho(\varepsilon_n, \hat{\varepsilon}_n) = -1$ (perfect negative correlation). We will see that it is common for $|\rho(\varepsilon_n, \hat{\varepsilon}_n)|$ to be small. If this is so and the true error ε_n does not change significantly for different samples, then $\text{Var}(\varepsilon_n) \approx 0$ and $\text{Var}_{\text{dev}}(\hat{\varepsilon}_n) \approx \text{Var}(\hat{\varepsilon}_n)$ so that the deviation variance is dominated by the variance of the error estimator itself. One should be forewarned that it is not uncommon for $\text{Var}(\varepsilon_n)$ to be large, especially for small samples.

The RMS is generally considered the most important error estimation performance metric. The other performance metrics appear within the computation of the RMS; indeed, all of the five basic moments — the expectations $E[\varepsilon_n]$ and $E[\hat{\varepsilon}_n]$, the variances $\text{Var}(\varepsilon_n)$ and $\text{Var}(\hat{\varepsilon}_n)$, and the covariance $\text{Cov}(\varepsilon_n, \hat{\varepsilon}_n)$ — appear within the RMS. In addition, with $X = |Z|^2$ and $a = \tau^2$ in Markov's inequality (A.45), it follows that

$P(|Z| \geq \tau) \leq E(|Z|^2)/\tau^2$, for any random variable Z. By letting $Z = |\hat{\varepsilon}_n - \varepsilon_n|$, one obtains

$$P(|\hat{\varepsilon}_n - \varepsilon_n| \geq \tau) \leq \frac{\text{MSE}(\hat{\varepsilon}_n)}{\tau^2} = \frac{\text{RMS}(\hat{\varepsilon}_n)^2}{\tau^2}. \tag{2.11}$$

Therefore, a small RMS implies small tail probabilities.

2.2.2 Consistency

One can also consider asymptotic performance as sample size increases. In particular, an error estimator is said to be *consistent* if $\hat{\varepsilon}_n \to \varepsilon_n$ in probability as $n \to \infty$, and *strongly consistent* if convergence is with probability 1. It turns out that consistency is related to the tail probabilities of the deviation distribution: by definition of convergence in probability, $\hat{\varepsilon}_n$ is consistent if and only if, for all $\tau > 0$,

$$\lim_{n\to\infty} P(|\hat{\varepsilon}_n - \varepsilon_n| \geq \tau) = 0. \tag{2.12}$$

By an application of the First Borel-Cantelli lemma (c.f. Lemma A.1), we obtain that $\hat{\varepsilon}_n$ is, in addition, strongly consistent if the following stronger condition holds:

$$\sum_{n=1}^{\infty} P(|\hat{\varepsilon}_n - \varepsilon_n| \geq \tau) < \infty, \tag{2.13}$$

for all $\tau > 0$, that is, the tail probabilities vanish sufficiently fast for the summation to converge. If an estimator is consistent regardless of the underlying feature–label distribution, one says it is *universally* consistent.

If the given classification rule is consistent and the error estimator is consistent, then $\varepsilon_n \to \varepsilon_{\text{bay}}$ and $\hat{\varepsilon}_n \to \varepsilon_n$, implying $\hat{\varepsilon}_n \to \varepsilon_{\text{bay}}$. Hence, the error estimator provides a good estimate of the Bayes error, provided one has a large sample S_n. The question of course is how large the sample size needs to be. While the rate of convergence of $\hat{\varepsilon}_n$ to ε_n can be bounded for some classification rules and some error estimators regardless of the distribution, there will always be distributions for which ε_n converges to ε_{bay} arbitrarily slowly. Hence, one cannot guarantee that $\hat{\varepsilon}_n$ is close to ε_{bay} for a given n, unless one has additional information about the distribution.

An important caveat is that, in small-sample cases, consistency does not play a significant role, the key issue being accuracy as measured by the RMS.

2.2.3 Conditional Expectation

RMS, bias, and deviation variance are global quantities in that they relate to the full joint distribution of the true error and error estimator. They do not relate directly to the knowledge obtained from a given error estimate. The knowledge derived from a specific estimate is embodied in the the conditional distribution of the true error given the estimated error. In particular, the conditional expectation (nonlinear regression) $E[\varepsilon_n|\hat{\varepsilon}_n]$ gives the average value of the true error given the estimate. As the best mean-square-error estimate of the true error given the estimate, it provides a useful measure of performance (Xu et al., 2006). In practice, $E[\varepsilon_n|\hat{\varepsilon}_n]$ is not known; however, when evaluating the performance of an error estimator for a specific feature–label distribution, it is known. Since $E[\varepsilon_n|\hat{\varepsilon}_n]$ is the best mean-square estimate of the true error, we would like to have $E[\varepsilon_n|\hat{\varepsilon}_n] \approx \hat{\varepsilon}_n$, since then the true error is expected to be close to the error estimate. Since the accuracy of the conditional expectation as a representation of the overall conditional distribution depends on the conditional variance, $\mathrm{Var}(\varepsilon_n|\hat{\varepsilon}_n)$ is also of interest.

2.2.4 Linear Regression

The correlation between ε_n and $\hat{\varepsilon}_n$ is directly related to the linear regression between the true and estimated errors, where a linear model for the conditional expectation is assumed:

$$E[\varepsilon_n|\hat{\varepsilon}_n] = a\hat{\varepsilon}_n + b. \tag{2.14}$$

Based on the sample data, the least-squares estimate of the regression coefficient a is given by

$$\hat{a} = \frac{\hat{\sigma}_{\varepsilon_n}}{\hat{\sigma}_{\hat{\varepsilon}_n}}\hat{\rho}(\varepsilon_n, \hat{\varepsilon}_n), \tag{2.15}$$

where $\hat{\sigma}_{\varepsilon_n}$ and $\hat{\sigma}_{\hat{\varepsilon}_n}$ are the sample standard deviations of the true and estimated errors, respectively, and $\hat{\rho}(\varepsilon_n, \hat{\varepsilon}_n)$ is the sample correlation coefficient between the true and estimated errors. The closer $\hat{a}$ is to 0, and thus the closer $\hat{\rho}(\varepsilon_n, \hat{\varepsilon}_n)$ is to 0, the weaker the regression. With small-sample data, it is actually not unusual to have $\hat{\sigma}_{\varepsilon_n}/\hat{\sigma}_{\hat{\varepsilon}_n} \geq 1$, in which case a small regression coefficient is solely due to lack of correlation.

2.2.5 Confidence Intervals

If one is concerned with having the true error below some tolerance given the estimate, then interest centers on finding a conditional bound for the true error given the estimate. This leads us to consider a one-sided confidence interval of the form $[0, \tau]$ conditioned on the error estimate. Specifically, given $0 < \alpha < 1$ and the estimate $\hat{\varepsilon}_n$, we would like to find an upper bound $\tau(\alpha, \hat{\varepsilon}_n)$ on ε_n, such that

$$P(\varepsilon_n < \tau(\alpha, \hat{\varepsilon}_n) \mid \hat{\varepsilon}_n) = 1 - \alpha. \tag{2.16}$$

Since

$$P(\varepsilon_n < \tau(\alpha, \hat{\varepsilon}_n)|\hat{\varepsilon}_n) = \int_0^{\tau(\alpha,\hat{\varepsilon}_n)} dF(\varepsilon_n|\hat{\varepsilon}_n), \tag{2.17}$$

where $F(\varepsilon_n|\hat{\varepsilon}_n)$ is the conditional distribution for ε_n given $\hat{\varepsilon}_n$, determining the upper bound depends on obtaining the required conditional distributions, which can be derived from the joint distribution of the true and estimated errors.

A confidence bound is related both to the conditional mean $E[\varepsilon_n|\hat{\varepsilon}_n]$ and conditional variance $\text{Var}[\varepsilon_n|\hat{\varepsilon}_n]$. If $\text{Var}[\varepsilon_n|\hat{\varepsilon}_n]$ is large, this increases $\tau(\alpha, \hat{\varepsilon}_n)$; if $\text{Var}[\varepsilon_n|\hat{\varepsilon}_n]$ is small, this decreases $\tau(\alpha, \hat{\varepsilon}_n)$.

Whereas the one-sided bound in (2.16) provides a conservative parameter (given the estimate we want to be confident that the true error is not too great), a two-sided interval involves accuracy parameters (we want to be confident that the true error lies in some interval). Such a confidence interval is given by

$$P(\hat{\varepsilon}_n - \tau(\alpha, \hat{\varepsilon}_n) < \varepsilon_n < \hat{\varepsilon}_n + \tau(\alpha, \hat{\varepsilon}_n) \mid \hat{\varepsilon}_n) = 1 - \alpha. \tag{2.18}$$

Equations (2.16) and (2.18) involve conditional probabilities. A different perspective is to consider an unconditioned interval. Since

$$P(\varepsilon_n \in (\hat{\varepsilon}_n - \tau, \hat{\varepsilon}_n + \tau)) = P(|\hat{\varepsilon}_n - \varepsilon_n| < \tau) = 1 - P(|\hat{\varepsilon}_n - \varepsilon_n| \geq \tau), \tag{2.19}$$

the tail probabilities of the deviation distribution yield a "confidence interval" for the true error in terms of the estimated error. Note, however, that here the sample data are not given and both the true and estimated errors are random. Hence, this is an unconditional (average) measure of confidence and is therefore less informative about the particular problem than the conditional confidence intervals specified in (2.16) and (2.18).

2.3 TEST-SET ERROR ESTIMATION

This error estimation procedure assumes the availability of a second random sample $S_m = \{(\mathbf{X}_i^t, Y_i^t); i = 1, \ldots, m\}$ from the joint feature–label distribution $F_{\mathbf{X}Y}$, independent from the original sample S_n, which is set aside and not used in classifier design. In this context, S_n is sometimes called the *training data*, whereas S_m is called the *test data*. The classification rule is applied on the training data and the error of the resulting classifier is estimated to be the proportion of errors it makes on the test data. Formally, this *test-set* or *holdout* error estimation rule is given by:

$$\Xi_{n,m}(\Psi_n, S_n, S_m) = \frac{1}{m} \sum_{i=1}^{m} |Y_i^t - \Psi_n(S_n)\left(\mathbf{X}_i^t\right)|. \tag{2.20}$$

This is a randomized error estimation rule, with the test data themselves as the internal random factors, $\xi = S_m$.

For any given classification rule Ψ_n, the *test-set error estimator* $\hat{\varepsilon}_{n,m} = \Xi_{n,m}(\Psi_n, S_n, S_m)$ has a few remarkable properties. First of all, the test-set error estimator is unbiased in the sense that

$$E[\hat{\varepsilon}_{n,m} \mid S_n] = \varepsilon_n, \tag{2.21}$$

for all m. From this and the Law of Large Numbers (c.f. Theorem A.2) it follows that, given the training data S_n, $\hat{\varepsilon}_{n,m} \to \varepsilon_n$ with probability 1 as $m \to \infty$, regardless of the feature–label distribution. In addition, it follows from (2.21) that $E[\hat{\varepsilon}_{n,m}] = E[\varepsilon_n]$. Hence, the test-set error estimator is also unbiased in the sense of Section 2.2.1:

$$\text{Bias}(\hat{\varepsilon}_{n,m}) = E[\hat{\varepsilon}_{n,m}] - E[\varepsilon_n] = 0. \tag{2.22}$$

We now show that the internal variance of this randomized estimator vanishes as the number of testing samples increases to infinity. Using (2.21) allows one to obtain

$$V_{\text{int}} = E\left[(\hat{\varepsilon}_{n,m} - E[\hat{\varepsilon}_{n,m} \mid S_n])^2 \mid S_n\right] = E\left[(\hat{\varepsilon}_{n,m} - \varepsilon_n)^2 \mid S_n\right]. \tag{2.23}$$

Moreover, given S_n, $m\hat{\varepsilon}_{n,m}$ is binomially distributed with parameters (m, ε_n):

$$P(m\hat{\varepsilon}_{n,m} = k \mid S_n) = \binom{m}{k} \varepsilon_n^k (1 - \varepsilon_n)^{m-k}, \tag{2.24}$$

for $k = 0, \dots, m$. From the formula for the variance of a binomial random variable, it follows that

$$\begin{aligned} V_{\text{int}} = E\left[(\hat{\varepsilon}_{n,m} - \varepsilon_n)^2 | S_n\right] &= \frac{1}{m^2} E\left[(m\hat{\varepsilon}_{n,m} - m\varepsilon_n)^2 | S_n\right] \\ &= \frac{1}{m^2}\text{Var}(m\hat{\varepsilon}_{n,m} \mid S_n) = \frac{1}{m^2}\, m\varepsilon_n(1 - \varepsilon_n) = \frac{\varepsilon_n(1 - \varepsilon_n)}{m}. \end{aligned} \tag{2.25}$$

The internal variance of this estimator can therefore be bounded as follows:

$$V_{\text{int}} \leq \frac{1}{4m}. \tag{2.26}$$

Hence, $V_{\text{int}} \to 0$ as $m \to \infty$. Furthermore, by using (2.3), we get that the full variance of the holdout estimator is simply

$$\text{Var}(\hat{\varepsilon}_{n,m}) = E[V_{\text{int}}] + \text{Var}[\varepsilon_n]. \tag{2.27}$$

Thus, provided that m is large, so that V_{int} is small — this is guaranteed by (2.26) for large enough m — the variance of the holdout estimator is approximately equal to the variance of the true error itself, which is typically small if n is large.

It follows from (2.23) and (2.26) that

$$\begin{aligned}\text{MSE}(\hat{\varepsilon}_{n,m}) &= E\left[(\hat{\varepsilon}_{n,m} - \varepsilon_n)^2\right] \\ &= E\left[E\left[(\hat{\varepsilon}_{n,m} - \varepsilon_n)^2 \mid S_n\right]\right] = E[V_{\text{int}}] \le \frac{1}{4m}.\end{aligned} \tag{2.28}$$

Thus, we have the distribution-free RMS bound

$$\text{RMS}(\hat{\varepsilon}_{n,m}) \le \frac{1}{2\sqrt{m}}. \tag{2.29}$$

The holdout error estimator is reducible, in the sense defined in Section 2.1, that is, $\hat{\varepsilon}^d_{n,m} = \hat{\varepsilon}^D_{n,m}$, if we reduce the test sample to the same feature space selected in the feature selection step.

Despite its many favorable properties, the holdout estimator has a considerable drawback. In practice, one has N total labeled sample points that must be split into training and testing sets S_n and S_m, with $N = n + m$. For large samples, $E[\varepsilon_n] \approx E[\varepsilon_N]$, while m is still large enough to obtain a good (small-variance) test-set estimator $\hat{\varepsilon}_{n,m}$. However, for small samples, this is not the case: $E[\varepsilon_n]$ may be too large with respect to $E[\varepsilon_N]$ (i.e., the loss of performance may be intolerable), m may be too small (leading to a poor test-set error estimator), or both. Simply put, setting aside a part of the sample for testing means that there are fewer data available for design, thereby typically resulting in a poorer performing classifier, and if fewer points are set aside to reduce this undesirable effect, there are insufficient data for testing. As seen in Figure 2.2, the negative impact on the expected classification error of reducing the

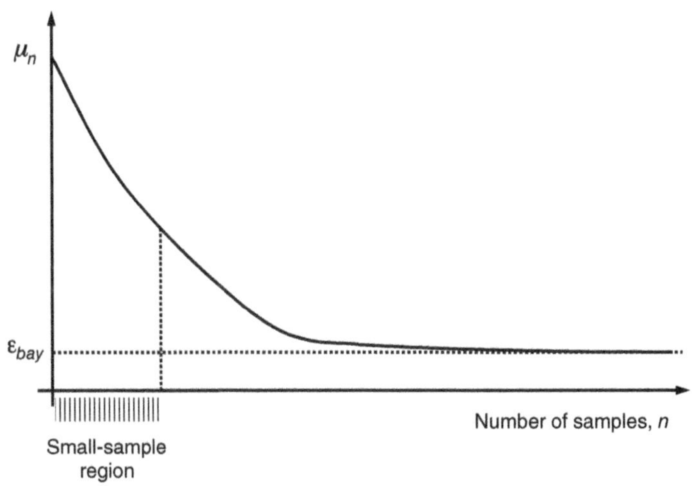

FIGURE 2.2 Expected error as a function of training sample size, demonstrating the problem of higher rate of classification performance deterioration with a reduced training sample size in the small-sample region.

training sample size by a substantial margin to obtain enough testing points is typically negligible when there is an abundance of data, but can be significant when sample size is small.

If the number m of testing samples is small, then the variance of $\hat{\varepsilon}_{n,m}$ is usually large and this is reflected in the RMS bound of (2.29). For example, to get the RMS bound down to 0.05, it would be necessary to use 100 test points, and data sets containing fewer than 100 overall sample points are common. Moreover, the bound of (2.26) is tight, it being achieved by letting $\varepsilon_n = 0.5$ in (2.25).

In conclusion, splitting the available data into training and test data is problematic in practice due to the unavailability of enough data to be able to make both n and m large enough. Therefore, for a large class of practical problems, where data are at a premium, test-set error estimation is effectively ruled out. In such cases, one must use the same data for training and testing.

2.4 RESUBSTITUTION

The simplest and fastest error estimation rule based solely on the training data is the resubstitution estimation rule, defined previously in (2.1). Given the classification rule Ψ_n, and the classifier $\psi_n = \Psi_n(S_n)$ designed on S_n, the resubstitution error estimator is given by

$$\hat{\varepsilon}_n^r = \frac{1}{n} \sum_{i=1}^{n} |Y_i - \psi_n(\mathbf{X}_i)|. \tag{2.30}$$

An alternative way to look at resubstitution is as the classification error according to the *empirical distribution*. The feature–label empirical distribution $F^*_{\mathbf{X}Y}$ for the pair $(\mathbf{X}, Y)$ is given by the probability mass function $P(\mathbf{X} = \mathbf{x}_i, Y = y_i) = \frac{1}{n}$, for $i = 1, \ldots, n$. It is easy to see that the resubstitution estimator is given by

$$\hat{\varepsilon}_n^r = E_{F^*}[|Y - \psi_n(\mathbf{X})|]. \tag{2.31}$$

The concern with resubstitution is that it is generally, but not always, optimistically biased; that is, typically, $\text{Bias}(\hat{\varepsilon}_n^r) < 0$. The optimistic bias of resubstitution may not be of great concern if it is small; however, it tends to become unacceptably large for overfitting classification rules, especially in small-sample situations. An extreme example is given by the 1-nearest neighbor classification rule, where $\hat{\varepsilon}_n^r \equiv 0$ for all sample sizes, classification rules, and feature–label distributions. Although resubstitution may be optimistically biased, the bias will often disappear as sample size increases, as we show below. In addition, resubstitution is typically a low-variance error estimator. This often translates to good asymptotic RMS properties. In the presence of feature selection, resubstitution is a reducible error estimation rule, as mentioned in Section 2.1.

Resubstitution is a universally strongly consistent error estimator for any classification rule with a finite VC dimension. In addition, the absolute bias of resubstitution is $O(\sqrt{V_C \log(n)/n})$, regardless of the feature–label distribution. We begin the proof of these facts by dropping the supremum in the Vapnik–Chervonenkis theorem (c.f. Theorem B.1) and letting $\psi = \psi_n$, in which case $\varepsilon[\psi] = \varepsilon_n$, the designed classifier error, and $\hat{\varepsilon}[\psi] = \hat{\varepsilon}_n^r$, the apparent error (i.e., the resubstitution error). This leads to:

$$P\left(|\hat{\varepsilon}_n^r - \varepsilon_n| > \tau\right) \leq 8\mathcal{S}(\mathcal{C}, n)e^{-n\tau^2/32}, \tag{2.32}$$

for all $\tau > 0$. Applying Lemma A.3 to the previous inequality produces the following bound on the bias of resubstitution:

$$|\text{Bias}\left(\hat{\varepsilon}_n^r\right)| = |E\left[\hat{\varepsilon}_n^r - \varepsilon_n\right]| \leq E\left[|\hat{\varepsilon}_n^r - \varepsilon_n|\right] \leq 8\sqrt{\frac{\log \mathcal{S}(\mathcal{C}, n) + 4}{2n}}. \tag{2.33}$$

Now, if $V_C < \infty$, then we can use inequality (B.10) to replace $\mathcal{S}(\mathcal{C}, n)$ by $(n+1)^{V_C}$ in the previous equations. From (2.32), we obtain

$$P\left(|\hat{\varepsilon}_n^r - \varepsilon_n| > \tau\right) \leq 8(n+1)^{V_C} e^{-n\tau^2/32}, \tag{2.34}$$

for all $\tau > 0$. Note that in this case $\sum_{n=1}^{\infty} P(|\hat{\varepsilon}_n^r - \varepsilon_n| \geq \tau) < \infty$, for all $\tau > 0$ and all distributions, so that, by the First Borel-Cantelli lemma (c.f. Lemma A.1), $\hat{\varepsilon}_n^r \to \varepsilon_n$ with probability 1 for all distributions, that is, resubstitution is a universally strongly consistent error estimator. In addition, from (2.33) we get

$$|\text{Bias}(\hat{\varepsilon}_n^r)| \leq 8\sqrt{\frac{V_C \log(n+1) + 4}{2n}}, \tag{2.35}$$

and the absolute value of the bias is $O(\sqrt{V_C \log(n)/n})$. Vanishingly small resubstitution bias is guaranteed if $n \gg V_C$, regardless of the feature–label distribution. For fixed V_C, the behavior of the bias is $O(\sqrt{\log(n)/n})$, which goes to 0 as $n \to \infty$, albeit slowly. This is a distribution-free result; better bounds can be obtained if distributional assumptions are made. For example, it is proved in McLachlan (1976) that if the distributions are Gaussian with the same covariance matrix, and the classification rule is LDA, then $|\text{Bias}(\hat{\varepsilon}_n^r)| = O(1/n)$.

Another use of the resubstitution estimator is in a model selection method known as *structural risk minimization* (SRM) (Vapnik, 1998). Here, the problem is to select a classifier in a given set of options in such a way that a balance between small apparent error and small complexity of the classification rule is achieved in order to avoid overfitting. We will assume throughout that the VC dimension is finite and begin by rewriting the bound in the VC Theorem, doing away with the supremum and the absolute value, and letting $\psi = \psi_n$, which allows us to write

$$P\left(\varepsilon_n - \hat{\varepsilon}_n^r > \tau\right) \leq 8(n+1)^{V_C} e^{n\tau^2/32}, \tag{2.36}$$

for all $\tau > 0$. Let the right-hand side of the previous equation be equal to ξ, where $0 \leq \xi \leq 1$. Solving for τ gives

$$\tau(\xi) = \sqrt{\frac{32}{n}\left[V_C \log(n+1) - \log\left(\frac{\xi}{8}\right)\right]}. \tag{2.37}$$

Thus, for a given ξ,

$$P\left(\varepsilon_n - \hat{\varepsilon}_n^r > \tau(\xi)\right) \leq \xi \;\Rightarrow\; P\left(\varepsilon_n - \hat{\varepsilon}_n^r \leq \tau(\xi)\right) \geq 1 - \xi; \tag{2.38}$$

that is, the inequality $\varepsilon_n \leq \hat{\varepsilon}_n^r + \tau(\xi)$ holds with probability at least $1 - \xi$. In other words,

$$\varepsilon_n \leq \hat{\varepsilon}_n^r + \sqrt{\frac{32}{n}\left[V_C \log(n+1) - \log\left(\frac{\xi}{8}\right)\right]} \tag{2.39}$$

holds with probability at least $1 - \xi$. The second term on the right-hand side is the so-called *VC confidence*. The actual form of the VC confidence may vary according to the derivation, but in any case: (1) it depends only on V_C and n, for a given ξ; (2) it is small if $n \gg V_C$ and large otherwise.

Now let C^k be a sequence of classes associated with classification rules Ψ_n^k, for $k = 1, 2, \ldots$. For example, k might be the number of neurons in the hidden layer of a neural network, or the number of levels in a classification tree. Denote the classification error and the resubstitution estimator under each classification rule by ε_{n,C^k} and $\hat{\varepsilon}^r_{n,C^k}$, respectively. We want to be able to pick a classification rule such that the classification error is minimal. From (2.39), this can be done by picking a ξ sufficiently close to 1 (the value $\xi = 0.95$ is typical), computing

$$\hat{\varepsilon}^r_{n,C^k} + \sqrt{\frac{32}{n}\left[V_{C^k} \log(n+1) - \log\left(\frac{\xi}{8}\right)\right]}$$

for each k, and then picking k such that this is minimal. In other words, we want to minimize $\hat{\varepsilon}^r_{n,C^k}$, while penalizing large V_C compared to n. This comprises the SRM method for classifier selection. Figure 2.3 illustrates the compromise between small resubstitution error and small complexity achieved by the optimal classifier selected by SRM.

2.5 CROSS-VALIDATION

Cross-validation, mentioned briefly in Section 2.1, improves upon the bias of resubstitution by using a resampling strategy in which classifiers are designed from

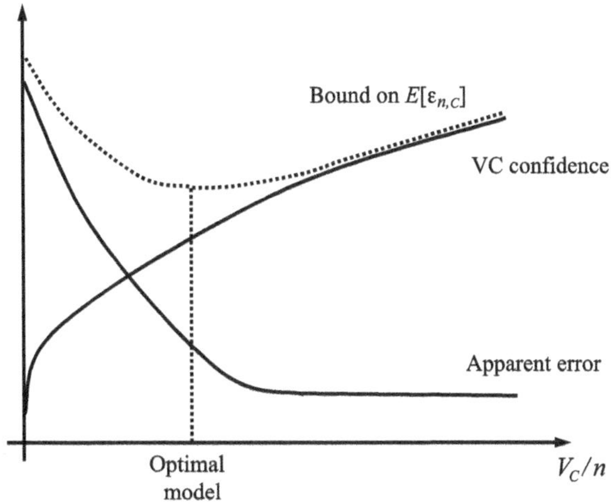

FIGURE 2.3 Model selection by structural risk minimization procedure, displaying the compromise between small apparent error and small complexity.

parts of the sample, each is tested on the remaining data, and the error estimate is given by the average of the errors (Cover, 1969; Lachenbruch and Mickey, 1968; Stone, 1974; Toussaint and Donaldson, 1970). In *k-fold cross-validation*, where it is assumed that k divides n, the sample S_n is randomly partitioned into k equal-size *folds* $S_{(i)} = \{(\mathbf{X}_j^{(i)}, Y_j^{(i)}); j = 1, \dots, n/k\}$, for $i = 1, \dots, k$. The procedure is as follows: each fold $S_{(i)}$ is left out of the design process to obtain a deleted sample $S_n - S_{(i)}$, a classification rule $\Psi_{n-n/k}$ is applied to $S_n - S_{(i)}$, and the error of the resulting classifier is estimated as the proportion of errors it makes on $S_{(i)}$. The average of these k error rates is the cross-validated error estimate. The *k-fold cross-validation error estimation rule* can thus be formulated as

$$\Xi_n^{\mathrm{cv}(k)}(\Psi_n, S_n, \xi) = \frac{1}{n}\sum_{i=1}^{k}\sum_{j=1}^{n/k}\left|Y_j^{(i)} - \Psi_{n-n/k}(S_n - S_{(i)})\left(\mathbf{X}_j^{(i)}\right)\right|. \tag{2.40}$$

The random factors ξ specify the random partition of S_n into the folds; therefore, k-fold cross-validation is, in general, a randomized error estimation rule. However, if $k = n$, then each fold contains a single point, and randomness is removed as there is only one possible partition of the sample. As a result, the error estimation rule is nonrandomized. This is called the *leave-one-out error estimation rule*.

Several variants of the basic k-fold cross-validation estimation rule are possible. In *stratified* cross-validation, the classes are represented in each fold in the same proportion as in the original data, in an attempt to reduce estimation variance. As with all randomized error estimation rules, there is concern about the internal variance of cross-validation; this can be reduced by repeating the process of random selection of the folds several times and averaging the results. This is called *repeated* k-fold cross-validation. In the limit, one will be using all possible folds of size n/k, and there will be no internal variance left, the estimation rule becoming nonrandomized. This is called *complete* k-fold cross-validation (Kohavi, 1995).

The most well-known property of the *k-fold cross-validation error estimator* $\hat{\varepsilon}_n^{\,\mathrm{cv}(k)} = \Xi_n^{\mathrm{cv}(k)}(\Psi_n, S_n, \xi)$, for any given classification rule Ψ_n, is the "near-unbiasedness" property

$$E\left[\hat{\varepsilon}_n^{\,\mathrm{cv}(k)}\right] = E[\varepsilon_{n-n/k}]. \tag{2.41}$$

This is a distribution-free property, which will be proved in Section 5.5. It is often and mistakenly taken to imply that the k-fold cross-validation error estimator is unbiased, but this is not true, since $E[\varepsilon_{n-n/k}] \neq E[\varepsilon_n]$, in general. In fact, $E[\varepsilon_{n-n/k}]$ is typically larger than $E[\varepsilon_n]$, since the former error rate is based on a smaller sample size, and so the k-fold cross-validation error estimator tends to be pessimistically biased. To reduce the bias, it is advisable to increase the number of folds (which in turn will tend to increase estimator variance). The maximum value $k = n$ corresponds to the *leave-one-out error estimator* $\hat{\varepsilon}_n^{\,l}$, which is typically the least biased cross-validation estimator, with

$$E\left[\hat{\varepsilon}_n^{\,l}\right] = E[\varepsilon_{n-1}]. \tag{2.42}$$

In the presence of feature selection, cross-validation is not a reducible error estimation rule, in the sense defined in Section 2.1. Consider leave-one-out: the estimator $\hat{\varepsilon}_n^{\,l,D}$ is computed by leaving out a point, applying the feature selection rule to the deleted sample, then applying Ψ_n^d in the reduced feature space, testing it on the deleted sample point (after projecting it into the reduced feature space), repeating the process n times, and computing the average error rate. This requires application of the feature selection rule anew at each iteration, for a total of n times. This process, called by some authors "external cross-validation," ensures that (2.42) is satisfied: $E[\hat{\varepsilon}_n^{\,l,D}] = E[\varepsilon_{n-1}^D]$. On the other hand, if feature selection is applied once at the beginning of the process and then leave-one-out proceeds in the reduced feature space while ignoring the original data, then one obtains an estimator $\hat{\varepsilon}_n^{\,l,d}$ that not only differs from $\hat{\varepsilon}_n^{\,l,D}$, but also does not satisfy any version of (2.42): neither $E[\hat{\varepsilon}_n^{\,l,d}] = E[\varepsilon_{n-1}^D]$ nor $E[\hat{\varepsilon}_n^{\,l,d}] = E[\varepsilon_{n-1}^d]$, in general. While the fact that the former identity does not hold is intuitively clear, the case of the latter identity is more subtle. This is also the identity that is mistakenly sometimes assumed to hold. The reason it does not hold is that the feature selection process "biases" the reduced sample S_n^d, making it have a different sampling distribution than data independently generated in the reduced feature space. In fact, $\hat{\varepsilon}_n^{\,l,d}$ can be substantially optimistically biased. This phenomenon is called "selection bias" in Ambroise and McLachlan (2002).

The major concern regarding the leave-one-out error estimator, and cross-validation in general, is not bias but estimation variance. Despite the fact that, as mentioned previously, leave-one-out cross-validation is a nonrandomized error estimation rule, the leave-one-out error estimator can still exhibit large variance. Small internal variance (in this case, zero) by itself is not sufficient to guarantee a low-variance estimator. Following (2.3), the variance of the cross-validation error estimator also depends on the term $\text{Var}(E[\hat{\varepsilon}_n^{\text{cv}(k)} | S_n])$, which is the variance corresponding to the randomness of the sample S_n. This term tends to be large when sample size is small, which makes the sample unstable. Even though our main concern is not with the variance itself, but with the RMS, which is a more complete measure of the accuracy of the error estimator, it is clear from (2.5) and (2.6) that, unless the variance is small, the RMS cannot be small.

The variance problem of cross-validation has been known for a long time. In a 1978 paper, Glick considered LDA classification for one-dimensional Gaussian class-conditional distributions possessing unit variance, with means μ_0 and μ_1, and a sample size of $n = 20$ with an equal number of sample points from each distribution (Glick, 1978). Figure 2.4 is based on Glick's paper; however, with the number of Monte-Carlo repetitions raised from 400 to 20,000 for increased accuracy (Dougherty and Bittner, 2011) — analytical methods that allow the exact computation of these curves will be presented in Chapter 6. The x-axis is labeled with $m = |\mu_1 - \mu_0|$, with the parentheses containing the corresponding Bayes error. The y-axis plots the standard deviation corresponding to the true error, resubstitution estimator, and leave-one-out estimator. With small Bayes error, the standard deviations of the leave-one-out error and the resubstitution error are close, but when the Bayes error is large, the leave-one-out error has a much greater standard deviation. Thus, Glick's warning was not to use leave-one-out with small samples. Further discussion on the large variance of cross-validation can be found in Braga-Neto and Dougherty (2004a), Devroye et al., (1996), Toussaint (1974).

In small sample cases, cross-validation not only tends to display large variance, but can also display large bias, and even negative correlation with the true error. An extreme example of this for the leave-one-out estimator is afforded by considering two equally-likely Gaussian populations $\Pi_0 \sim N_d(\boldsymbol{\mu}_0, \Sigma)$ and $\Pi_1 \sim N_d(\boldsymbol{\mu}_1, \Sigma)$, where the vector means $\boldsymbol{\mu}_0$ and $\boldsymbol{\mu}_1$ are far enough apart to make $\delta = \sqrt{(\boldsymbol{\mu}_1 - \boldsymbol{\mu}_0)^T \Sigma^{-1} (\boldsymbol{\mu}_1 - \boldsymbol{\mu}_0)} \gg 0$. It follows from (1.23) that the Bayes error is $\varepsilon_{\text{bay}} = \Phi(-\frac{\delta}{2}) \approx 0$. Let the classification rule Ψ_n be the 3-nearest neighbor rule (c.f. Section 1.6.1), with $n = 4$. For any sample S_n with $N_0 = N_1 = 2$, which occurs $P(N_0 = 2) = \binom{4}{2} 2^{-4} = 37.5\%$ of the time, it is easy to see that $\varepsilon_n \approx 0$ but $\hat{\varepsilon}_n^l = 1$! It is straightforward to calculate (see Exercise 2.2) that $E[\varepsilon_n] \approx 5/16 = 0.3125$, but $E[\hat{\varepsilon}_n^l] = 0.5$, so that $\text{Bias}(\hat{\varepsilon}_n^l) \approx 3/16 = 0.1875$, and the leave-one-out estimator is far from unbiased. In addition, one can show that $\text{Var}_{\text{dev}}(\hat{\varepsilon}_n^l) \approx 103/256 \approx 0.402$, which corresponds to a standard deviation of $\sqrt{0.402} = 0.634$. The leave-one-out estimator is therefore highly-biased and highly-variable in this case. But even more surprising is the fact that $\rho(\varepsilon_n, \hat{\varepsilon}_n^l) \approx -0.98$, that is, the leave-one-out estimator is almost perfectly negatively correlated with the true classification error. For comparison,

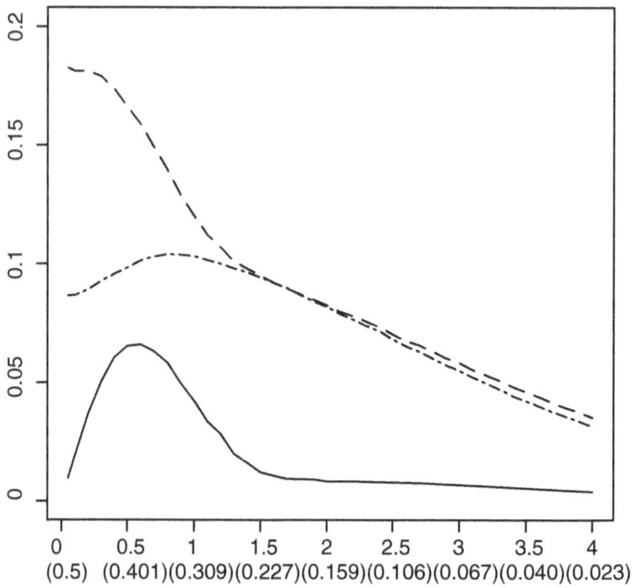

FIGURE 2.4 Standard deviation as a function of the distance between the means (and Bayes error) for LDA discrimination in a one-dimensional Gaussian model: true error (solid), resubstitution error (dots), and leave-one-out error (dashes). Reproduced from Dougherty and Bittner (2011) with permission from Wiley-IEEE Press.

the numbers for the resubstitution estimator in this case are $E[\hat{\varepsilon}_n^r] = 1/8 = 0.125$, so that $\text{Bias}(\hat{\varepsilon}_n^r) \approx -3/16 = -0.1875$, which is exactly the negative of the bias of leave-one-out; but $\text{Var}_{\text{dev}}(\hat{\varepsilon}_n^r) \approx 7/256 \approx 0.027$, for a standard deviation of $\sqrt{7}/16 \approx 0.165$, which is several times smaller than the leave-one-out variance. Finally, we obtain $\rho(\varepsilon_n, \hat{\varepsilon}_n^r) \approx \sqrt{3/5} \approx 0.775$, showing that the resubstitution estimator is highly positively correlated with the true error. The stark results in this example are a product of the large number of neighbors in the *k*NN classification rule compared to the small sample size: $k = n - 1$ with $n = 4$. This confirms the principle that small sample sizes require very simple classification rules.[1]

One element that can increase the variance of cross-validation is the fact that it attempts to estimate the error of a classifier $\psi_n = \Psi_n(S_n)$ via estimates of the error of *surrogate classifiers* $\psi_{n,i} = \Psi_{n-n/k}(S_n - S_{(i)})$, for $i = 1, \ldots, k$. If the classification rule is unstable (e.g. overfitting), then deleting different sets of points from the sample may produce wildly disparate surrogate classifiers. Instability of classification rules is illustrated in Figure 2.5, which shows different classifiers resulting from a single point left out from the original sample. We can conclude from these plots that LDA

[1]The peculiar behavior of leave-one-out error estimation for the *k*NN classification rule, especially for $k = n - 1$, is discussed in Devroye et al. (1996).

is the most stable rule among the three and CART (Classification and Regression Trees) is the most unstable, with 3NN displaying moderate stability.

Incidentally, the fact that k-fold cross-validation relies on an average of error rates of surrogate classifiers designed on samples of size $n - n/k$ implies that it is more properly an estimator of $E[\varepsilon_{n-n/k}]$ than of ε_n. This is a reasonable approach if both the cross-validation and true error variances are small, since the rationale here follows the chain $\hat{\varepsilon}_n^{\,\mathrm{cv}(k)} \approx E[\hat{\varepsilon}_n^{\,\mathrm{cv}(k)}] = E[\varepsilon_{n-n/k}] \approx E[\varepsilon_n] \approx \varepsilon_n$. However, the large variance typically displayed by cross-validation at small samples implies that the first approximation $\hat{\varepsilon}_n^{\,\mathrm{cv}(k)} \approx E[\hat{\varepsilon}_n^{\,\mathrm{cv}(k)}]$ in the chain is not valid, so that cross-validation cannot provide accurate error estimation in small-sample settings.

Returning to the impact of instability of the classification rule to sample deletion on the accuracy of cross-validation, this issue is beautifully quantified in a result given in Devroye et al. (1996), which proves very useful in generating distribution-free bounds on the RMS of the leave-one-out estimator for a given classification rule, provided that it be *symmetric*, that is, a rule for which the designed classifier is not affected by arbitrarily permuting the pairs of the sample. Most classification rules encountered in practice are symmetric, but in a few cases one must be careful. For instance, with k even, the kNN rule with ties broken according to the label of the first sample point (or any sample point of a fixed index) is not symmetric; this would apply to any classification rule that employs such a tie-breaking rule. However, the kNN rule is symmetric if ties are broken randomly. The next theorem is equivalent to the result in Devroye et al. (1996, Theorem. 24.2). See that reference for a proof.

Theorem 2.1 *(Devroye et al., 1996) For a symmetric classification rule* Ψ_n,

$$\mathrm{RMS}\left(\hat{\varepsilon}_n^{\,l}\right) \leq \sqrt{\frac{E[\varepsilon_n]}{n} + 6P(\Psi_n(S_n)(\mathbf{X}) \neq \Psi_{n-1}(S_{n-1})(\mathbf{X}))}. \tag{2.43}$$

The theorem relates the RMS of leave-one-out with the stability of the classification rule: the smaller the term $P(\Psi_n(S_n)(\mathbf{X}) \neq \Psi_{n-1}(S_{n-1})(\mathbf{X}))$ is, that is, the more stable the classification rule is (with respect to deletion of a point), the smaller the RMS of leave-one-out is guaranteed to be. For large n, the term $P(\Psi_n(S_n)(\mathbf{X}) \neq \Psi_{n-1}(S_{n-1})(\mathbf{X}))$ will be small for most classification rules; however, with small n, the situation changes and unstable classification rules become common. There is a direct relationship between the complexity of a classification rule and its stability under small-sample settings. For example, in connection with Figure 2.5, it is expected that the term $P(\Psi_n(S_n)(\mathbf{X}) \neq \Psi_{n-1}(S_{n-1})(\mathbf{X}))$ will be smallest for LDA, guaranteeing a smaller RMS for leave-one-out, and largest for CART. Even though Theorem 2.1 does not allow us to derive any firm conclusions as to the comparison among classification rules, since it gives an upper bound only, empirical evidence suggests that, in small-sample settings, the RMS of leave-one-out is better for LDA than for 3NN, and it is better for 3NN than for CART.

Simple bounds for $P(\Psi_n(S_n)(\mathbf{X}) \neq \Psi_{n-1}(S_{n-1})(\mathbf{X}))$ are known for a few classification rules. Together with the fact that $E[\varepsilon_n]/n < 1/n$, this leads to simple bounds

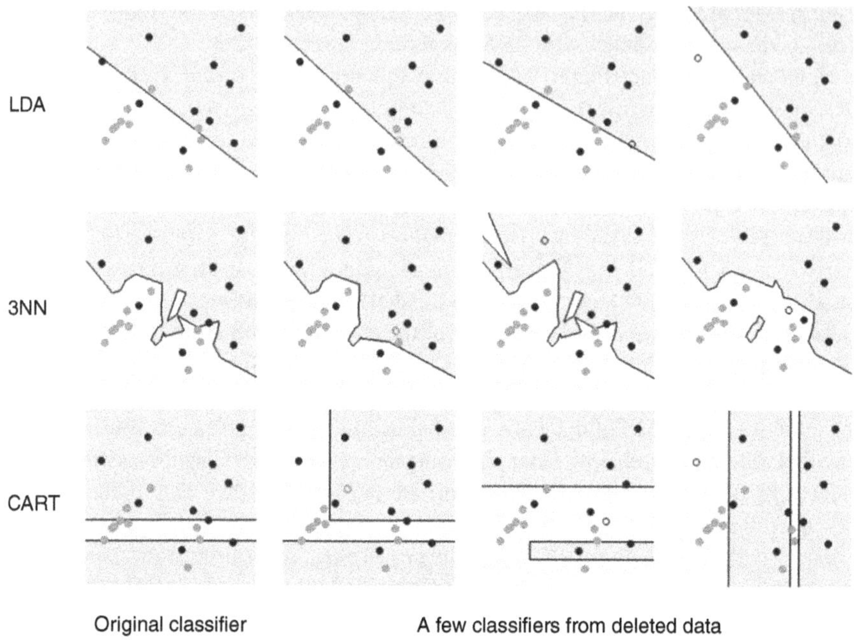

FIGURE 2.5 Illustration of instability of different classification rules. Original classifier and a few classifiers obtained after one of the sample points is deleted. Deleted sample points are represented by unfilled circles. Reproduced from Braga-Neto (2007) with permission from IEEE.

for the RMS of leave-one-out. For example, consider the kNN rule with randomized tie-breaking, in which case

$$\text{RMS}\left(\hat{\varepsilon}_n^l\right) \leq \sqrt{\frac{6k+1}{n}}. \tag{2.44}$$

This follows from the fact that the inequality $\Psi_n(S_n)(\mathbf{X}) \neq \Psi_{n-1}(S_{n-1})(\mathbf{X})$ is only possible if the deleted point is one of the k-nearest neighbors of $\mathbf{X}$. Owing to symmetry, the probability of this occurring is k/n, as all sample points are equally likely to be among the k-nearest neighbors of $\mathbf{X}$. For another example, the uniform kernel rule (c.f. Section 1.4.3) yields

$$\text{RMS}\left(\hat{\varepsilon}_n^l\right) \leq \sqrt{\frac{1}{2n} + \frac{24}{n^{1/2}}}. \tag{2.45}$$

A proof of this inequality is given in Devroye and Wagner (1979). Notice that in both examples, $\text{RMS}(\hat{\varepsilon}_n^l) \to 0$ as $n \to \infty$.

Via (2.11), such results lead to corresponding bounds on the tail probabilities $P(|\hat{\varepsilon}_n^l - \varepsilon_n| \geq \tau)$. For example, from (2.44) it follows that

$$P\left(|\hat{\varepsilon}_n^l - \varepsilon_n| \geq \tau\right) \leq \frac{6k+1}{n\tau^2}. \tag{2.46}$$

Clearly, $P(|\hat{\varepsilon}_n^l - \varepsilon_n| \geq \tau) \to 0$, as $n \to \infty$, that is, leave-one-out cross-validation is universally consistent (c.f. Section 2.2.2) for kNN classification. The same would be true for the uniform kernel rule, via (2.45).

As previously remarked in the case of resubstitution, these RMS bounds are typically not useful in small-sample cases. By being distribution-free, they are worst-case, and tend to be loose for small n. They should be viewed as asymptotic bounds, which characterize error estimator behavior for large sample sizes. Attempts to ignore this often lead to meaningless results. For example, with $k = 3$ and $n = 100$, the bound in (2.44) only insures that the RMS is less than 0.436, which is virtually useless. To be assured that the RMS is less than 0.1, one must have $n \geq 1900$.

2.6 BOOTSTRAP

The bootstrap methodology is a resampling strategy that can be viewed as a kind of smoothed cross-validation, with reduced variance (Efron, 1979). As stated in Efron and Tibshirani (1997), bootstrap smoothing is a way to reduce the variance of cross-validation by averaging. The bootstrap methodology employs the notion of the feature–label empirical distribution $F^*_{\mathbf{X}Y}$ introduced at the beginning of Section 2.4. A *bootstrap sample* S_n^* from $F^*_{\mathbf{X}Y}$ consists of n equally-likely draws with replacement from the original sample S_n. Some sample points will appear multiple times, whereas others will not appear at all. The probability that any given sample point will not appear in S_n^* is $(1 - 1/n)^n \approx e^{-1}$. It follows that a bootstrap sample of size n contains on average $(1 - e^{-1})n \approx 0.632n$ of the original sample points.

The basic bootstrap error estimation procedure is as follows: given a sample S_n, independent bootstrap samples $S_n^{*,j}$, for $j = 1, \ldots, B$, are generated; in Efron (1983), B has been recommended to be between 25 and 200. Then a classification rule Ψ_n is applied to each bootstrap sample $S_n^{*,j}$, and the number of errors the resulting classifier makes on $S_n - S_n^{*,j}$ is recorded. The average number of errors made over all B bootstrap samples is the bootstrap estimate of the classification error. This *zero bootstrap error estimation rule* (Efron, 1983) can thus be formulated as

$$\Xi_n^{\text{boot}}(\Psi_n, S_n, \xi) = \frac{\sum_{j=1}^{B} \sum_{i=1}^{n} |Y_i - \Psi_n\left(S_n^{*,j}\right)(\mathbf{X}_i)|\, I_{(\mathbf{X}_i, Y_i) \notin S_n^{*,j}}}{\sum_{j=1}^{B} \sum_{i=1}^{n} I_{(\mathbf{X}_i, Y_i) \notin S_n^{*,j}}}. \tag{2.47}$$

The random factors ξ govern the resampling of S_n to create the bootstrap sample $S_n^{*,j}$, for $j = 1, \ldots, B$; hence, the zero bootstrap is a randomized error estimation rule. Given a classification rule Ψ_n, the *zero bootstrap error estimator* $\hat{\varepsilon}_n^{\,\text{boot}} = \Xi_n^{\text{boot}}(\Psi_n, S_n, \xi)$ tends to be a pessimistically biased estimator of $E[\varepsilon_n]$, since the number of points available for classifier design is on average only $0.632n$.

As in the case of cross-validation, the bootstrap is not reducible. In the presence of feature selection, bootstrap resampling must be applied to the original data rather than the reduced data. Also as in the case of cross-validation, several variants of the basic bootstrap scheme exist. In the *balanced* bootstrap, each sample point is made to appear exactly B times in the computation (Chernick, 1999); this is supposed to reduce estimation variance, by reducing the internal variance associated with bootstrap sampling. This variance can also be reduced by increasing the value of B. In the limit as B increases, all bootstrap samples will be used up, and there will be no internal variance left, the estimation rule becoming nonrandomized. This is the *complete* zero bootstrap error estimator, which corresponds to the expectation of the $\hat{\varepsilon}_n^{\,\text{boot}}$ over the bootstrap sampling mechanism:

$$\hat{\varepsilon}_n^{\,\text{zero}} = E\left[\hat{\varepsilon}_n^{\,\text{boot}} \mid S_n\right]. \tag{2.48}$$

The ordinary zero bootstrap $\hat{\varepsilon}_n^{\,\text{boot}}$ is therefore a randomized Monte-Carlo approximation of the nonrandomized complete zero bootstrap $\hat{\varepsilon}_n^{\,\text{zero}}$. By the Law of Large Numbers, $\hat{\varepsilon}_n^{\,\text{boot}}$ converges to $\hat{\varepsilon}_n^{\,\text{zero}}$ with probability 1 as the number of bootstrap samples B grows without bound. Note that $E[\hat{\varepsilon}_n^{\,\text{boot}}] = E[\hat{\varepsilon}_n^{\,\text{zero}}]$.

A practical way of implementing the complete zero bootstrap estimator is to generate beforehand all distinct bootstrap samples (this is possible if n is not too large) and form a weighted average of the error rates based on each bootstrap sample, the weight for each bootstrap sample being its probability; for a formal definition of this process, see Section 5.4.4. As in the case of cross-validation, for all bootstrap variants there is a trade-off between computational cost and estimation variance.

The *.632 bootstrap error estimator* is a convex combination of the (typically optimistic) resubstitution and the (typically pessimistic) zero bootstrap estimators (Efron, 1983):

$$\hat{\varepsilon}_n^{\,632} = 0.368\,\hat{\varepsilon}_n^{\,r} + 0.632\,\hat{\varepsilon}_n^{\,\text{boot}}. \tag{2.49}$$

Alternatively, the complete zero bootstrap $\hat{\varepsilon}_n^{\,\text{zero}}$ can be substituted for $\hat{\varepsilon}_n^{\,\text{boot}}$. The weights in (2.49) are set heuristically. We will see in the next section, as well as in Sections 6.3.4 and 7.2.2.1, that they can be far from optimal for a given feature–label distribution and classification rule. A simple example where the weights are off is afforded by the 1NN rule, for which $\hat{\varepsilon}_n^{\,r} \equiv 0$.

In an effort to overcome these issues, an adaptive approach to finding the weights has been proposed (Efron and Tibshirani, 1997). The basic idea is to adjust the weight of the resubstitution estimator when resubstitution is exceptionally optimistically biased, which indicates strong overfitting. Given the sample S_n, let

$$\hat{\gamma} = \frac{1}{n^2} \sum_{i=1}^{n} \sum_{j=1}^{n} I_{Y_i \neq \Psi_n(S_n)(\mathbf{X}_j)}, \tag{2.50}$$

and define the relative overfitting rate by

$$\hat{R} = \frac{\hat{\varepsilon}_n^{\text{boot}} - \hat{\varepsilon}_n^r}{\hat{\gamma} - \hat{\varepsilon}_n^r}. \tag{2.51}$$

To be certain that $\hat{R} \in [0, 1]$, one sets $\hat{R} = 0$ if $\hat{R} < 0$ and $\hat{R} = 1$ if $\hat{R} > 1$. The parameter $\hat{R}$ indicates the degree of overfitting via the relative difference between the zero bootstrap and resubstitution estimates, a larger difference indicating more overfitting. Although we shall not go into detail, $\hat{\gamma} - \hat{\varepsilon}_n^r$ approximately represents the largest degree to which resubstitution can be optimistically biased, so that usually $\hat{\varepsilon}_n^{\text{boot}} - \hat{\varepsilon}_n^r \leq \hat{\gamma} - \hat{\varepsilon}_n^r$. The weight

$$\hat{w} = \frac{0.632}{1 - 0.368\hat{R}} \tag{2.52}$$

replaces 0.632 in (2.49) to produce the *.632+ bootstrap error estimator,*

$$\hat{\varepsilon}_n^{632+} = (1 - \hat{w})\,\hat{\varepsilon}_n^r + \hat{w}\,\hat{\varepsilon}_n^{\text{boot}}, \tag{2.53}$$

except when $\hat{R} = 1$, in which case $\hat{\gamma}$ replaces $\hat{\varepsilon}_n^{\text{boot}}$ in (2.53). If $\hat{R} = 0$, then there is no overfitting, $\hat{w} = 0.632$, and the .632+ bootstrap estimate is equal to the plain .632 bootstrap estimate. If $\hat{R} = 1$, then there is maximum overfit, $\hat{w} = 1$, the .632+ bootstrap estimate is equal to $\hat{\gamma}$, and the resubstitution estimate does not contribute, as in this case it is not to be trusted. As $\hat{R}$ changes between 0 and 1, the .632+ bootstrap estimate ranges between these two extremes. In particular, in all cases we have $\hat{\varepsilon}_n^{632+} \geq \hat{\varepsilon}_n^{632}$. The nonnegative difference $\hat{\varepsilon}_n^{632+} - \hat{\varepsilon}_n^{632}$ corresponds to the amount of "correction" introduced to compensate for too much resubstitution bias.

2.7 CONVEX ERROR ESTIMATION

Given any two error estimation rules, a new one (in fact, an infinite number of new ones) can be obtained via convex combination. Usually, this is done in an attempt to reduce bias. The .632 bootstrap error estimator discussed in the previous section is an example of convex error estimator.

Formally, given an optimistically biased estimator $\hat{\varepsilon}_n^{\text{opm}}$ and a pessimistically biased estimator $\hat{\varepsilon}_n^{\text{psm}}$, we define the *convex error estimator*

$$\hat{\varepsilon}_n^{\text{conv}} = w\,\hat{\varepsilon}_n^{\text{opm}} + (1 - w)\,\hat{\varepsilon}_n^{\text{psm}}, \tag{2.54}$$

where $0 \leq w \leq 1$, $\text{Bias}(\hat{\varepsilon}_n^{\text{opm}}) < 0$, and $\text{Bias}(\hat{\varepsilon}_n^{\text{psm}}) > 0$ (Sima and Dougherty, 2006a). The .632 bootstrap error estimator is an example of a convex error estimator, with $\hat{\varepsilon}_n^{\text{opm}} = \hat{\varepsilon}_n^{\text{boot}}$, $\hat{\varepsilon}_n^{\text{psm}} = \hat{\varepsilon}_n^r$, and $w = 0.632$. However, according to this definition, the .632+ bootstrap is not, because its weights are not constant.

The first convex estimator was apparently proposed in Toussaint and Sharpe (1974) (see also Raudys and Jain (1991)). It employed the typically optimistically biased resubstitution estimator and the pessimistically biased cross-validation estimator with equal weights:

$$\hat{\varepsilon}_n^{\mathrm{r,\,cv}(k)} = 0.5\,\hat{\varepsilon}_n^{\mathrm{r}} + 0.5\,\hat{\varepsilon}_n^{\mathrm{cv}(k)}. \tag{2.55}$$

This estimator can be bad owing to the large variance of cross-validation. As will be shown below in Theorem 2.2, for unbiased convex error estimation, one should look for component estimators with low variances. The .632 bootstrap estimator addresses this by replacing cross-validation with the lower-variance estimator $\hat{\varepsilon}_n^{\mathrm{boot}}$.

The weights for the aforementioned convex error estimators are chosen heuristically, without reference to the classification rule, or the feature–label distribution. Given the component estimators, a principled convex estimator is obtained by optimizing w to minimize $\mathrm{RMS}(\hat{\varepsilon}_n^{\mathrm{conv}})$ or, equivalently, to minimize $\mathrm{MSE}(\hat{\varepsilon}_n^{\mathrm{conv}})$. Rather than finding the weights to minimize the RMS of a convex estimator, irrespective of bias, one can impose the additional constraint that the estimator be unbiased. In general, the added constraint can have the effect of increased RMS; however, not only is unbiasedness intuitively desirable, the increase in RMS owing to unbiasedness is often negligible. Moreover, as is often the case with estimators, the unbiasedness assumption makes for a directly appreciable analysis. The next theorem shows that the key to good unbiased convex estimation is choosing component estimators possessing low variances.

Theorem 2.2 *(Sima and Dougherty, 2006a) For an unbiased convex error estimator,*

$$\mathrm{MSE}(\hat{\varepsilon}_n^{\mathrm{conv}}) \le \max\left\{\mathrm{Var}_{\mathrm{dev}}\left(\hat{\varepsilon}_n^{\mathrm{opm}}\right), \mathrm{Var}_{\mathrm{dev}}\left(\hat{\varepsilon}_n^{\mathrm{psm}}\right)\right\}. \tag{2.56}$$

Proof:

$$\begin{aligned}
\mathrm{MSE}\left(\hat{\varepsilon}_n^{\mathrm{conv}}\right) &= \mathrm{Var}_{\mathrm{dev}}\left(\hat{\varepsilon}_n^{\mathrm{conv}}\right) = w^2\mathrm{Var}_{\mathrm{dev}}\left(\hat{\varepsilon}_n^{\mathrm{opm}}\right) + (1-w)^2\mathrm{Var}_{\mathrm{dev}}\left(\hat{\varepsilon}_n^{\mathrm{psm}}\right)\\
&\quad + 2w(1-w)\,\rho\left(\hat{\varepsilon}_n^{\mathrm{opm}} - \varepsilon_n, \hat{\varepsilon}_n^{\mathrm{psm}} - \varepsilon_n\right)\sqrt{\mathrm{Var}_{\mathrm{dev}}\left(\hat{\varepsilon}_n^{\mathrm{opm}}\right)\mathrm{Var}_{\mathrm{dev}}\left(\hat{\varepsilon}_n^{\mathrm{psm}}\right)}\\
&\le w^2\mathrm{Var}_{\mathrm{dev}}\left(\hat{\varepsilon}_n^{\mathrm{opm}}\right) + (1-w)^2\mathrm{Var}_{\mathrm{dev}}\left(\hat{\varepsilon}_n^{\mathrm{psm}}\right)\\
&\quad + 2w(1-w)\sqrt{\mathrm{Var}_{\mathrm{dev}}\left(\hat{\varepsilon}_n^{\mathrm{opm}}\right)\mathrm{Var}_{\mathrm{dev}}\left(\hat{\varepsilon}_n^{\mathrm{psm}}\right)}\\
&= \left(w\sqrt{\mathrm{Var}_{\mathrm{dev}}\left(\hat{\varepsilon}_n^{\mathrm{opm}}\right)} + (1-w)\sqrt{\mathrm{Var}_{\mathrm{dev}}\left(\hat{\varepsilon}_n^{\mathrm{psm}}\right)}\right)^2\\
&\le \max\left\{\mathrm{Var}_{\mathrm{dev}}\left(\hat{\varepsilon}_n^{\mathrm{opm}}\right), \mathrm{Var}_{\mathrm{dev}}\left(\hat{\varepsilon}_n^{\mathrm{psm}}\right)\right\},
\end{aligned} \tag{2.57}$$

where the last inequality follows from the relation

$$(wx + (1 - w)y)^2 \leq \max\{x^2, y^2\}, \quad 0 \leq w \leq 1. \tag{2.58}$$

■

The following theorem provides a sufficient condition for an unbiased convex estimator to be more accurate than either component estimator.

Theorem 2.3 *(Sima and Dougherty, 2006a) If $\hat{\varepsilon}_n^{\text{conv}}$ is an unbiased convex estimator and*

$$|\text{Var}_{\text{dev}}\left(\hat{\varepsilon}_n^{\text{psm}}\right) - \text{Var}_{\text{dev}}\left(\hat{\varepsilon}_n^{\text{opm}}\right)| \leq \min\left\{\text{Bias}\left(\hat{\varepsilon}_n^{\text{opm}}\right)^2, \text{Bias}\left(\hat{\varepsilon}_n^{\text{psm}}\right)^2\right\}, \tag{2.59}$$

then

$$\text{MSE}\left(\hat{\varepsilon}_n^{\text{conv}}\right) \leq \min\left\{\text{MSE}\left(\hat{\varepsilon}_n^{\text{opm}}\right), \text{MSE}\left(\hat{\varepsilon}_n^{\text{psm}}\right)\right\}. \tag{2.60}$$

Proof: If $\text{Var}_{\text{dev}}(\hat{\varepsilon}_n^{\text{opm}}) \leq \text{Var}_{\text{dev}}(\hat{\varepsilon}_n^{\text{psm}})$, then we use Theorem 2.2 to get

$$\text{MSE}\left(\hat{\varepsilon}_n^{\text{conv}}\right) \leq \text{Var}_{\text{dev}}\left(\hat{\varepsilon}_n^{\text{psm}}\right) \leq \text{MSE}\left(\hat{\varepsilon}_n^{\text{psm}}\right). \tag{2.61}$$

In addition, if $\text{Var}_{\text{dev}}(\hat{\varepsilon}_n^{\text{psm}}) - \text{Var}_{\text{dev}}(\hat{\varepsilon}_n^{\text{opm}}) \leq \text{Bias}(\hat{\varepsilon}_n^{\text{opm}})^2$, we have

$$\begin{aligned}
\text{MSE}\left(\hat{\varepsilon}_n^{\text{conv}}\right) &\leq \text{Var}_{\text{dev}}\left(\hat{\varepsilon}_n^{\text{psm}}\right) \\
&= \text{MSE}\left(\hat{\varepsilon}_n^{\text{opm}}\right) - \text{Bias}\left(\hat{\varepsilon}_n^{\text{opm}}\right)^2 + \text{Var}_{\text{dev}}\left(\hat{\varepsilon}_n^{\text{psm}}\right) - \text{Var}_{\text{dev}}\left(\hat{\varepsilon}_n^{\text{opm}}\right) \\
&\leq \text{MSE}\left(\hat{\varepsilon}_n^{\text{opm}}\right).
\end{aligned} \tag{2.62}$$

The two previous equations result in (2.60). If $\text{Var}_{\text{dev}}(\hat{\varepsilon}_n^{\text{psm}}) \leq \text{Var}_{\text{dev}}(\hat{\varepsilon}_n^{\text{opm}})$, an analogous derivation shows that (2.60) holds if $\text{Var}_{\text{dev}}(\hat{\varepsilon}_n^{\text{opm}}) - \text{Var}_{\text{dev}}(\hat{\varepsilon}_n^{\text{psm}}) \leq \text{Bias}^2(\hat{\varepsilon}_n^{\text{psm}})$. Hence, (2.60) holds in general if (2.59) is satisfied. ■

While it may be possible to achieve a lower MSE by not requiring unbiasedness, these inequalities regarding unbiased convex estimators carry a good deal of insight. For instance, since the bootstrap variance is generally smaller than the cross-validation variance, (2.56) implies that convex combinations involving bootstrap error estimators are likely to be more accurate than those involving cross-validation error estimators when the weights are chosen in such a way that the resulting bias is very small.

In the general, biased case, the optimal convex estimator requires finding an optimal weight w_{opt} that minimizes $\text{MSE}(\hat{\varepsilon}_n^{\text{conv}})$ (note that the optimal weight is a function of the sample size n, but we omit denoting this dependency for simplicity of notation). From (2.8) and (2.54),

$$\mathrm{MSE}\left(\hat{\varepsilon}_n^{\,\mathrm{conv}}\right) = E\left[\left(w\hat{\varepsilon}_n^{\,\mathrm{opm}} + (1-w)\hat{\varepsilon}_n^{\,\mathrm{psm}} - \varepsilon_n\right)^2\right], \tag{2.63}$$

so that

$$w_{\mathrm{opt}} = \arg\min_{0\le w\le 1} \; E\left[\left(w\hat{\varepsilon}_n^{\,\mathrm{opm}} + (1-w)\hat{\varepsilon}_n^{\,\mathrm{psm}} - \varepsilon_n\right)^2\right]. \tag{2.64}$$

The optimal weight can be computed to arbitrary accuracy if the feature–label distribution is known, as follows. Consider M independent sample training sets of size n drawn from the feature–label distribution. Let $\hat{\varepsilon}_n^{\,\mathrm{opm}} = (\hat{\varepsilon}_n^{\,\mathrm{opm}1}, \ldots, \hat{\varepsilon}_n^{\,\mathrm{opm}M})$ and $\hat{\varepsilon}_n^{\,\mathrm{psm}} = (\hat{\varepsilon}_n^{\,\mathrm{psm}1}, \ldots, \hat{\varepsilon}_n^{\,\mathrm{psm}M})$ be the vectors of error estimates for the classifiers $\psi_n^1, \ldots, \psi_n^M$ designed on each of the M sample sets, and let $\varepsilon_n = (\varepsilon_n^1, \ldots, \varepsilon_n^M)$ be the vector of true errors of these classifiers (which can be computed with the knowledge of the feature–label distribution). Our problem is to find a weight w_{opt} that minimizes

$$\frac{1}{M}\left\|w\hat{\varepsilon}_n^{\,\mathrm{opm}} + (1-w)\hat{\varepsilon}_n^{\,\mathrm{psm}} - \varepsilon_n\right\|_2^2 = \frac{1}{M}\sum_{k=1}^{M}\left(w\hat{\varepsilon}_n^{\,\mathrm{opm}k} + (1-w)\hat{\varepsilon}_n^{\,\mathrm{psm}k} - \varepsilon_n^k\right)^2, \tag{2.65}$$

subject to the constraint $0 \le w \le 1$. This problem can be solved explicitly using the method of Lagrange multipliers, with or without the additional constraint of unbiasedness (see (Sima and Dougherty, 2006a) for details).

An important application of the preceding theory is the optimization of the weight for bootstrap convex error estimation. We have that

$$\hat{\varepsilon}_n^{\,\mathrm{bopt}} = w_{\mathrm{bopt}}\hat{\varepsilon}_n^{\,r} + (1 - w_{\mathrm{bopt}})\hat{\varepsilon}_n^{\,\mathrm{boot}}, \tag{2.66}$$

where $0 \le w_{\mathrm{bopt}} \le 1$ is the optimal weight to be determined. Given a feature–label distribution, classification rule, and sample size, w_{bopt} can be determined by minimizing the functional (2.65). Figure 2.6a shows how far the weights of the optimized bootstrap can deviate from those of the .632 bootstrap, for a Gaussian model with class covariance matrices I and $(1.5)^2 I$ and the 3NN classification rule. The .632 bootstrap is closest to optimal when the Bayes error is 0.1 but it can be far from optimal for smaller or larger Bayes errors. Using the same Gaussian model and the 3NN classification rule, Figure 2.6b displays the RMS errors for the .632, .632+, and optimized bootstrap estimators for increasing sample size when the Bayes error is 0.2. Notice that, as the sample size increases, the .632+ bootstrap outperforms the .632 bootstrap, but both are substantially inferior to the optimized bootstrap. For very small sample sizes, the .632 bootstrap outperforms the .632+ bootstrap.

In Sections 6.3.4 and 7.2.2.1, this problem will be solved analytically in the univariate and multivariate cases, respectively, for the LDA classification rule and Gaussian distributions — we will see once again that the best weight can deviate significantly from the 0.632 weight in (2.49).

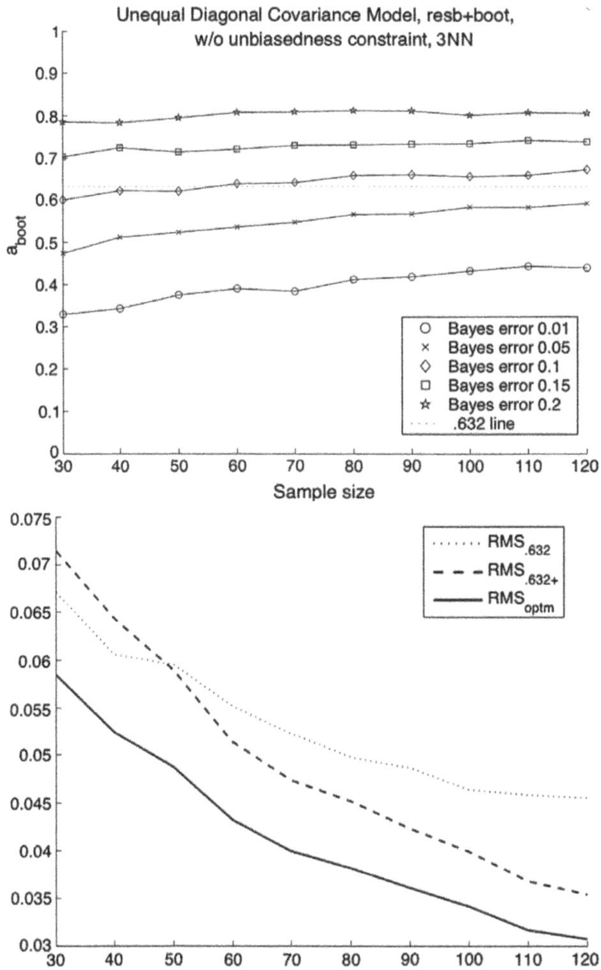

FIGURE 2.6 Bootstrap convex error estimation performance as a function of sample size for 3NN classification. (a) Optimal zero-bootstrap weight for different Bayes errors; the dotted horizontal line indicates the constant 0.632 weight. (b) RMS for different bootstrap estimators. Reproduced from Sima and Dougherty (2006a) with permission from Elsevier.

2.8 SMOOTHED ERROR ESTIMATION

In this and the next section we will discuss two broad classes of error estimators based on the idea of smoothing the error count.

We begin by noting that the resubstitution estimator can be rewritten as

$$\hat{\varepsilon}_n^r = \frac{1}{n}\sum_{i=1}^{n}\left(\psi_n(\mathbf{X}_i)I_{Y_i=0} + (1-\psi_n(\mathbf{X}_i))I_{Y_i=1}\right), \tag{2.67}$$

where $\psi_n = \Psi_n(S_n)$ is the designed classifier. The classifier ψ_n is a sharp 0-1 step function that can introduce variance by the fact that a point near the decision boundary can change its contribution from 0 to $\frac{1}{n}$ (and vice versa) via a slight change in the training data, even if the corresponding change in the decision boundary is small, and hence so is the change in the true error. In small-sample settings, $\frac{1}{n}$ is relatively large.

The idea behind smoothed error estimation (Glick, 1978) is to replace ψ_n in (2.67) by a suitably chosen "smooth" function taking values in the interval $[0, 1]$, thereby reducing the variance of the original estimator. In Glick (1978), Hirst (1996), Snapinn and Knoke (1985, 1989), this idea is applied to LDA classifiers (the same basic idea can be easily applied to any linear classification rule). Consider the LDA discriminant W_L in (1.48). While the sign of W_L gives the decision region to which a point belongs, its magnitude measures robustness of that decision: it can be shown that $|W_L(\mathbf{x})|$ is proportional to the Euclidean distance from a point $\mathbf{x}$ to the separating hyperplane (see Exercise 1.11). In addition, W_L is not a step function of the data, but varies in a linear fashion. To achieve smoothing of the error count, the idea is to use a monotone increasing function $r : R \to [0, 1]$ applied on W_L. The function r should be such that $r(-u) = 1 - r(u)$, $\lim_{u \to -\infty} r(u) = 0$, and $\lim_{u \to \infty} r(u) = 1$. The *smoothed resubstitution* estimator is given by

$$\hat{\epsilon}_n^{\mathrm{rs}} = \frac{1}{n} \sum_{i=1}^{n} \left((1 - r(W_L(\mathbf{X}_i))) I_{Y_i=0} + r(W_L(\mathbf{X}_i)) I_{Y_i=1} \right). \tag{2.68}$$

For instance, in Snapinn and Knoke (1985) the Gaussian smoothing function

$$r(u) = \Phi\left(\frac{u}{b\,\hat{\Delta}} \right) \tag{2.69}$$

is proposed, where Φ is the cumulative distribution function of a standard Gaussian variable,

$$\hat{\Delta} = \sqrt{(\bar{\mathbf{X}}_1 - \bar{\mathbf{X}}_0)^T \hat{\Sigma}^{-1} (\bar{\mathbf{X}}_1 - \bar{\mathbf{X}}_0)} \tag{2.70}$$

is the estimated Mahalanobis distance between the classes, and b is a free parameter that must be provided. Another example is the windowed linear function $r(u) = 0$ on $(-\infty, -b)$, $r(u) = 1$ on (b, ∞), and $r(u) = (u + b)/2b$ on $[-b, b]$. Generally, a choice of function r depends on tunable parameters, such as b in the previous examples. The choice of parameter is a major issue, which affects the bias and variance of the resulting estimator. A few approaches have been tried, namely, arbitrary choice (Glick, 1978; Tutz, 1985), arbitrary function of the separation between classes (Tutz, 1985), parametric estimation assuming normal populations (Snapinn and Knoke, 1985, 1989), and simulation-based methods (Hirst, 1996).

Extension of smoothing to nonlinear classification rules is not straightforward, since a suitable discriminant is not generally available. This problem has received little attention in the literature. In Devroye et al. (1996), and under a different but equivalent guise in Tutz (1985), it is suggested that one should use an estimate $\hat{\eta}_1(\mathbf{x})$ of the posterior probability $\eta_1(\mathbf{x}) = P(Y = 1|\mathbf{X} = \mathbf{x})$ as the discriminant (with the usual threshold of $1/2$). For example, for a kNN classifier, the estimate of the posterior probability is given by

$$\hat{\eta}_1(\mathbf{x}) = \frac{1}{k}\sum_{i=1}^{n} Y_{(i)}(\mathbf{x}), \tag{2.71}$$

where $Y_{(i)}(\mathbf{x})$ denotes the label of the ith closest observation to $\mathbf{x}$. A monotone increasing function $r : [0, 1] \to [0, 1]$, such that $r(u) = 1 - r(1 - u)$, $r(0) = 0$, and $r(1) = 1$ (this may simply be the identity function $r(u) = u$), can then be applied to $\hat{\eta}_1$, and the smoothed resubstitution estimator is given as before by (2.68), with W_L replaced by $\hat{\eta}_0 = 1 - \hat{\eta}_1$. In the special case $r(u) = u$, one has the *posterior-probability error estimator* of (Lugosi and Pawlak, 1994). However, this approach is problematic. It is not clear how to choose the estimator $\hat{\eta}_1$ and it will not, in general, possess a geometric interpretation. In Tutz (1985), it is suggested that one estimate the actual class-conditional probabilities from the data to arrive at $\hat{\eta}_1$, but this method has serious issues in small-sample settings. To illustrate the difficulty of the problem, consider the case of 1NN classification, for which smoothed resubstitution clearly is identically zero with the choice of $\hat{\eta}_1$ in (2.71). In the next section we discuss an alternative, kernel-based approach to smoothing the error count that is applicable in principle to any classification rule.

2.9 BOLSTERED ERROR ESTIMATION

The resubstitution estimator is written in (2.31) in terms of the empirical distribution $F^*_{\mathbf{X}Y}$, which is confined to the original data points, so that no distinction is made between points near or far from the decision boundary. If one spreads out the probability mass put on each point by the empirical distribution, variation is reduced in (2.31) because points near the decision boundary will have more mass go to the other side than will points far from the decision boundary. Another way of looking at this is that more confidence is attributed to points far from the decision boundary than points near it. For $i = 1, \ldots, n$, consider a d-variate probability density function $f_i^\diamond$, called a *bolstering kernel*. Given the sample S_n, the *bolstered empirical distribution* $F^\diamond_{\mathbf{X}Y}$ has probability density function

$$f^\diamond_{\mathbf{X}Y}(\mathbf{x}, y) = \frac{1}{n}\sum_{i=1}^{n} f_i^\diamond(\mathbf{x} - \mathbf{X}_i) I_{y=Y_i}. \tag{2.72}$$

Given a classification rule Ψ_n, and the designed classifier $\psi_n = \Psi_n(S_n)$, the *bolstered resubstitution* error estimator (Braga-Neto and Dougherty, 2004a) is obtained by replacing $F^*_{\mathbf{X}Y}$ by $F^\diamond_{\mathbf{X}Y}$ in (2.31):

$$\hat{\varepsilon}_n^{\,\mathrm{br}} = E_{F^\diamond_{\mathbf{X}Y}}\left[|Y - \psi_n(\mathbf{X})|\right]. \tag{2.73}$$

The following result gives a computational expression for the bolstered resubstitution error estimate.

Theorem 2.4 *(Braga-Neto and Dougherty, 2004a) Let $A_j = \{\mathbf{x} \in R^d \mid \psi_n(\mathbf{x}) = j\}$, for $j = 0, 1,$ be the decision regions for the designed classifier. Then the bolstered resubstitution error estimator can be written as*

$$\hat{\varepsilon}_n^{\,\mathrm{br}} = \frac{1}{n}\sum_{i=1}^{n}\left(\int_{A_1} f_i^\diamond(\mathbf{x} - \mathbf{X}_i)\, d\mathbf{x}\, I_{Y_i=0} + \int_{A_0} f_i^\diamond(\mathbf{x} - \mathbf{X}_i)\, d\mathbf{x}\, I_{Y_i=1}\right). \tag{2.74}$$

Proof: From (2.73),

$$\begin{aligned}
\hat{\varepsilon}_n^{\,\mathrm{br}} &= \int |y - \psi_n(\mathbf{x})|\, dF^\diamond(\mathbf{x}, y) \\
&= \sum_{y=0}^{1} \int_{R^d} |y - \psi_n(\mathbf{x})| f^\diamond(\mathbf{x}, y)\, d\mathbf{x} \\
&= \frac{1}{n}\sum_{y=0}^{1}\sum_{i=1}^{n} \int_{R^d} |y - \psi_n(\mathbf{x})| f_i^\diamond(\mathbf{x} - \mathbf{X}_i) I_{y=Y_i}\, d\mathbf{x} \\
&= \frac{1}{n}\sum_{i=1}^{n} \int_{R^d} \psi_n(\mathbf{x}) f_i^\diamond(\mathbf{x} - \mathbf{X}_i)\, d\mathbf{x}\, I_{Y_i=0} \\
&\quad + \int_{R^d} (1 - \psi_n(\mathbf{x})) f_i^\diamond(\mathbf{x} - \mathbf{X}_i)\, d\mathbf{x}\, I_{Y_i=1}.
\end{aligned} \tag{2.75}$$

But $\psi_n(\mathbf{x}) = 0$ over A_0 and $\psi_n(\mathbf{x}) = 1$ over A_1, from which (2.74) follows. ■

Equation (2.74) extends a similar expression proposed in Kim et al. (2002) in the context of LDA. The integrals in (2.74) are the error contributions made by the data points, according to the label $Y_i = 0$ or $Y_i = 1$. The bolstered resubstitution estimate is equal to the sum of all error contributions divided by the number of points (see Figure 2.7 for an illustration, where the bolstering kernels are given by uniform circular distributions).

When the classifier is linear (i.e., the decision boundary is a hyperplane), then it is usually possible to find analytic expressions for the integrals in (2.74) — see

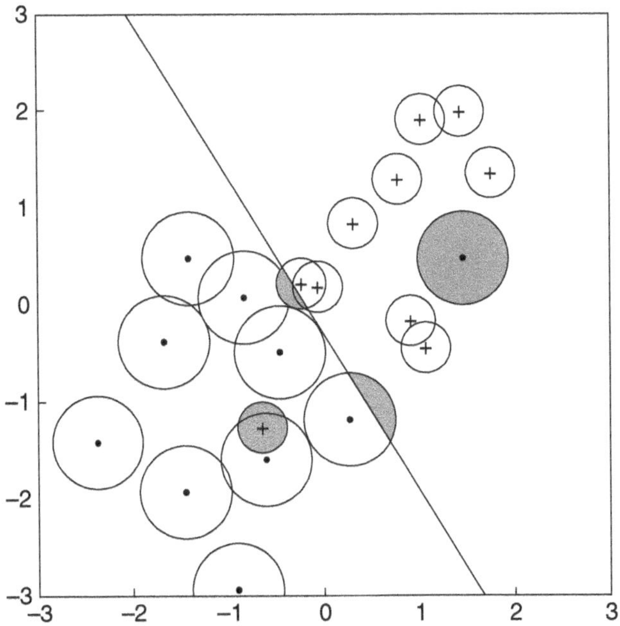

FIGURE 2.7 Bolstered resubstitution for LDA, assuming uniform circular bolstering kernels. The area of each shaded region divided by the area of the associated circle is the error contribution made by a point. The bolstered resubstitution error is the sum of all contributions divided by the number of points. Reproduced from Braga-Neto and Dougherty (2004) with permission from Elsevier.

the next subsection for an important case. Otherwise, one has to apply Monte-Carlo integration:

$$\hat{\varepsilon}_n^{\,\mathrm{br}} \approx \frac{1}{n} \sum_{i=1}^{n} \left(\sum_{j=1}^{M} I_{X_{ij} \in A_1} I_{Y_i = 0} + \sum_{j=1}^{M} I_{X_{ij} \in A_0} I_{Y_i = 1} \right), \tag{2.76}$$

where $\{X_{ij}; j = 1, \ldots, M\}$ are random points drawn from the density $f_i^{\diamond}$.

In the presence of feature selection, the integrals necessary to find the bolstered error estimate in (2.74) in the original feature space R^D can be equivalently carried out in the reduced space R^d, if the kernel densities are comprised of independent components, in such a way that

$$f_i^{\diamond,D}(\mathbf{x}) = f_i^{\diamond,d}(\mathbf{x}')f_i^{\diamond,D-d}(\mathbf{x}''), \quad \text{for } i = 1, \ldots, n, \tag{2.77}$$

where $\mathbf{x} = (\mathbf{x}', \mathbf{x}'') \in R^D$, $\mathbf{x}' \in R^d$, $\mathbf{x}'' \in R^{D-d}$, and $f_i^{\diamond,D}(\mathbf{x})$, $f_i^{\diamond,d}(\mathbf{x}')$, and $f_i^{\diamond,D-d}(\mathbf{x}'')$ denote the densities in the original, reduced, and difference feature spaces, respectively. One example would be Gaussian kernel densities with spherical or diagonal covariance matrices, described in fuller detail in Section 2.9.1. For each training point, write $\mathbf{X}_i = (\mathbf{X}_i', \mathbf{X}_i'') \in R^D$, where $\mathbf{X}_i' \in R^d$ and $\mathbf{X}_i'' \in R^{D-d}$ for $i = 1, \ldots, n$. For a given set of kernel densities satisfying (2.77), we have

$$\begin{aligned}
\hat{\varepsilon}_n^{\,\mathrm{br},D} &= \frac{1}{n}\sum_{i=1}^{n}\left(I_{Y_i=0}\int_{A_1} f_i^{\diamond,d}(\mathbf{x}'-\mathbf{X}_i')f_i^{\diamond,D-d}(\mathbf{x}''-\mathbf{X}_i'')\,d\mathbf{x}\right.\\
&\qquad\left.+\,I_{Y_i=1}\int_{A_0} f_i^{\diamond,d}(\mathbf{x}'-\mathbf{X}_i')f_i^{\diamond,D-d}(\mathbf{x}''-\mathbf{X}_i'')\,d\mathbf{x}\right)\\
&= \frac{1}{n}\sum_{i=1}^{n}\left(I_{Y_i=0}\int_{A_1^d} f_i^{\diamond,d}(\mathbf{x}'-\mathbf{X}_i')\,d\mathbf{x}'\int_{R^{D-d}} f_i^{\diamond,D-d}(\mathbf{x}''-\mathbf{X}_i'')\,d\mathbf{x}''\right.\\
&\qquad\left.+\,I_{Y_i=1}\int_{A_0^d} f_i^{\diamond,d}(\mathbf{x}'-\mathbf{X}_i')\,d\mathbf{x}'\int_{R^{D-d}} f_i^{\diamond,D-d}(\mathbf{x}''-\mathbf{X}_i'')\,d\mathbf{x}''\right)\\
&= \frac{1}{n}\sum_{i=1}^{n}\left(I_{Y_i=0}\int_{A_1^d} f_i^{\diamond,d}(\mathbf{x}'-\mathbf{X}_i')\,d\mathbf{x}' + I_{Y_i=1}\int_{A_0^d} f_i^{\diamond,d}(\mathbf{x}'-\mathbf{X}_i')\,d\mathbf{x}'\right),
\end{aligned} \tag{2.78}$$

where $A_j^d = \{\mathbf{x}' \in R^d \mid \psi_n^d(\mathbf{x}') = j\}$, for $j = 0, 1$, are the decision regions for the classifier designed in the reduced feature space R^d. While being an important computation-saving device, this identity does not imply that bolstered resubstitution is always reducible, in the sense defined in Section 2.1, when the kernel densities decompose as in (2.77). This will depend on the way that the kernel densities are adjusted to the sample data. In Section 2.9.2, we will see an example where bolstered resubstitution with independent-component kernel densities is reducible, and one where it is not.

When resubstitution is too optimistically biased due to overfitting classification rules, it is not a good idea to bolster incorrectly classified data points because that increases the optimistic bias of the error estimator. Bias is reduced if one assigns no bolstering kernels to incorrectly classified points. This decreases bias but increases variance because there is less bolstering. This variant is called *semi-bolstered resubstitution* (Braga-Neto and Dougherty, 2004a).

Bolstering can be applied to any error-counting error estimation method. For example, consider leave-one-out estimation. Let $A_j^{(i)} = \{\mathbf{x} \in R^d \mid \Psi_{n-1}(S_n - \{(\mathbf{X}_i, Y_i)\})(\mathbf{x}) = j\}$, for $j = 0, 1$, be the the decision regions for the classifier designed on the deleted sample $S_n - \{(\mathbf{X}_i, Y_i)\}$. The *bolstered leave-one-out* estimator can be computed via

$$\hat{\varepsilon}_n^{\,\mathrm{bloo}} = \frac{1}{n}\sum_{i=1}^{n}\left(\int_{A_1^{(i)}} f_i^{\diamond}(\mathbf{x}-\mathbf{X}_i)\,d\mathbf{x}\,I_{Y_i=0} + \int_{A_0^{(i)}} f_i^{\diamond}(\mathbf{x}-\mathbf{X}_i)\,d\mathbf{x}\,I_{Y_i=1}\right). \tag{2.79}$$

When the integrals cannot be computed exactly, a Monte-Carlo expression similar to (2.76) can be employed instead.

2.9.1 Gaussian-Bolstered Error Estimation

If the bolstering kernels are zero-mean multivariate Gaussian densities (c.f. Section A.11)

$$f_i^{\diamond}(\mathbf{x}) = \frac{1}{\sqrt{(2\pi)^d \det(C_i)}} \exp\left(-\frac{1}{2}\mathbf{x}^T C_i^{-1}\mathbf{x}\right), \tag{2.80}$$

where the kernel covariance matrix C_i can in principle be distinct at each training point $\mathbf{X}_i$, and the classification rule is produced by a linear discriminant as in Section 1.4.2 (e.g. this includes LDA, linear SVMs, perceptrons), then simple analytic solutions of the integrals in (2.74) and (2.79) become possible.

Theorem 2.5 *Let Ψ_n be a linear classification rule, defined by*

$$\Psi_n(S_n)(\mathbf{x}) = \begin{cases} 1, & \mathbf{a}_n^T x + b_n \leq 0, \\ 0, & \textit{otherwise}, \end{cases} \tag{2.81}$$

where $\mathbf{a}_n \in R^d$ and $b_n \in R$ are sample-based coefficients. Then the Gaussian-bolstered resubstitution error estimator can be written as

$$\hat{\varepsilon}_n^{\,\mathrm{br}} = \frac{1}{n}\sum_{i=1}^{n}\left(\Phi\left(-\frac{\mathbf{a}_n^T\mathbf{X}_i + b_n}{\sqrt{\mathbf{a}_n^T C_i \mathbf{a}_n}}\right) I_{Y_i=0} + \Phi\left(\frac{\mathbf{a}_n^T\mathbf{X}_i + b_n}{\sqrt{\mathbf{a}_n^T C_i \mathbf{a}_n}}\right) I_{Y_i=1}\right), \tag{2.82}$$

where $\Phi(x)$ is the cumulative distribution function of a standard $N(0,1)$ Gaussian random variable.

Proof: By comparing (2.74) and (2.82), it suffices to show that

$$\int_{A_j} f_i^{\diamond}(\mathbf{x} - \mathbf{X}_i)\, d\mathbf{x} = \Phi\left((-1)^j \frac{\mathbf{a}_n^T\mathbf{X}_i + b_n}{\sqrt{\mathbf{a}_n^T C_i \mathbf{a}_n}}\right), \quad j = 0, 1. \tag{2.83}$$

We show the case $j = 1$, the other case being entirely similar. Let $f_i^{\diamond}(\mathbf{a} - \mathbf{a}_i)$ be the density of a random vector $\mathbf{Z}$. We have that

$$\int_{A_1} f_i^{\diamond}(\mathbf{x} - \mathbf{X}_i)\, d\mathbf{x} = P(\mathbf{Z} \in A_1) = P\left(\mathbf{a}_n^T\mathbf{Z} + b_n \leq 0\right). \tag{2.84}$$

Now, by hypothesis, $\mathbf{Z} \sim N_d(\mathbf{X}_i, C_i)$. From the properties of the multivariate Gaussian (c.f. Section A.11), it follows that $\mathbf{a}_n^T\mathbf{Z} + b_n \sim N(\mathbf{a}_n^T\mathbf{X}_i + b_n, \mathbf{a}_n^T C_i \mathbf{a}_n)$. The result follows from the fact that $P(U \le 0) = \Phi(-E[U]/\sqrt{\text{Var}(U)})$ for a Gaussian random variable U. ■

Note that for any training point $\mathbf{X}_i$ exactly on the decision hyperplane, we have $\mathbf{a}_n^T\mathbf{X}_i + b_n = 0$, and thus its contribution to the bolstered resubstitution error estimate is exactly $\Phi(0) = 1/2$, regardless of the label Y_i.

2.9.2 Choosing the Amount of Bolstering

Selecting the correct amount of bolstering, that is, the "size" of the bolstering kernels, is critical for estimator performance. We now describe simple non-parametric procedures for adjusting the kernel size to the sample data, which are particularly suited to small-sample settings.

The kernel covariance matrices C_i for each training point $\mathbf{X}_i$, for $i = 1, \dots, n$, will be estimated from the sample data. Under small sample sizes, restrictions have to be imposed on the kernel covariances. First, a natural assumption is to make all kernel densities, and thus covariance matrices, equal for training points with the same class label: $C_i = D_0$ if $Y_i = 0$ or $C_i = D_1$ if $Y_i = 1$. This reduces the number of parameters to be estimated to $2d(d+1)$. Furthermore, it could be assumed that the covariance matrices D_0 and D_1 are diagonal, with diagonal elements, or *kernel variances*, $\sigma_{01}^2, \dots, \sigma_{0d}^2$ and $\sigma_{11}^2, \dots, \sigma_{1d}^2$. This further reduces the number of parameters to $2d$. Finally, it could be assumed that $D_0 = \sigma_0^2 I_d$ and $D_1 = \sigma_1^2 I_d$, which correspond to spherical kernels with variances σ_0^2 and σ_1^2, respectively, in which case there are only two parameters to be estimated.

In the diagonal and spherical cases, there is a simple interpretation of the relationship between the bolstered and plain resubstitution estimators. As the kernel variances all shrink to zero, it is easy to see that the bolstered resubstitution error estimator reduces to the plain resubstitution estimator. On the other hand, as the kernel variances increase, the bolstered estimator becomes more and more distinct from plain resubstitution. As resubstitution is usually optimistically biased, increasing the kernel variances will initially reduce the bias, but increasing them too much will eventually make the estimator pessimistic. Hence, there are typically optimal values for the kernel variances that make the estimator unbiased. In addition, increasing the kernel variances generally decreases variance of the error estimator. Therefore, by properly selecting the kernel variances, one can simultaneously reduce estimator bias and variance.

If the kernel densities are multivariate Gaussian, then (2.77) and (2.78) apply in both the spherical and diagonal cases, which presents a computational advantage in the presence of feature selection.

Let us consider first a method to fit spherical kernels, in which case we need to determine the kernel variances σ_0 and σ_1 for the kernel covariance matrices $D_0 = \sigma_0^2 I_d$ and $D_1 = \sigma_1^2 I_d$. One reason why plain resubstitution is optimistically biased is that the test points in (2.30) are all at distance zero from the training points (i.e., the test points are the training points). Bolstered estimators compensate for that by spreading the test point probability mass over the training points. Following a distance argument used in Efron (1983), we propose to find the amount of spreading that makes the test points to be as close as possible to the true mean distance to the training data points. The true mean distances d_0 and d_1 among points from populations Π_0 and Π_1, respectively, are estimated here by the sample-based mean minimum distance among points from each population:

$$\hat{d}_j = \frac{1}{n_j} \sum_{i=1}^{n} \left(\min_{\substack{i'=1,\ldots,n \\ i' \neq i, Y_{i'}=j}} \{||\mathbf{X}_i - \mathbf{X}_{i'}||\} \right) I_{Y_i=j}, \quad j = 0, 1. \tag{2.85}$$

The basic idea is to set the kernel variance σ_j^2 in such a way that the median distance of a point randomly drawn from a kernel to the origin matches the estimated mean distance $\hat{d}_j$, for $j = 0, 1$ (Braga-Neto and Dougherty, 2004a). This implies that half of the probability mass (i.e., half of the test points) of the bolstering kernel will be farther from the center than the estimated mean distance and the other half will be nearer. Assume for the moment a unit-variance kernel covariance matrix $D_0 = I_d$ (to fix ideas, we consider class label 0). Let R_0 be the random variable corresponding to the distance of a point randomly selected from the kernel density to the origin with cumulative distribution function $F_{R_0}(x)$. The median distance of a point randomly drawn from a kernel to the origin is given by $\alpha_{d,0} = F_{R_0}^{-1}(1/2)$, where the subscript d indicates explicitly that $\alpha_{d,0}$ depends on the dimensionality. For a kernel covariance matrix $D_0 = \sigma_0^2 I_d$, all distances get multiplied by σ_0. Hence, σ_0 is the solution of the equation $\sigma_0 F_{R_0}^{-1}(1/2) = \sigma \alpha_{d,0} = \hat{d}_0$. The argument for class label 1 is obtained by replacing 0 by 1 throughout; hence, the kernel standard deviations are set to

$$\sigma_j = \frac{\hat{d}_j}{\alpha_{d,j}}, \quad j = 0, 1. \tag{2.86}$$

As sample size increases, it is easy to see that $\hat{d}_0$ and $\hat{d}_1$ decrease, so that the kernel variances σ_0^2 and σ_1^2 also decrease, and there is less bias correction introduced by bolstered resubstitution. This is in accordance with the fact that resubstitution tends to be less optimistically biased as the sample size increases. In the limit, as sample size grows to infinity, the kernel variances tend to zero, and the bolstered resubstitution converges to the plain resubstitution. A similar approach to estimating the kernel variances can be employed for the bolstered leave-one-out estimator (see Braga-Neto and Dougherty (2004a) for the details).

For the specific case where all bolstering kernels are multivariate spherical Gaussian densities, the distance variables R_0 and R_1 are both distributed as a *chi* random variable with d degrees of freedom, with density given by (Evans et al., 2000)

$$f_{R_0}(r) = f_{R_1}(r) = \frac{2^{1-d/2} r^{d-1} e^{-r^2/2}}{\Gamma(\frac{d}{2})}, \tag{2.87}$$

where Γ denotes the gamma function. For $d = 2$, this becomes the well-known Rayleigh density. The cumulative distribution function F_{R_0} or F_{R_1} can be computed by numerical integration of (2.87) and the inverse at point $1/2$ can be found by a simple binary search procedure (using the fact that cumulative distribution functions are monotonically increasing), yielding the dimensional constant α_d. There is no subscript "j" in this case because the constant is the same for both class labels. The kernel standard deviations are then computed as in (2.86), with α_d in place of $\alpha_{d,j}$. The constant α_d depends only on the dimensionality, and in fact can be interpreted as a type of "dimensionality correction," which adjusts the value of the estimated mean distance to account for the feature space dimensionality. In the spherical Gaussian case, the values of the dimensional constant up to five dimensions are $\alpha_1 = 0.674$, $\alpha_2 = 1.177$, $\alpha_3 = 1.538$, $\alpha_4 = 1.832$, $\alpha_5 = 2.086$.

In the presence of feature selection, the use of spherical kernel densities with the previous method of estimating the kernel variances results in a bolstered resubstitution estimation rule that is not reducible, even if the kernel densities are Gaussian. This is clear, since both the mean distance estimate and dimensional constant change between the original and reduced feature spaces, rendering $\hat{\varepsilon}_n^{\text{br},d} \neq \hat{\varepsilon}_n^{\text{br},D}$, in general.

To fit diagonal kernels, we need to determine the kernel variances $\sigma_{01}^2, \dots, \sigma_{0d}^2$ and $\sigma_{11}^2, \dots, \sigma_{1d}^2$. A simple approach proposed in Jiang and Braga-Neto (2014) is to estimate the kernel variances σ_{0k}^2 and σ_{1k}^2 separately for each k, using the univariate data $S_{n,k} = \{(X_{1k}, Y_1), \dots, (X_{nk}, Y_n)\}$, for $k = 1, \dots, d$, where X_{ik} is the kth feature (component) in vector $\mathbf{X}_i$. This can be seen as an application of the "Naive Bayes" principle (Dudoit et al., 2002). We define the mean minimum distance along feature k as

$$\hat{d}_{jk} = \frac{1}{n_j} \sum_{i=1}^{n} \left(\min_{\substack{i'=1,\dots,n \\ i' \neq i, Y_{i'}=j}} \{ ||X_{ik} - X_{i'k}|| \} \right) I_{Y_i=j}, \quad j = 0, 1. \tag{2.88}$$

The kernel standard deviations are set to

$$\sigma_{jk} = \frac{\hat{d}_{jk}}{\alpha_{1,j}}, \quad j = 0, 1, \ k = 1, \dots, d. \tag{2.89}$$

In the Gaussian case, one has $\sigma_{jk} = \hat{d}_{jk}/\alpha_1 = \hat{d}_{jk}/0.674$, for $j = 0, 1$, $k = 1, \dots, d$.

In the presence of feature selection, the previous "Naive Bayes" method of fitting kernel densities yields a reducible bolstered resubstitution error estimation rule, if the kernel densities are Gaussian. This is clear because (2.77) and (2.78) hold for diagonal Gaussian kernels, and the "Naive Bayes" method produces the same kernel variances in both the original and reduced feature spaces, so that $\hat{\varepsilon}_n^{\text{br},d} = \hat{\varepsilon}_n^{\text{br},D}$.

2.9.3 Calibrating the Amount of Bolstering

The method of choosing the amount of bolstering for spherical Gaussian kernels has proved to work very well with a small number of features d (Braga-Neto and Dougherty, 2004a; Sima et al., 2005a,b) (see also the performance results in Chapter 3). However, experimental evidence exists that this may not be the case in high-dimensional spaces (Sima et al., 2011; Vu and Braga-Neto, 2008) — there is, however, some experimental evidence that diagonal Gaussian kernels with the "Naive Bayes" method described in the previous section may work well in high-dimensional spaces (Jiang and Braga-Neto, 2014). In Sima et al. (2011), a method is provided for calibrating the kernel variances for spherical Gaussian kernels in such a way that the estimator may have minimum RMS regardless of the dimensionality. We assume that the classification rule in the high-dimensional space R^D is constrained by feature selection to a reduced space R^d. The bolstered resubstitution error estimator $\hat{\varepsilon}_n^{\mathrm{br},D}$ is computed as in (2.78), using the spherical kernel variance fitting method described in the previous section — note that all quantities are computed in the high-dimensional space R^D. The only change is the introduction of a multiplicative correction factor k_D in (2.86):

$$\sigma_j = k_D \times \frac{\hat{d}_j^D}{\alpha_D}, \quad j = 0, 1. \tag{2.90}$$

The correction factor k_D is determined by the dimensionality D. The choice $k_D = 1$ naturally yields the previously proposed kernel variances.

Figure 2.8 shows the results of a simulation experiment, with synthetic data generated from four different Gaussian models with different class-conditional covariance matrices Σ_0 and Σ_1, ranging from spherical and equal (M1), spherical and unequal (M2), blocked (a generalization of a diagonal matrix, where there are blocks of nonzero submatrices along the diagonal, specifying subsets of correlated variables) and equal (M3), and blocked and unequal (M4). For performance comparison, we also compute the bolstered resubstitution error estimator $\hat{\varepsilon}_n^{\mathrm{br},d}$ with kernel variances estimated using only the low-dimensional data S_n^d and no correction ($k_D = 1$), as well as an "external" 10-fold cross-validation error estimator $\hat{\varepsilon}_n^{\mathrm{cv}(10)}$ on the high-dimensional feature space. For feature selection, we use sequential forward floating search. We plot the RMS versus kernel scaling factor k_D in Figure 2.8, for LDA, 3NN, and CART for $n = 50$, selecting $d = 3$ out of $D = 200$ features for 3NN and CART, $d = 3$ out of $D = 500$ features for LDA, and $E[\varepsilon_n] = 0.10$. The horizontal lines show the RMS for $\hat{\varepsilon}_n^{\mathrm{br},d}$ and $\hat{\varepsilon}_n^{\mathrm{cv}(10)}$. The salient point is that the RMS curve for optimal bolstering is well below that for cross-validation across a large interval of k_D values, indicating robustness with respect to choosing $k_D^{\min}$, which in this case is 0.8 for all models. Putting this together with the observation that the curves are close for the four models and for the different classification rules indicates that $k_D^{\min}$ for one model and classification rule can be used for others while remaining close to optimal. This property is key to developing a method to use optimal bolstering in real-data

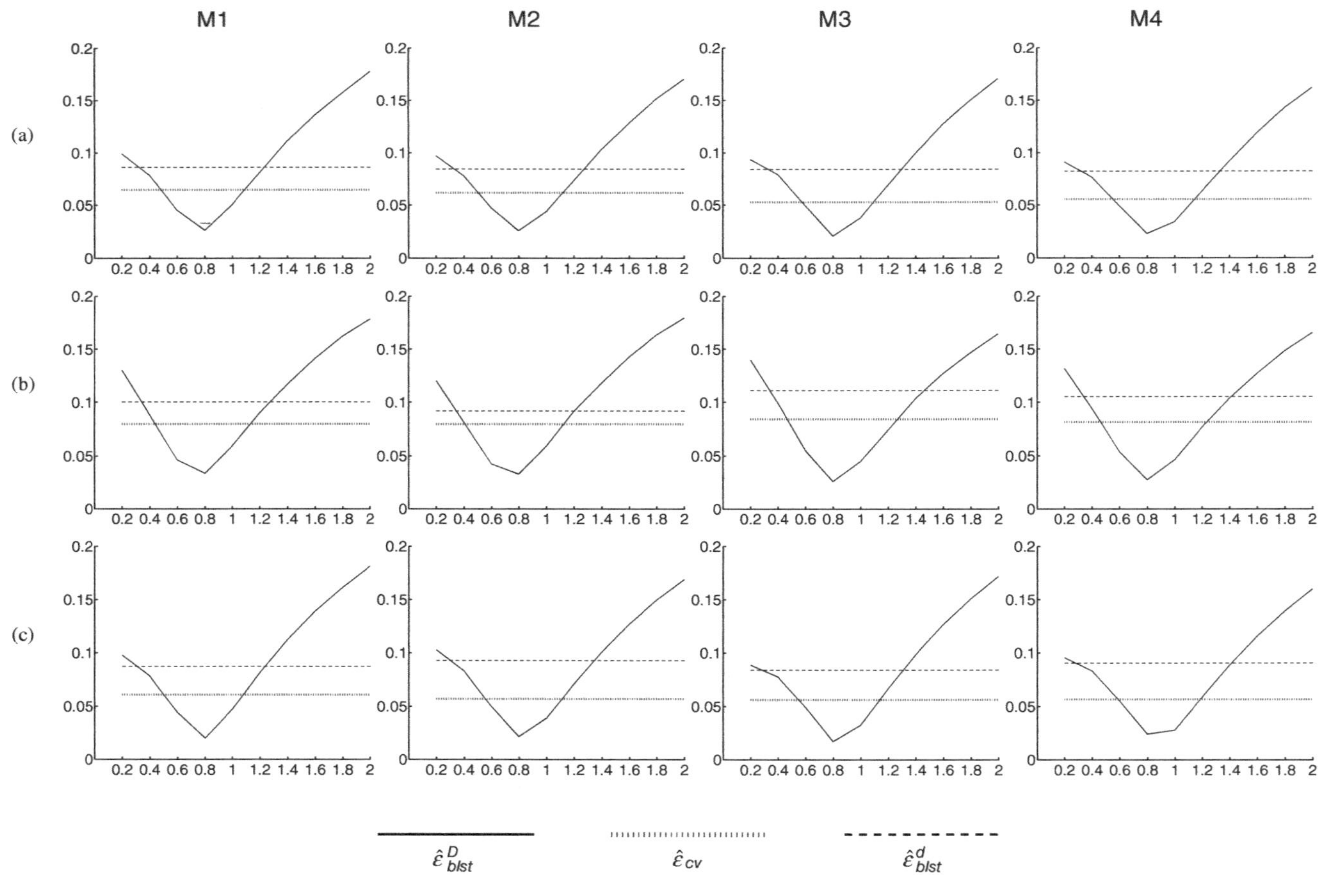

FIGURE 2.8 RMS versus scaling factor k_D for sample size $n = 50$ and selected feature size $d = 3$, for (a) 3NN with total feature size $D = 200$, (b) CART with $D = 200$, and (c) LDA with $D = 500$. Reproduced from Sima et al. (2011) with permission from Oxford University Press.

applications, for which we refer to Sima et al. (2011), where it is also demonstrated that $k_D^{\min}$ is relatively stable with respect to $E[\varepsilon_n^D]$ so long as $E[\varepsilon_n^D] \leq 0.25$, beyond which it grows with increasing $E[\varepsilon_n^D]$.

EXERCISES

2.1 You are given that the classification error ε_n and an error estimator $\hat{\varepsilon}_n$ are distributed, as a function of the random sample S_n, as two Gaussians

$$\varepsilon_n \sim N(\varepsilon_{\text{bay}} + 1/n, 1/n^2), \quad \hat{\varepsilon}_n \sim N(\varepsilon_{\text{bay}} - 1/n, 1/n^2).$$

Find the bias of $\hat{\varepsilon}_n$ as a function of n and plot it; is this estimator optimistically or pessimistically biased? If $\rho(\varepsilon, \hat{\varepsilon}_n) = (n-1)/n$, compute the deviation variance, RMS, and a bound on the tail probabilities, and plot them as a function of n.

2.2 You are given that an error estimator $\hat{\varepsilon}_n$ is related to the classification error ε_n through the simple model

$$\hat{\varepsilon}_n = \varepsilon_n + Z,$$

where the conditional distribution of the random variable Z given the training data S_n is Gaussian, $Z \sim N(0, 1/n^2)$. Is $\hat{\varepsilon}_n$ randomized or nonrandomized? Find the internal variance and variance of $\hat{\varepsilon}_n$. What happens as the sample size grows without bound?

2.3 Obtain a sufficient condition for an error estimator to be consistent in terms of the RMS. Repeat for strong consistency. Hint: Consider (2.12) and (2.13), and then use (2.11).

2.4 You are given that $\text{Var}(\varepsilon_n) \leq 5 \times 10^{-5}$. Find the minimum number of test points m that will guarantee that the standard deviation of the test-set error estimator $\hat{\varepsilon}_{n,m}$ will be at most 1%.

2.5 You are given that the error of a given classifier is $\varepsilon_n = 0.1$. Find the probability that the test-set error estimate $\hat{\varepsilon}_{n,m}$ will be exactly equal to ε_n, if $m = 20$ testing points are available.

2.6 This problem concerns additional properties of test-set error estimation.

(a) Show that

$$\text{Var}(\hat{\varepsilon}_{n,m}) = \frac{E[\varepsilon_n](1 - E[\varepsilon_n])}{m} + \frac{m-1}{m}\,\text{Var}(\varepsilon_n).$$

From this, show that $\text{Var}(\hat{\varepsilon}_{n,m}) \to \text{Var}(\varepsilon_n)$, as the number of testing points $m \to \infty$.

(b) Using the result of part (a), show that

$$\mathrm{Var}(\varepsilon_n) \le \mathrm{Var}(\hat{\varepsilon}_{n,m}) \le E[\varepsilon_n](1 - E[\varepsilon_n])$$

In particular, this shows that when $E[\varepsilon_n]$ is small, so is $\mathrm{Var}(\hat{\varepsilon}_{n,m})$. Hint: For any random variable X such that $0 \le X \le 1$ with probability 1, one has $\mathrm{Var}(X) \le E[X](1 - E[X])$.

(c) Show that the tail probabilities given the training data S_n satisfy:

$$P(|\hat{\varepsilon}_{n,m} - \varepsilon_n| \ge \tau \mid S_n) \le e^{-2m\tau^2}, \quad \text{for all } \tau > 0.$$

Hint: Use Hoeffding's inequality (c.f. Theorem A.4).

(d) By using the Law of Large Numbers (c.f. Theorem A.2), show that, given the training data S_n, $\hat{\varepsilon}_{n,m} \to \varepsilon_n$ with probability 1.

(e) Repeat item (d), but this time using the result from item (c).

(f) The bound on $\mathrm{RMS}(\hat{\varepsilon}_{n,m})$ in (2.29) is distribution-free. Prove the distribution-based bound

$$\mathrm{RMS}(\hat{\varepsilon}_{n,m}) \le \frac{1}{2\sqrt{m}} \min(1, 2\sqrt{E[\varepsilon_n]}). \tag{2.91}$$

2.7 Concerning the example in Section 2.5 regarding the 3NN classification rule with $n = 4$, under Gaussian populations $\Pi_0 \sim N_d(\boldsymbol{\mu}_0, \Sigma)$ and $\Pi_1 \sim N_d(\boldsymbol{\mu}_1, \Sigma)$, where $\boldsymbol{\mu}_0$ and $\boldsymbol{\mu}_1$ are far enough apart to make $\delta = \sqrt{(\boldsymbol{\mu}_1 - \boldsymbol{\mu}_0)^T \Sigma^{-1} (\boldsymbol{\mu}_1 - \boldsymbol{\mu}_0)} \gg 0$, compute

(a) $E[\varepsilon_n]$, $E[\hat{\varepsilon}_n^r]$, and $E[\hat{\varepsilon}_n^l]$.

(b) $E[(\hat{\varepsilon}_n^r)^2]$, $E[(\hat{\varepsilon}_n^l)^2]$, and $E[\hat{\varepsilon}_n^r \hat{\varepsilon}_n^l]$.

(c) $\mathrm{Var}_{\mathrm{dev}}(\hat{\varepsilon}_n^r)$, $\mathrm{Var}_{\mathrm{dev}}(\hat{\varepsilon}_n^l)$, $\mathrm{RMS}(\hat{\varepsilon}_n^r)$, and $\mathrm{RMS}(\hat{\varepsilon}_n^l)$.

(d) $\rho(\varepsilon_n, \hat{\varepsilon}_n^r)$ and $\rho(\varepsilon_n, \hat{\varepsilon}_n^l)$.

Hint: Reason that the true error, the resubstitution estimator, and the leave-one-out estimator are functions only of the sample sizes N_0 and N_1 and then write down their values for each possible case $N_0 = 0, 1, 2, 3, 4$, with the associated probabilities.

2.8 For the models M1 and M2 of Exercise 1.8, for the LDA, QDA, 3NN, and linear SVM classification rules, and for $d = 6$, $\sigma = 1, 2$, $\rho = 0, 0.2, 0.8$, and $n = 25, 50, 200$, plot the performance metrics of (2.4) through (2.7), the latter with $\tau = 0.05$, as a function of the Bayes errors $0.05, 0.15, 0.25, 0.35$ for resubstitution, leave-one-out, 5-fold cross-validation, and .632 bootstrap, with $c_0 = 0.1, 0.3, 0.5$.

2.9 Find the weight w_{bopt} in (2.66) for model M1 of Exercise 1.8, classification rules LDA and 3NN, $d = 6$, $\sigma = 1, 2$, $\rho = 0, 0.2, 0.8$, $n = 25, 50, 200$, and the Bayes errors $0.05, 0.15, 0.25, 0.35$. In all cases compute the bias and the RMS.

2.10 The weights for the convex combination between resubstitution and k-fold cross-validation in (2.55) have been set heuristically to 0.5. Redo Exercise 2.9 except that now find the weights that minimize the RMS for the convex combination in (2.55), with $k = 5$. Again find the bias and RMS in all cases.

2.11 Redo Exercise 2.9 under the constraint of unbiasedness.

2.12 Redo Exercise 2.10 under the constraint of unbiasedness.

2.13 Using spherical Gaussian bolstering kernels and the kernel variance estimation technique of Section 2.9.2, redo Exercise 2.8 for bolstered resubstitution with $d = 2, 3, 4, 5$.

2.14 Redo Exercise 2.13 for diagonal Gaussian bolstering kernels, using the Naive Bayes method of Section 2.9.2.

CHAPTER 3

PERFORMANCE ANALYSIS

In this chapter we evaluate the small-sample performance of several error estimators under a broad variety of experimental conditions, using the performance metrics introduced in the previous chapter. The results are based on simulations using synthetic data generated from known models, so that the true classification error is known exactly.

3.1 EMPIRICAL DEVIATION DISTRIBUTION

The experiments in this section assess the empirical distribution of $\hat{\varepsilon}_n - \varepsilon_n$. The error estimator $\hat{\varepsilon}_n$ is one of: resubstitution (resb), leave-one-out cross-validation (loo), k-fold cross-validation (cv$k$), k-fold cross-validation repeated r times ($r$cv$k$), bolstered resubstitution (bresb), semi-bolstered resubstitution (sresb), calibrated bolstering (opbolst), .632 bootstrap (b632), .632+ bootstrap (b632+), or optimized bootstrap (optboot).[1] Bolstering kernels are spherical Gaussian.

The model consists of equally likely Gaussian class-conditional densities with unequal covariance matrices. Class 0 has mean at $(0, 0, \dots, 0)$, with covariance matrix $\Sigma_0 = \sigma_0^2 I$, and class 1 has mean at $(a_1, a_2, \dots, a_D)$, with covariance matrix $\Sigma_1 = \sigma_1^2 I$. The parameters $a_1, a_2, \dots, a_D$ have been independently sampled from a beta distribution, Beta(0.75, 2). With the assumed covariance matrices, the features are

[1] Please consult Chapter 2 for the definition of these error estimators.

Error Estimation for Pattern Recognition, First Edition. Ulisses M. Braga-Neto and Edward R. Dougherty.

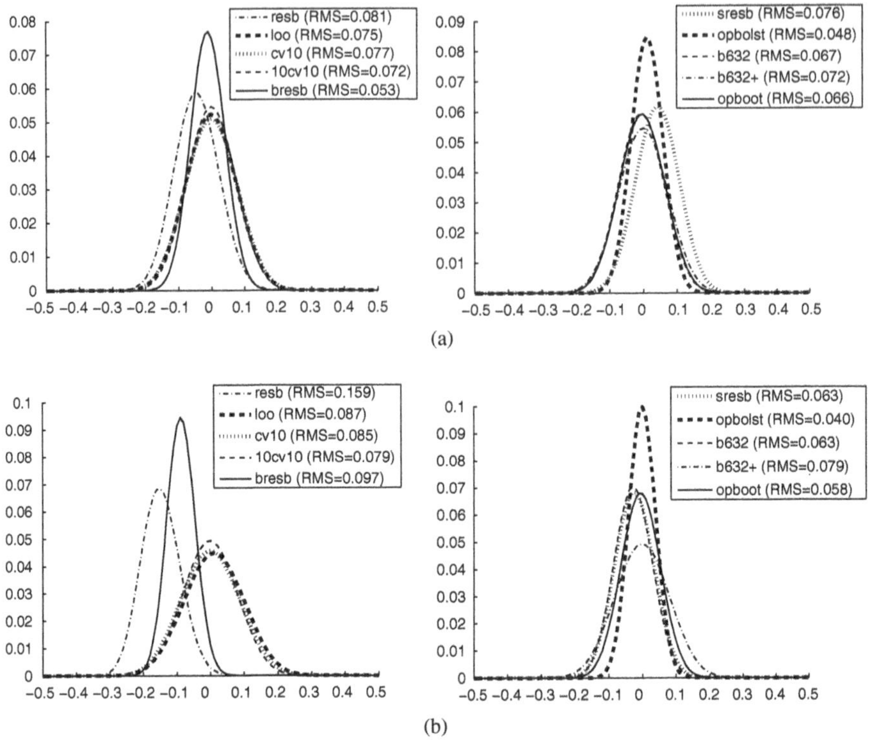

FIGURE 3.1 Beta-distribution fits (1000 points) for the empirical deviation distributions and estimated RMS values for several error estimators using three features with $\sigma_0 = 1$, $\sigma_1 = 1.5$, and sample size $n = 40$. (a) LDA and (b) 3NN classification rules.

independent and can be ranked according to the values of $a_1, a_2, \ldots, a_D$, the better (i.e., more discriminatory) features resulting from larger values.

Figure 3.1 shows beta-distribution fits (1000 points) for the empirical deviation distributions arising from (a) Linear Discriminant Analysis (LDA) and (b) 3NN classification using $d = 3$ features with $\sigma_0 = 1$, $\sigma_1 = 1.5$, and sample size $n = 40$. Also indicated are the estimated RMS values of each estimator. A centered distribution indicates low bias and a narrow density indicates low deviation variance, leading to small RMS values. Notice the large dispersions for cross-validation. Cross-validation generally suffers from a large deviation variance. Bolstered resubstitution has the minimum deviation variance among the rules but in both cases it is low-biased. Semi-bolstered resubstitution mitigates the low bias, even being slightly high-based in the LDA case. The correction is sufficiently good for LDA that semi-bolstered resubstitution has the minimum RMS among all the classification rules considered. For LDA, the .632+ bootstrap does not outperform the .632 bootstrap on account of increased variance, even though it is less biased. For 3NN, the .632+ bootstrap outperforms the .632 bootstrap, basically eliminating the low bias of the

.632 bootstrap. This occurs even though the .632+ bootstrap shows increased variance in comparison to the .632 bootstrap. Both the .632 and .632+ bootstrap are significantly outperformed by the optimized bootstrap. Note the consistently poor performance of cross-validation. One must refrain from drawing general conclusions concerning performance because it depends heavily on the feature–label distribution and classification rule. More extensive studies can be found in Braga-Neto and Dougherty (2004a,b). Whereas the dimensionality in these experiments is only $d = 3$, performance degrades significantly in high-dimensional spaces, even if feature selection is applied (Molinaro et al., 2005; Xiao et al., 2007).

Computation time varies widely among the different error estimators. Resubstitution is the fastest estimator, with its bolstered versions coming just behind. Leave-one-out and its bolstered version are fast for small sample size, but performance quickly degrades with increasing sample size. The 10-fold cross-validation and the bootstrap estimators are much slower. Resubstitution and its bolstered version can be hundreds of times faster than the bootstrap estimator (see Braga-Neto and Dougherty (2004a) for a listing of timings). When there is feature selection, cross-validation and bootstrap resampling must be carried out with respect to the original set of features, with feature selection being part of the classification rule, as explained in Chapter 2. This greatly increases the computational complexity of these resampling procedures.

3.2 REGRESSION

To illustrate regression relating the true and estimated errors, we consider the QDA classification rule using the t-test to reduce the number of features from $D = 200$ to $d = 5$ in a Gaussian model with sample size $n = 50$. Class 0 has mean $(0, 0, \ldots, 0)$ with covariance matrix $\Sigma_0 = \sigma_0^2 I$ and $\sigma_0 = 1$. For class 1, a vector $\mathbf{a} = (a_1, a_2, \ldots, a_{200})$, where a_i is distributed as Beta(0.75, 1), is found and fixed for the entire simulation. The mean for class 1 is $\delta\mathbf{a}$, where δ is tuned to obtain the desired expected true error, which is 0.30. The covariance matrix for class 1 is $\Sigma_1 = \sigma_1^2 I$ and $\sigma_1 = 1.5$.

Figure 3.2 displays scatter plots of the true error versus estimated error, which include the regression, that is, the estimated conditional expectation $E[\varepsilon_n \mid \hat{\varepsilon}_n]$, of the true error on the estimated error and 95% confidence-interval (upper bound) curves, for the QDA classification rule and various error estimators (the ends of the curves have been truncated in some cases because, owing to very little data in that part of the distribution, the line becomes erratic). Notice the lack of regression (flatness of the curves), especially for the cross-validation methods. Bolstered resubstitution and bootstrap show more regression. Lack of regression, and therefore poor error estimation, is commonplace when the sample size is small, especially for randomized error estimators (Hanczar et al., 2007). Notice the large variance of cross-validation and the much smaller variances of bolstered resubstitution, bootstrap, and resubstitution. Notice also the optimistic bias of resubstitution. Figure 3.3 is similar to the previous one, except that the roles of true and estimated errors are reversed, and the estimated conditional expectation $E[\hat{\varepsilon}_n \mid \varepsilon_n]$ and 95% confidence-interval (upper bound) curves are displayed.

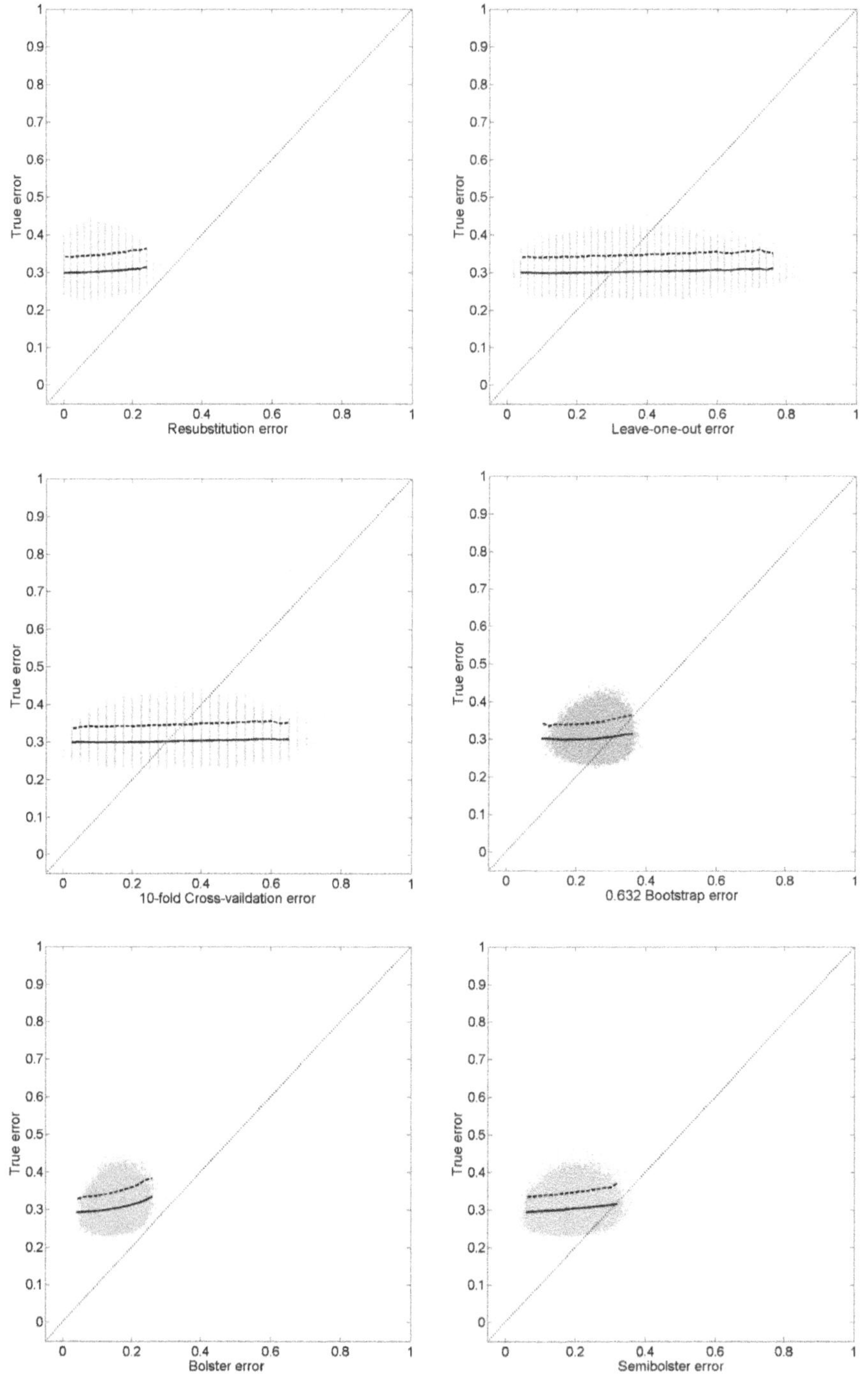

FIGURE 3.2 Scatter plots of the true error versus estimated error, including the conditional expectation of the true error on the estimated error (solid line) and 95% confidence-interval (upper bound) curves (dashed line), for the QDA classification rule, $n = 50$, and various error estimators.

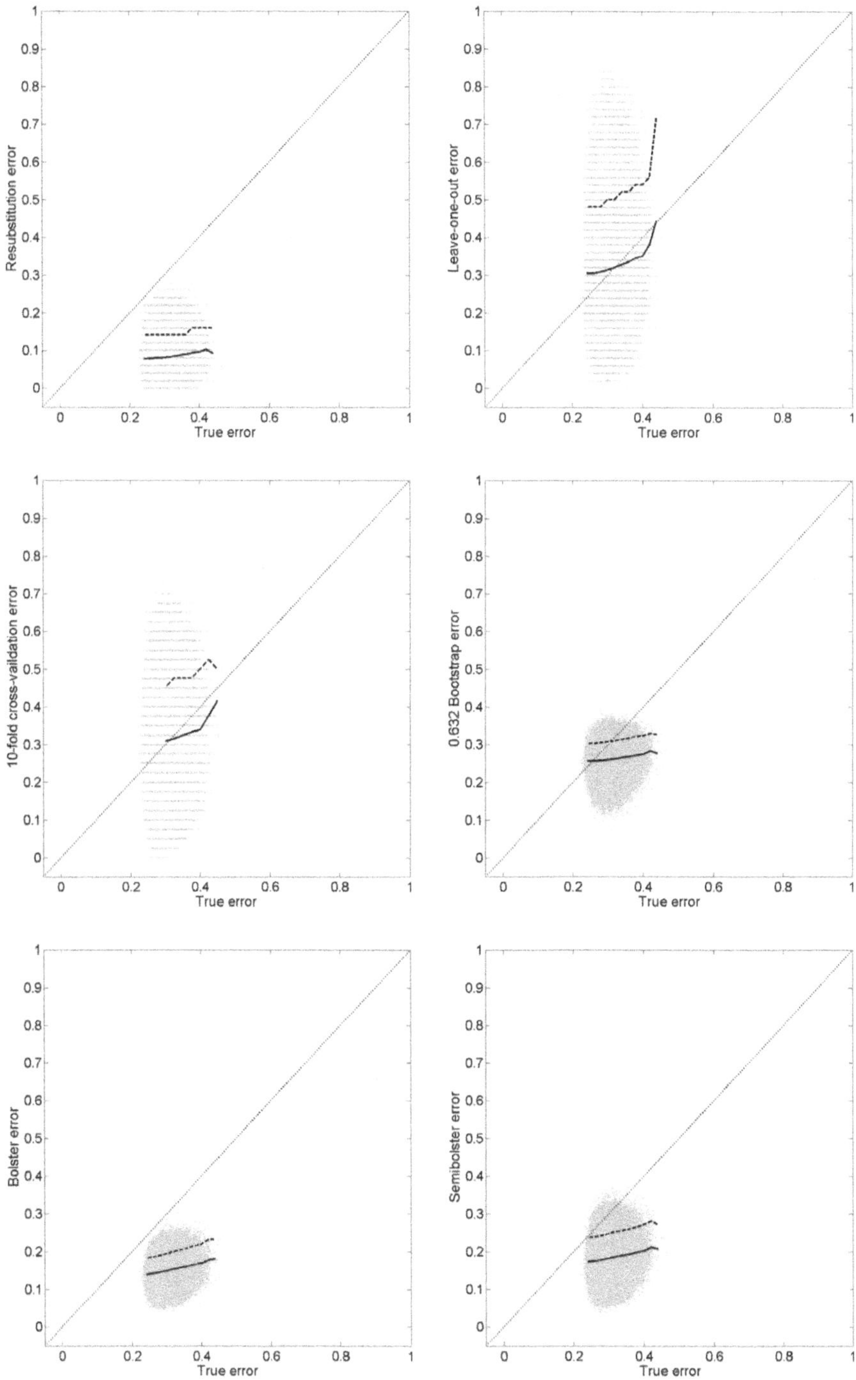

FIGURE 3.3 Scatter plots of estimated error versus the true error, including the conditional expectation of the estimated error on the true error (solid line) and 95% confidence-interval (upper bound) curves (dashed line), for the QDA classification rule, $n = 50$, and various error estimators.

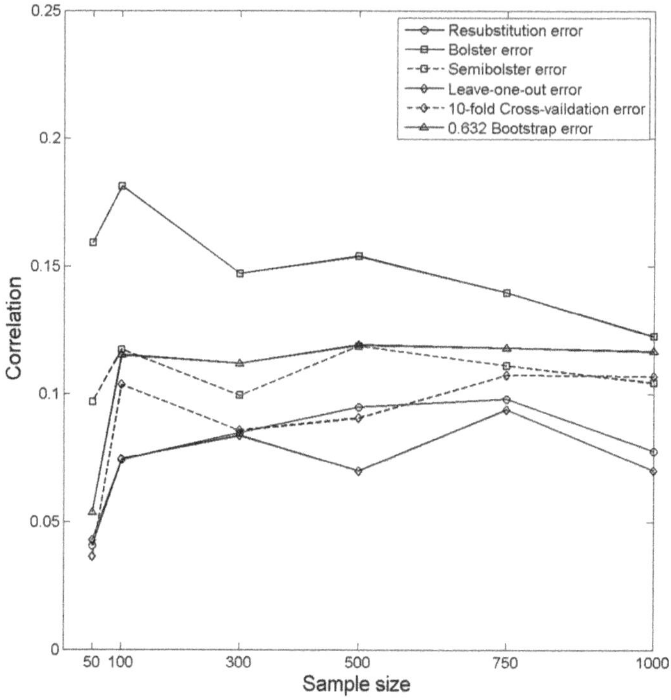

FIGURE 3.4 Correlation coefficient versus sample size for the QDA classification rule, $n = 50$, and various error estimators.

Figure 3.4 displays the correlation coefficient between true and estimated errors as a function of the sample size. Increasing the sample size up to 1000 does not help. Keep in mind that with a large sample, the variation in both the true and estimated errors will get very small, so that for large samples there is good error estimation even with a small correlation coefficient between true and estimated errors, if the estimator is close to being unbiased. Of course, if one has a large sample, one could simply use test-set error estimation and obtain superior results.

For comparison purposes, Figure 3.5 displays the regression for the test-set error estimator, displaying the estimated conditional expectations $E[\varepsilon_n \mid \hat{\varepsilon}_{n,m}]$ and $E[\hat{\varepsilon}_{n,m} \mid \varepsilon_n]$ in the left and right plots, respectively. Here the training set size is $n = 50$, as before, and the test set size is $m = 100$. We observe that, in comparison with the previous error estimators, there is less variance and there is better regression. Note that the better performance of the test-set error comes at the cost of a significantly larger overall sample size, $m + n = 150$.

3.3 IMPACT ON FEATURE SELECTION

Error estimation plays a role in feature selection. In this section, we consider two measures of error estimation performance in feature selection, one for ranking features

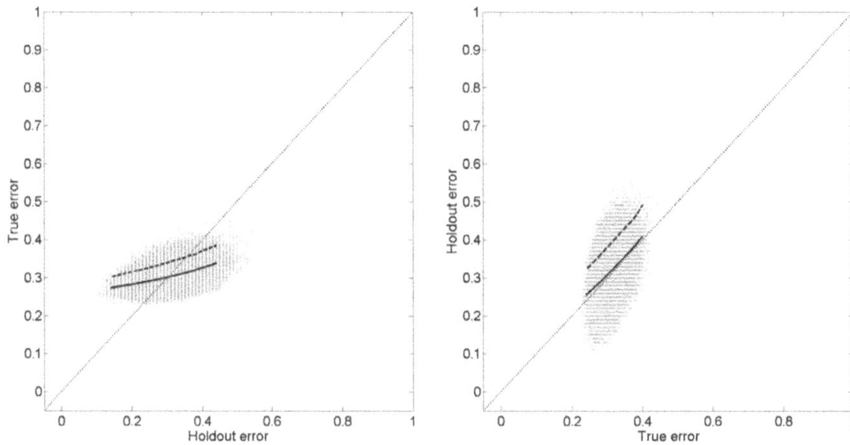

FIGURE 3.5 Scatter plots of the true error versus test-set error estimator with regression line superimposed, for the QDA classification rule, $n = 50$, and $m = 100$.

and the other for determining feature sets. The model is similar to the one used in Section 3.2, except that $a_i \sim \text{Beta}(0.75,2)$ and the expected true error is 0.27.

A natural measure of worth for an error estimator is its ranking accuracy for feature sets. Two figures of merit have been considered for an error estimator in its ranking accuracy for feature sets (Sima et al., 2005b). Each compares ranking based on true and estimated errors under the condition that the true error is less than t. The first is the number $R_1^K(t)$ of feature sets in the truly top K feature sets that are also among the top K feature sets based on error estimation. It measures how well the error estimator finds top feature sets. The second is the mean-absolute rank deviation $R_2^K(t)$ for the K best feature sets. Figure 3.6 shows plots obtained by averaging these measures over 10,000 trials with $K = 40$, sample size $n = 40$, and four error estimators: resubstitution, 10-fold cross-validation, bolstered resubstitution, and .632+ bootstrap. The four parts of the figure illustrate the following scenarios: (a) LDA and $R_1^K(t)$, (b) LDA and $R_2^K(t)$, (c) 3NN and $R_1^K(t)$, and (d) 3NN and $R_2^K(t)$. An exhaustive search of all feature sets has been performed for the true ranking with each feature set being of size $d = 3$ out of the $D = 20$ total number of features. As one can see, cross-validation generally performs poorer than the bootstrap and bolstered estimators.

When selecting features via a wrapper algorithm such as sequential forward floating selection (SFFS), which employs error estimation within it, the choice of error estimator obviously impacts feature selection performance, the degree of impact depending on the classification rule and feature–label distribution (Sima et al., 2005a). SFFS can perform close to using an exhaustive search when the true error is used within the SFFS algorithm; however, this is not possible in practice, where error estimates must be used within the SFFS algorithm. It has been observed that the choice of error estimator can make a greater difference than the choice of feature-selection algorithm — indeed, the classification error can be smaller using SFFS with bolstering or bootstrap than performing a full search using leave-one-out cross-validation

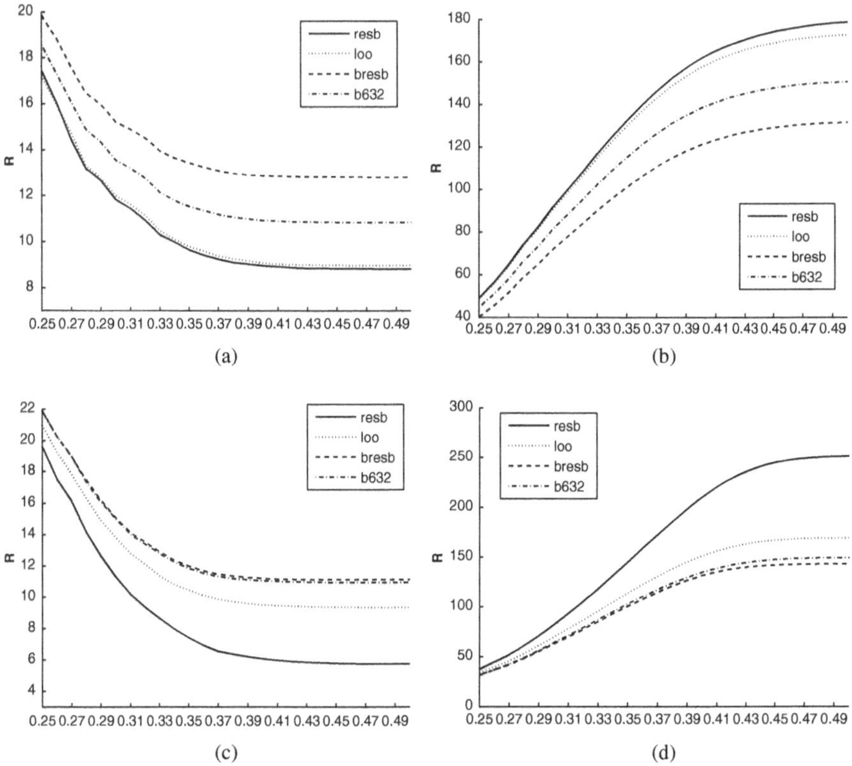

FIGURE 3.6 Feature-set ranking measures: (a) LDA, $R_1^{40}(t)$; (b) LDA, $R_2^{40}(t)$; (c) 3NN, $R_1^{40}(t)$; (d) 3NN, $R_2^{40}(t)$.

(Sima et al., 2005a). Figure 3.7 illustrates the impact of error estimation within SFFS feature selection to select $d = 3$ among $D = 50$ features with sample size $n = 40$. The errors in the figure are the average true errors of the resulting designed classifiers for 1000 replications. Parts (a) and (b) are from LDA and 3NN, respectively. In both cases, bolstered resubstitution performs the best, with semi-bolstered resubstitution performing second best.

3.4 MULTIPLE-DATA-SET REPORTING BIAS

Suppose rather than drawing a single sample from a feature–label distribution, one draws m samples independently, applies a pattern recognition rule on each, and then takes the minimum of the resulting error estimates

$$\hat{\varepsilon}_n^{\min} = \min\left\{\hat{\varepsilon}_n^1, \dots, \hat{\varepsilon}_n^m\right\}. \tag{3.1}$$

Furthermore, suppose that this minimum is then taken as an estimator of the classification error. Not only does this estimator suffer from the variances of the individual

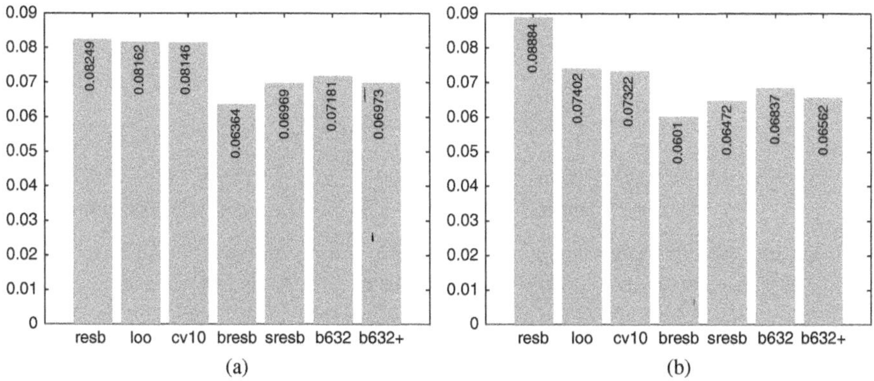

FIGURE 3.7 Average true error rates obtained from different estimation methods inside the SFFS feature selection algorithm: (a) LDA; (b) 3NN.

estimators, the minimum tends to create a strong optimistic tendency, which increases with m, even if the individual biases may be small. The potential for optimism is seen in the probability distribution function for this estimator, which is given by

$$P\left(\hat{\varepsilon}_n^{\min} \le x\right) = 1 - \left[1 - P(\hat{\varepsilon}_n \le x)\right]^m. \tag{3.2}$$

It follows from (3.2) that $P(\hat{\varepsilon}_n^{\min} \le x) \to 1$, as $m \to \infty$, for all $x \ge 0$ such that $P(\hat{\varepsilon}_n \le x) > 0$.

One might at first dismiss these considerations regarding $\hat{\varepsilon}_n^{\min}$ as outlandish, arguing that no one would use such an estimator; however, they are indeed used, perhaps unknowingly. Suppose a researcher takes m independent samples of size n, computes $\hat{\varepsilon}_n^i$ for $i = 1, 2, \ldots, m$, and then reports the "best" result, that is, the minimum in (3.1), as the estimate of the classification error. We show below that the reported estimate can be strongly optimistic (depending on m). This kind of bias has been termed "reporting bias" (Yousefi et al., 2010) and "fishing for significance" (Boulesteix, 2010). There are different ways of looking at the bias resulting from taking the minimum of the error estimates as an estimate of the expected error (Yousefi et al., 2010). We discuss two of these. To do so, let ε_n^i denote the true error of the classifier designed on a random sample S_n^i of size n, for $i = 1, 2, \ldots, m$. Let $S_n = \{S_n^1, \ldots, S_n^m\}$. Assuming the samples to be independent, we have that $E_{S_n}[\varepsilon_n^1] = \cdots = E_{S_n}[\varepsilon_n^m] = E[\varepsilon_n]$, the expected classification error in a single experiment, where "E_{S_n}" denotes expectation being taken across the collection of samples S_n. Suppose the minimum in (3.1) occurs on $i = i_{\min}$, so that the reported error estimate is $\hat{\varepsilon}_n^{\min} = \hat{\varepsilon}_n^{i_{\min}}$, and the true classification error of the selected model is $\varepsilon_n^{i_{\min}}$. Note that $E_{S_n}[\varepsilon_n^{i_{\min}}]$ is not equal to $E[\varepsilon_n^{i_{\min}}] = E[\varepsilon_n]$. We consider two kinds of bias:

$$E_{S_n}\left[\hat{\varepsilon}_n^{i_{\min}} - \varepsilon_n\right] = E_{S_n}\left[\hat{\varepsilon}_n^{i_{\min}}\right] - E[\varepsilon_n] \tag{3.3}$$

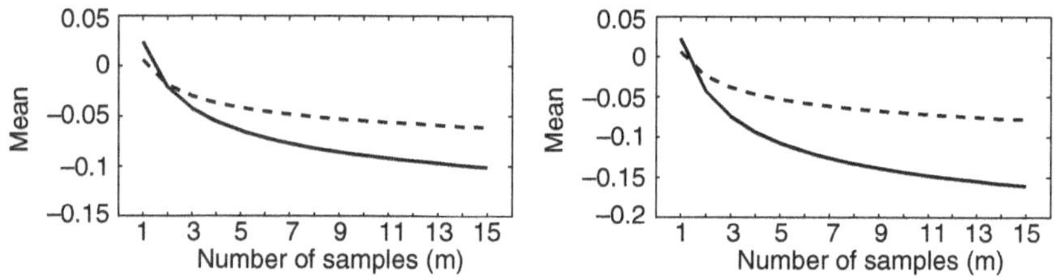

FIGURE 3.8 Multiple data-set (reporting) bias. Left panel: average deviation of the estimated and true errors for the sample showing the least-estimated error. Right panel: average deviation of the estimated error for the sample showing the least-estimated error and the expected true error. Solid line: $n = 60$; dashed line: $n = 120$. Reproduced from Dougherty and Bittner (2011) with permission from Wiley-IEEE Press.

and

$$E_{S_n}\left[\hat{\varepsilon}_n^{i_{\min}} - \varepsilon_n^{i_{\min}}\right] = E_{S_n}\left[\hat{\varepsilon}_n^{i_{\min}}\right] - E_{S_n}\left[\varepsilon_n^{i_{\min}}\right]. \tag{3.4}$$

The first bias provides the average discrepancy between the reported error estimate $\hat{\varepsilon}_n^{i_{\min}}$ and the true error ε_n. This is the most significant kind of bias from a practical perspective and reveals the true extent of the problem. The second bias describes the effect of taking $\hat{\varepsilon}_n^{\min}$ as an error estimate of $\varepsilon_n^{i_{\min}}$. Figure 3.8 plots the two kinds of bias (first bias on the right, second bias on the left) as a function of m, for samples of size $n = 60$ (solid line) and $n = 120$ (dashed line) generated from a 20,000-dimensional model described in Hua et al. (2009), Yousefi et al. (2010). The pattern recognition rule consists of LDA classification, t-test feature selection and the 10-fold cross-validation error estimator (applied externally to the feature selection step, naturally).

We can see that the bias is slightly positive for $m = 1$ (that is a feature of cross-validation), but it immediately becomes negative, and remains so, for $m > 1$. The extent of the problem is exemplified by an optimistic bias exceeding 0.1 with as few as 5 samples of size 60 in Figure 3.8b. It is interesting that the bias in Figure 3.8a is also quite significant, even though it is smaller.

3.5 MULTIPLE-RULE BIAS

Whereas multiple-data-set bias results from using multiple data sets to evaluate a single classification rule and taking the minimum estimated error, multiple-rule bias results from applying multiple pattern recognition rules (i.e., multiple classification and error estimation rules) on the same data set. Following Yousefi et al. (2011), we consider r classification rules, $\Psi_1, \dots, \Psi_r$, and s error estimation rules, $\Xi_1, \dots, \Xi_s$, which are combined to form $m = rs$ pattern recognition rules: $(\Psi_1, \Xi_1), \dots, (\Psi_1, \Xi_s), \dots, (\Psi_r, \Xi_1), \dots, (\Psi_r, \Xi_s)$. Given a random sample S_n of size n drawn from a feature–label distribution $F_{\mathbf{X}Y}$, we obtain rs pattern recognition models, consisting of r designed classifiers $\psi_n^1, \dots, \psi_n^r$ (possessing respective true errors $\varepsilon_n^1, \dots, \varepsilon_n^r$) and s estimated errors $\hat{\varepsilon}_n^{i,1}, \dots, \hat{\varepsilon}_n^{i,s}$ for each classifier ψ_n^i, for $i = 1, \dots, r$.

Without loss of generality, we assume the classifier models are indexed so that $E[\varepsilon_n^1] \leq E[\varepsilon_n^2] \leq \cdots \leq E[\varepsilon_n^r]$. The reported estimated error is

$$\hat{\varepsilon}_n^{\min} = \min\left\{\hat{\varepsilon}_n^{1,1}, \ldots, \hat{\varepsilon}_n^{1,s}, \ldots, \hat{\varepsilon}_n^{r,1}, \ldots, \hat{\varepsilon}_n^{r,s}\right\}. \tag{3.5}$$

Let $i_{\min}$ and $j_{\min}$ be the classifier number and error estimator number, respectively, for which the error estimate is minimum. Note that $\hat{\varepsilon}_n^{\min} = \hat{\varepsilon}_n^{i_{\min}, j_{\min}}$. If, based on the single sample and the preceding minimum, $\Psi_{i_{\min}}$ is chosen as the best classification rule, this means that $\hat{\varepsilon}_n^{\min}$ is being used to estimate $\varepsilon_n^{i_{\min}}$. The goodness of this estimation is characterized by the distribution of the deviation $\Delta = \hat{\varepsilon}_n^{\min} - \varepsilon_n^{i_{\min}}$. The bias is given

$$\text{Bias}(m, n) = E[\Delta] = E\left[\hat{\varepsilon}_n^{\min}\right] - E\left[\varepsilon_n^{i_{\min}}\right]. \tag{3.6}$$

Estimation is optimistic if $\text{Bias}(m, n) < 0$. Owing to estimation-rule variance, this can happen even if none of the error estimation rules are optimistically biased. Optimistic bias results from considering multiple estimated errors among classifiers and from applying multiple error estimates for each classifier. Regarding the bias, as the number m of classifier models grows, the minimum of (3.5) is taken over more estimated errors, thereby increasing $|\text{Bias}(m, n)|$. Moreover, as the sample size n is increased, under the typical condition that error estimator variance decreases with increasing sample size, $E[\hat{\varepsilon}_n^{\min}]$ increases and $|\text{Bias}(m, n)|$ decreases. Bias and variance are combined into the RMS, given by

$$\text{RMS}(m, n) = \sqrt{E[\Delta^2]} = \sqrt{\text{Bias}(m, n)^2 + \text{Var}[\Delta]}. \tag{3.7}$$

We further generalize the experimental design by allowing a random choice of r classification rules out of a universe of R classification rules. There are $\binom{R}{r}$ possible collections of classification rules to be combined with s error estimators to form $\binom{R}{r}$ distinct collections of $m = rs$ pattern recognition rules. We denote the set of all such collections by Φ_m. To get the distribution of errors, one needs to generate independent samples from the same feature–label distribution and apply the procedure shown in Figure 3.9.

Before proceeding, we illustrate the foregoing process by considering $R = 4$ classification rules, $\Psi_1, \Psi_2, \Psi_3, \Psi_4$, and $s = 2$ error estimation rules, Ξ_1, Ξ_2. Suppose we wish to choose $r = 2$ classification rules. Then $m = rs = 4$ and the possible collections of $m = 4$ PR rules are listed below:

$$\begin{aligned}
&\{(\Psi_1, \Xi_1), (\Psi_1, \Xi_2), (\Psi_2, \Xi_1), (\Psi_2, \Xi_2)\},\\
&\{(\Psi_1, \Xi_1), (\Psi_1, \Xi_2), (\Psi_3, \Xi_1), (\Psi_3, \Xi_2)\},\\
&\{(\Psi_1, \Xi_1), (\Psi_1, \Xi_2), (\Psi_4, \Xi_1), (\Psi_4, \Xi_2)\},\\
&\{(\Psi_2, \Xi_1), (\Psi_2, \Xi_2), (\Psi_3, \Xi_1), (\Psi_3, \Xi_2)\},\\
&\{(\Psi_2, \Xi_1), (\Psi_2, \Xi_2), (\Psi_4, \Xi_1), (\Psi_4, \Xi_2)\},\\
&\{(\Psi_3, \Xi_1), (\Psi_3, \Xi_2), (\Psi_4, \Xi_1), (\Psi_4, \Xi_2)\}.
\end{aligned}$$

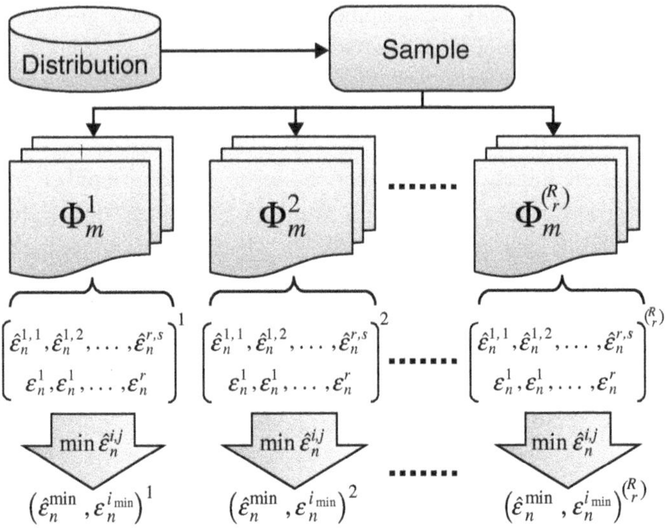

FIGURE 3.9 Multiple-rule testing procedure on a single sample.

The bias, variance, and RMS are adjusted to take model randomization into account by defining the *average bias*, *average variance*, and *average RMS* as

$$\mathrm{Bias}_{\mathrm{Av}}(m, n) = E_{\Phi_m}[E[\Delta \mid \Phi_m]], \tag{3.8}$$

$$\mathrm{Var}_{\mathrm{Av}}(m, n) = E_{\Phi_m}[\mathrm{Var}[\Delta \mid \Phi_m]], \tag{3.9}$$

$$\mathrm{RMS}_{\mathrm{Av}}(m, n) = E_{\Phi_m}\left[\sqrt{E[\Delta^2 \mid \Phi_m]}\right]. \tag{3.10}$$

In addition to the bias in estimating $\varepsilon_n^{i_{\min}}$ by $\hat{\varepsilon}_n^{\min}$, there is also the issue of the comparative advantage of the classification rule $\Psi_{i_{\min}}$. Its *true comparative advantage* with respect to the other considered classification rules is defined as

$$A_{\mathrm{true}} = E\left[\varepsilon_n^{i_{\min}}\right] - \frac{1}{r-1}\sum_{i \neq i_{\min}} E\left[\varepsilon_n^{i}\right]. \tag{3.11}$$

Its *estimated comparative advantage* is defined as

$$A_{\mathrm{est}} = \hat{\varepsilon}_n^{\min} - \frac{1}{r-1}\sum_{i \neq i_{\min}} \hat{\varepsilon}_n^{i, j_{\min}}. \tag{3.12}$$

The expectation $E[A_{\mathrm{est}}]$ of this distribution gives the *mean estimated comparative advantage* of $\Psi_{i_{\min}}$ with respect to the collection. An important bias to consider is the *comparative bias*

$$C_{\mathrm{bias}} = E[A_{\mathrm{est}}] - A_{\mathrm{true}}, \tag{3.13}$$

as it measures over-optimism in comparative performance. In the case of randomization, the *average comparative bias* is defined as

$$C_{\text{bias,Av}} = E_{\Phi_m}[C_{\text{bias}} \mid \Phi_m]\,. \tag{3.14}$$

We illustrate the analysis using the same 20,000-dimensional model employed in Section 3.4, with $n = 60$ and $n = 120$. However, here we consider two feature selection rules to reduce the number of features to 5: (1) t-test to reduce from 20,000 to 5, and (2) t-test to reduce from 20,000 to 500 followed by SFS to reduce to 5 (c.f. Section 1.7). Moreover, we consider equal variance and unequal variance models (see Yousefi et al. (2011) for full details). We consider LDA, diagonal LDA, NMC, 3NN, linear SVM, and radial basis function SVM. The combination of two feature selection rules and six classification rules on the reduced feature space make up a total of 12 classification rules on the original feature space. In addition, three error estimation rules (leave-one-out, 5-fold cross-validation, and 10-fold cross-validation) are employed, thereby leading to a total of 36 PR rules. As each classification rule is added (in an arbitrary order), the number m of PR rule models increments according to $3, 6, \ldots, 36$. We generate all $\binom{12}{r}$ possible collections of classification rules of size r, each associated with three error estimation rules, resulting in $\binom{12}{r}$ collections of m PR rules. The raw synthetic data consist of the true and estimated error pairs resulting from applying the 36 different classification schemes on 10,000 independent random samples drawn from the distribution models. For each sample we find the classifier model (including classification and error estimation) in the collection that gives the minimum estimated error. Then Bias_{Av}, Var_{Av}, RMS_{Av}, and $C_{\text{bias,Av}}$ are found. These are shown as functions of m in Figure 3.10.

Multiple-rule bias can be exacerbated by combining it with multiple data sets and then minimizing over both rules and data sets; however, multiple-rule bias can be mitigated by averaging over multiple data sets. In this case each sample is from a feature–label distribution $F^k_{\mathbf{X}Y}$, $k = 1, 2, \ldots, K$, and performance is averaged over the K feature–label distributions. Assuming multiple error estimators, there are m PR rules being considered over the K feature–label distributions and the estimator of interest is

$$\begin{aligned}\hat{\varepsilon}_n^{\min}(K) = \min\Bigg\{ &\frac{1}{K}\sum_{k=1}^{K} \hat{\varepsilon}_n^{1,1,k}, \ldots, \frac{1}{K}\sum_{k=1}^{K} \hat{\varepsilon}_n^{1,s,k}, \ldots, \\ &\frac{1}{K}\sum_{k=1}^{K} \hat{\varepsilon}_n^{r,1,k}, \ldots, \frac{1}{K}\sum_{k=1}^{K} \hat{\varepsilon}_n^{r,s,k} \Bigg\}\,, \end{aligned} \tag{3.15}$$

where $\hat{\varepsilon}_n^{i,j,k}$ is the estimated error of classifier ψ_i using the error estimation rule Ξ_j on the sample from feature–label distribution $F^k_{\mathbf{X}Y}$. The bias takes the form

$$\text{B}(m, n, K) = E\left[\hat{\varepsilon}_n^{\min}(K)\right] - \frac{1}{K}\sum_{k=1}^{K} E\left[\varepsilon_n^{i_{\min},k}\right]\,, \tag{3.16}$$

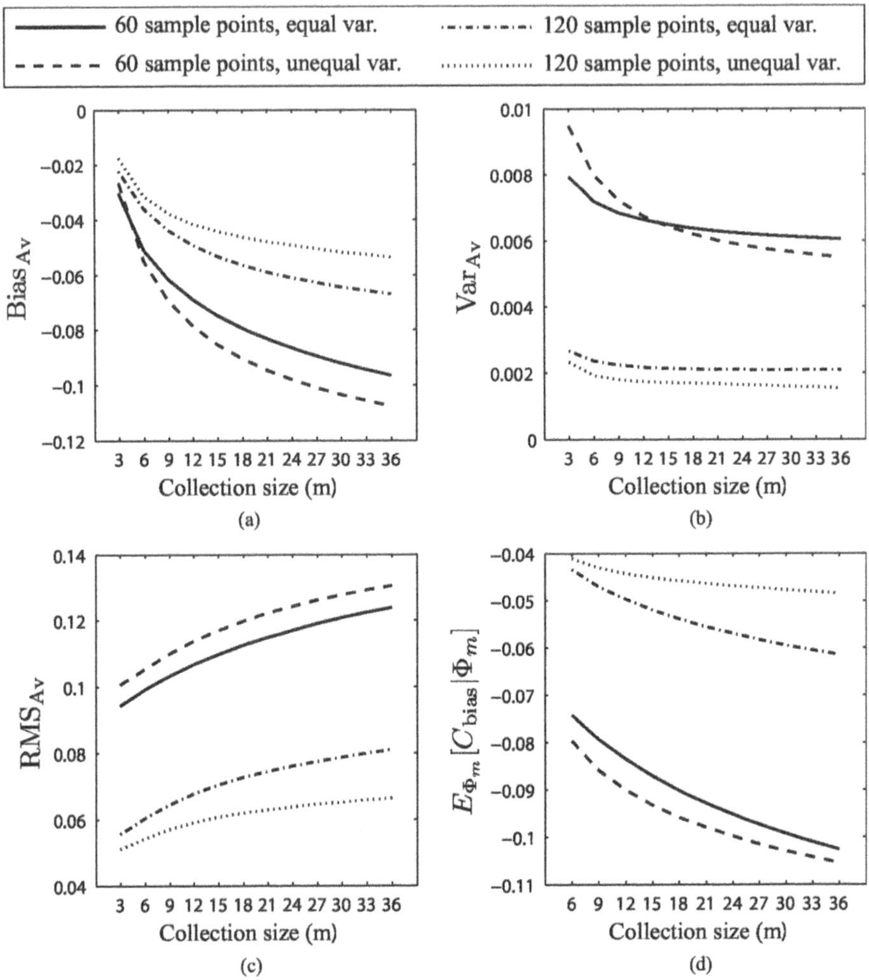

FIGURE 3.10 Multiple-rule bias: (a) Average bias, Bias_{Av}; (b) average variance, Var_{Av}; (c) average RMS, RMS_{Av}; (d) average comparative bias, $C_{\text{bias,Av}}$. Reproduced from Yousefi et al. with permission from Oxford University Press.

where $\varepsilon_n^{i_{\min},k}$ is the true error of classifier $\psi_{i_{\min}}$ on the sample from feature–label distribution $F_{\mathbf{X}Y}^k$.

Theorem 3.1 *(Yousefi et al., 2011) If all error estimators are unbiased, then* $\lim_{K\to\infty} \text{B}(m, n, K) = 0$.

Proof: The proof is given in the following two lemmas. The first shows that $\text{B}(m, n, K) \leq 0$ and the second that $\lim_{K\to\infty} \text{B}(m, n, K) \geq 0$. Together they prove the theorem. ■

Lemma 3.1 *If all error estimators are unbiased, then $B(m, n, K) \leq 0$.*

Proof: Define the set $S_n = \{S_n^1, S_n^2, \ldots, S_n^K\}$, where S_n^k is a random sample taken from the distribution $F_{\mathbf{X}Y}^k$ for $k = 1, 2, \ldots, K$. We can rewrite (3.15) as

$$\hat{\varepsilon}_n^{\min}(K) = \min_{i,j} \left\{ \frac{1}{K} \sum_{k=1}^{K} \hat{\varepsilon}_n^{i,j,k} \right\}, \tag{3.17}$$

where $i = 1, 2, \ldots, r$ and $j = 1, 2, \ldots, s$. Owing to the unbiasedness of the error estimators, $E_{S_n^k}[\hat{\varepsilon}_n^{i,j,k}] = E_{S_n^k}[\varepsilon_n^{i,k}]$. From (3.16) and (3.17),

$$\begin{aligned}
\mathrm{B}(m, n, k) &= E_{S_n}\left[\hat{\varepsilon}_n^{\min}(K)\right] - \frac{1}{K}\sum_{k=1}^{K} E_{S_n^k}\left[\varepsilon_n^{i_{\min},k}\right] \\
&= E_{S_n}\left[\min_{i,j}\left\{\frac{1}{K}\sum_{k=1}^{K}\hat{\varepsilon}_n^{i,j,k}\right\}\right] - \frac{1}{K}\sum_{k=1}^{K} E_{S_n^k}\left[\varepsilon_n^{i_{\min},k}\right] \\
&\leq \min_{i,j}\left\{E_{S_n}\left[\frac{1}{K}\sum_{k=1}^{K}\hat{\varepsilon}_n^{i,j,k}\right]\right\} - \frac{1}{K}\sum_{k=1}^{K} E_{S_n^k}\left[\varepsilon_n^{i_{\min},k}\right] \\
&= \min_{i,j}\left\{\frac{1}{K}\sum_{k=1}^{K} E_{S_n^k}\left[\hat{\varepsilon}_n^{i,j,k}\right]\right\} - \frac{1}{K}\sum_{k=1}^{K} E_{S_n^k}\left[\varepsilon_n^{i_{\min},k}\right] \\
&= \min_{i}\left\{\frac{1}{K}\sum_{k=1}^{K} E_{S_n^k}\left[\varepsilon_n^{i,k}\right]\right\} - \frac{1}{K}\sum_{k=1}^{K} E_{S_n^k}\left[\varepsilon_n^{i_{\min},k}\right] \leq 0.
\end{aligned}$$

where the relations in the third and sixth lines result from Jensen's inequality and unbiasedness of the error estimators, respectively. ■

Lemma 3.2 *If all error estimators are unbiased, then* $\lim_{K\to\infty} \mathrm{B}(m, n, K) \geq 0$.

Proof: Let

$$A^{i,j} = \frac{1}{K}\sum_{k=1}^{K}\hat{\varepsilon}_n^{i,j,k}, \qquad T^i = \frac{1}{K}\sum_{k=1}^{K} E_{S_n^k}\left[\varepsilon_n^{i,k}\right]. \tag{3.18}$$

Owing to the unbiasedness of the error estimators, $E_{S_n}[A^{i,j}] = T^i \leq 1$. Without loss of generality, we assume $T^1 \leq T^2 \leq \cdots \leq T^r$. To avoid cumbersome notation, we will further assume that $T^1 < T^2$ (with some adaptation, the proof goes through without this assumption). Let $2\delta = T^2 - T^1$ and

$$B_\delta = \left(\bigcap_{j=1}^{s}\left(T^1 - \delta \leq A^{1,j} \leq T^1 + \delta\right)\right) \bigcap \left(\min_{i \neq 1, j}\left\{A^{i,j}\right\} > T^1 + \delta\right).$$

Because $|\hat{\varepsilon}_n^{i,j,k}| \leq 1$, $\mathrm{Var}_{S_n}[A^{i,j}] \leq 1/K$. Hence, for $\tau > 0$, there exists $K_{\delta,\tau}$ such that $K \geq K_{\delta,\tau}$ implies $P(B_\delta(K)) > 1 - \tau$. Hence, referring to (3.16), for $K \geq K_{\delta,\tau}$,

$$\begin{aligned}
E_{S_n}\left[\hat{\varepsilon}_n^{\min}(K)\right] &= E_{S_n}\left[\hat{\varepsilon}_n^{\min}(K) \mid B_\delta\right] P(B_\delta) + E_{S_n}\left[\hat{\varepsilon}_n^{\min}(K) \mid B_\delta^c\right] P\left(B_\delta^c\right) \\
&\geq E_{S_n}\left[\hat{\varepsilon}_n^{\min}(K) \mid B_\delta\right] P(B_\delta) \\
&= E_{S_n}\left[\min_j A^{1,j}\right] P(B_\delta) \\
&\geq (T^1 - \delta)(1 - \tau).
\end{aligned} \tag{3.19}$$

Again referring to (3.16) and recognizing that $i_{\min} = 1$ in B_δ, for $K \geq K_{\delta,\tau}$,

$$\begin{aligned}
\frac{1}{K}\sum_{k=1}^{K} E_{S_n^k}\left[\varepsilon_n^{i_{\min},k}\right] &= \frac{1}{K}\sum_{k=1}^{K}\left(E_{S_n^k}\left[\varepsilon_n^{i_{\min},k} \mid B_\delta\right] P(B_\delta) + E_{S_n^k}\left[\varepsilon_n^{i_{\min},k} \mid B_\delta^c\right] P\left(B_\delta^c\right)\right) \\
&\leq \frac{1}{K}\sum_{k=1}^{K}\left(E_{S_n^k}\left[\varepsilon_n^{1,k} \mid B_\delta\right] P(B_\delta) + P\left(B_\delta^c\right)\right) \\
&\leq T^1 + \tau.
\end{aligned} \tag{3.20}$$

Putting (3.19) and (3.20) together and referring to (3.16) yields, for $K \geq K_{\delta,\tau}$,

$$\mathrm{B}(m, n, K) \geq (T^1 - \delta)(1 - \tau) - T^1 - \tau \geq -(2\tau + \delta). \tag{3.21}$$

Since δ and τ are arbitrary positive numbers, this implies that for any $\eta > 0$, there exists K_η such that $K \geq K_\eta$ implies $B(m, n, K) \geq -\eta$. ∎

Examining the proof of the theorem shows that the unbiasedness assumption can be relaxed in each of the lemmas. In the first lemma, we need only assume that none of the error estimators is pessimistically biased. In the second lemma, unbiasedness can be relaxed to weaker conditions regarding the expectations of the terms making up the minimum in (3.15). Moreover, since we are typically interested in close-to-unbiased error estimators, we can expect $|B(m, n, K)|$ to decrease with averaging, even if $B(m, n, K)$ does not actually converge to 0.

3.6 PERFORMANCE REPRODUCIBILITY

When data are expensive or difficult to obtain, it is common practice to run a preliminary experiment and then, if the results are good, to conduct a large experiment to verify the results of the preliminary experiment. In classification, a preliminary experiment might produce a classifier with an error estimate based on the training data and the follow-up study aims to validate the reported error rate from the first study. From (2.29), the RMS between the true and estimated errors for independent-test-data error estimation is bounded by $(2\sqrt{m})^{-1}$, where m is the size of the test sample; hence, a second study with 1000 data points will produce an RMS bounded by 0.016, so that the test-sample estimate can essentially be taken as the true error. There are two fundamental related questions (Dougherty, 2012): (1) Given the reported estimate from the preliminary study, is it prudent to commit large resources to the follow-on study? (2) Prior to that, is it possible that the preliminary study can obtain an error estimate that would warrant a decision to perform a follow-on study? A large follow-on study requires substantially more resources than those required for a preliminary study. If the preliminary study has a very low probability of producing reproducible results, then there is no purpose in doing it.

In Yousefi and Dougherty (2012) a reproducibility index is proposed to address these issues. Consider a small-sample preliminary study yielding a classifier ψ_n designed from a sample of size n with true error ε_n and estimated error $\hat{\varepsilon}_n$. The error estimate must be sufficiently small to motivate a large follow-on study. This means there is a threshold τ such that the second study occurs if and only if $\hat{\varepsilon}_n \leq \tau$. The original study is said to be *reproducible with accuracy* $\rho \geq 0$ if $\varepsilon_n \leq \hat{\varepsilon}_n + \rho$ (our concern being whether true performance is within a tolerable distance from the small-sample estimated error). These observations lead to the definition of the *reproducibility index*:

$$R_n(\rho, \tau) = P(\varepsilon_n \leq \hat{\varepsilon}_n + \rho \mid \hat{\varepsilon}_n \leq \tau). \tag{3.22}$$

The reproducibility index $R_n(\rho, \tau)$ depends on the classification rule, the error-estimation rule, and the feature–label distribution. Clearly, $\rho_1 \leq \rho_2$ implies $R_n(\rho_1, \tau) \leq R_n(\rho_2, \tau)$. If $|\varepsilon_n - \hat{\varepsilon}_n| \leq \rho$ almost surely, meaning $P(|\varepsilon_n - \hat{\varepsilon}_n| \leq \rho) = 1$, then $R_n(\rho, \tau) = 1$ for all τ. This means that, if the true and estimated errors are sufficiently close, then the reproducibility index is 1 for any τ. In practice, this ideal situation does not occur. Often, the true error greatly exceeds the estimated error when the latter is small, thereby driving down $R_n(\rho, \tau)$ for small τ.

In Yousefi and Dougherty (2012) the relationship between reproducibility and classification difficulty is examined. Fixing the classification rule and error estimator makes $R_n(\rho, \tau)$ dependent on the feature–label distribution $F_{\mathbf{X}Y}$. In a parametric setting, $R_n(\rho, \tau; \theta)$ depends on $F_{\mathbf{X}Y}(\theta)$. If there is a one-to-one monotone relation between θ and the Bayes error ε_{bay}, then there is a direct relationship between reproducibility and intrinsic classification difficulty.

To illustrate the reproducibility index we use a Gaussian model with common covariance matrix and uncorrelated features, $N_d(\boldsymbol{\mu}_y, \sigma^2 I)$, for $y = 0, 1$, where $\boldsymbol{\mu}_0 = (0, \dots, 0)$ and $\boldsymbol{\mu}_1 = (0, \dots, 0, \theta)$; note that there is a direct relationship between θ and ε_{bay}. A total of 10,000 random samples are generated. For each sample, the true and estimated error pairs are calculated. Figure 3.11 shows the reproducibility index as a function of $(\tau, \varepsilon_{\text{bay}})$ for LDA, 5-fold cross-validation, five features, $n = 60, 120$, and two values of ρ in (3.22) As the Bayes error increases, a larger reproducibility index is achieved only for larger τ; however, for $\rho = 0.0005 \approx 0$, the upper bound for the reproducibility index is about 0.5. We observe that the rate of change in the reproducibility index decreases for larger Bayes error and smaller sample size, which

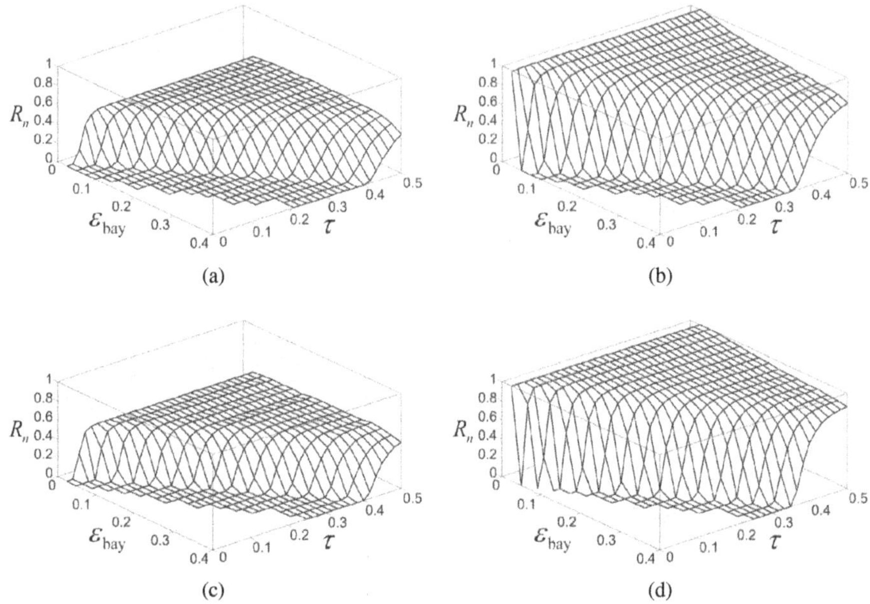

FIGURE 3.11 Reproducibility index for $n = 60, 120$, LDA classification rule and 5-fold cross-validation error estimation; $d = 5$ and the covariance matrices are equal with features uncorrelated. (a) $n = 60$, $\rho = 0.0005$; (b) $n = 60$, $\rho = 0.05$; (c) $n = 120$, $\rho = 0.0005$; (d) $n = 120$, $\rho = 0.05$. Reproduced from Yousefi and Dougherty (2012) with permission from Oxford University Press.

should be expected because these conditions increase the inaccuracy of error estimation. The figure reveals an irony. It seems prudent to only perform a large follow-on experiment if the estimated error is small in the preliminary study. This is accomplished by choosing a small value of τ. The figure reveals that the reproducibility index drops off once τ is chosen below a certain point. While small τ decreases the likelihood of a follow-on study, it increases the likelihood that the preliminary results are not reproducible. Seeming prudence is undermined by poor error estimation.

The impact of multiple-data-set and multiple-rule biases on the reproducibility index is also considered in Yousefi and Dougherty (2012). As might be expected, those biases have a detrimental effect on reproducibility. A method for application of the reproducibility index before any experiment is performed based on prior knowledge is also presented in the same reference.

EXERCISES

3.1 For the model M1 of Exercise 1.8, the LDA, QDA, and 3NN classification rules, $d = 5, \sigma = 1, \rho = 0, 0.8, n = 25, 50, 200$, and Bayes errors $0.05, 0.15, 0.25, 0.35$, plot the following curves for leave-one-out, 5-fold cross-validation, .632

bootstrap, and bolstered resubstitution: (a) conditional expectation of the true error on the estimated error; (b) 95% confidence bound for the true error as a function of the estimated error; (c) conditional expectation of the estimated error on the true error; (d) 95% confidence bound for the estimated error as a function of the true error; (e) linear regression of the true error on the estimated error; and (f) linear regression of the estimated error on the true error.

3.2 For the model M1 of Exercise 1.8, the LDA, QDA, and 3NN classification rules, $d = 5, \sigma = 1, \rho = 0, 0.8, n = 25, 50, 200$, and Bayes errors $0.05, 0.15, 0.25, 0.35$, consider (3.1) with leave-one-out, 5-fold cross-validation, and .632 bootstrap. Find $E_{S_n}[\hat{\varepsilon}_n^{i_{\min}}] - E_{S_n}[\varepsilon_n^{i_{\min}}]$ and $E_{S_n}[\hat{\varepsilon}_n^{\min}] - E_{S_n}[\varepsilon_n]$, for $m = 1, 2, \ldots, 15$.

3.3 For the model M1 of Exercise 1.8, $d = 5$, $\sigma = 1$, $\rho = 0, 0.8$, $n = 25, 50, 200$, and Bayes errors $0.05, 0.15, 0.25, 0.35$, consider (3.5) with the LDA, QDA, linear SVM, and 3NN classification rules and the leave-one-out, 5-fold cross-validation, .632 bootstrap and bolstered resubstitution error estimators. Find $\text{Bias}_{\text{Av}}, \text{Var}_{\text{Av}}, \text{RMS}_{\text{Av}}$, and $E_{\Phi_m}[C_{\text{bias}} | \Phi_m]$ as functions of m.

3.4 Redo Exercise 3.3 but now include multiple data sets.

3.5 For the model M1 of Exercise 1.8, the LDA, QDA, and 3NN classification rules, $d = 5, \sigma = 1, \rho = 0, 0.8$, and $n = 25, 50, 200$, compute the reproducibility index as a function of $(\tau, \varepsilon_{\text{bay}})$, for leave-one-out, 5-fold cross-validation, .632 bootstrap, and bolstered resubstitution, and $\rho = 0.05, 0.005$ in (3.22).

CHAPTER 4

ERROR ESTIMATION FOR DISCRETE CLASSIFICATION

Discrete classification was discussed in detail in Section 1.5. The study of error estimation for discrete classifiers is a fertile topic, as analytical characterizations of performance are often possible due to the simplicity of the problem. In fact, the entire joint distribution of the true and estimated errors can be computed exactly in this case by using computation-intensive complete enumeration methods.

The first comment on error estimation for discrete classification in the literature seems to have appeared in a 1961 paper by Cochran and Hopkins (1961), who argued that if the conditional probability mass functions are equal over any discretization bin (i.e., point in the discrete feature space), then the expected difference between the resubstitution error and the optimal error is proportional to $n^{-1/2}$. This observation was later proved and extended by Glick (1973), as will be discussed in this chapter. Another early paper on the subject was written by Hills (1966), who showed, for instance, that resubstitution is always optimistically biased for the discrete histogram rule. A few important results regarding error estimation for histogram classification (not necessarily discrete) were shown by Devroye et al. (1996), and will also be reviewed here. Exact expressions for the moments of resubstitution and leave-one-out, and their correlation with the true error, as well as efficient algorithms for the computation of the full marginal and joint distributions of the estimated and true errors, are found in Braga-Neto and Dougherty (2005, 2010), Chen and Braga-Neto (2010), Xu et al. (2006).

Error Estimation for Pattern Recognition, First Edition. Ulisses M. Braga-Neto and Edward R. Dougherty.

4.1 ERROR ESTIMATORS

Here we provide the definitions and simple properties of the main error estimators for the discrete histogram rule, which is the most important example of a discrete classification rule. General definitions and properties concerning the discrete histogram rule, including the true error rate, were given in Section 1.5.

4.1.1 Resubstitution Error

The resubstitution error estimator is the error rate of the designed discrete histogram classifier on the training data and in this case can be written as

$$\hat{\varepsilon}_n^r = \frac{1}{n} \sum_{i=1}^{b} \min\{U_i, V_i\} = \frac{1}{n} \sum_{i=1}^{b} [U_i I_{V_i > U_i} + V_i I_{U_i \geq V_i}], \tag{4.1}$$

where the variables $U_i, V_i, i = 1, \ldots, b$ are the bin counts associated with the sample data (c.f. Section 1.5). The resubstitution estimator is nonrandomized and very fast to compute. For a concrete example, the resubstitution estimate of the classification error in Figure 1.3 is $12/40 = 0.3$.

In the discrete case, the resubstitution estimator has the property of being the *plug-in* or nonparametric maximum-likelihood (NPMLE) estimator of the Bayes error. To see this, recall that the Bayes error for discrete classification is given in (1.56) and the maximum-likelihood estimators of the distributional parameters are given in (1.58). Plugging these estimators into the Bayes error results in expression (4.1).

4.1.2 Leave-One-Out Error

The leave-one-out error error estimator $\hat{\varepsilon}_n^l$ counts errors committed by n classifiers, each designed on $n - 1$ points and tested on the remaining left-out point, and divides the total count by n. A little reflection shows that this leads to

$$\hat{\varepsilon}_n^l = \frac{1}{n} \sum_{i=1}^{b} [U_i I_{V_i \geq U_i} + V_i I_{U_i \geq V_i - 1}]. \tag{4.2}$$

For example, in Figure 1.3 of Chapter 1, the leave-one-out estimate of the classification error is $15/40 = 0.375$. This is higher than the resubstitution estimate of 0.3. In fact, it is clear from (4.1) and (4.2) that

$$\hat{\varepsilon}_n^l = \hat{\varepsilon}_n^r + \frac{1}{n} \sum_{i=1}^{b} [U_i I_{V_i = U_i} + V_i I_{U_i = V_i - 1}]. \tag{4.3}$$

so that, in all cases, $\hat{\varepsilon}_n^l \geq \hat{\varepsilon}_n^r$; in particular, $E[\hat{\varepsilon}_n^l] \geq E[\hat{\varepsilon}_n^r]$, showing that the leave-one-out estimator must necessarily be less optimistic than the resubstitution estimator.

In fact, as was discussed in Section 2.5, it is always true that $E[\hat{\varepsilon}_n^l] = E[\varepsilon_{n-1}]$, making this estimator almost unbiased. This bias reduction is accomplished at the expense of an increase in variance (Braga-Neto and Dougherty, 2004b).

4.1.3 Cross-Validation Error

We consider in this chapter the non-randomized, complete cross-validation estimator of the overall error rate, also known as the "leave-k-out estimator" (Shao, 1993). This estimator averages the results of deleting each of $\binom{n}{k}$ possible folds from the data, designing the classifier on the remaining data points, and testing it on the data points in the removed fold. The resulting estimator can be written as

$$\hat{\varepsilon}_n^{l(k)} = \frac{1}{n\binom{n}{k}} \sum_{i=1}^{b} \sum_{\substack{k_1,k_2=0\\ k_1+k_2\le k}}^{k} \binom{n-U_i-V_i}{k-k_1-k_2} \left[k_1 \binom{U_i}{k_1} I_{V_i \ge U_i - k_1 + 1} + k_2 \binom{V_i}{k_2} I_{U_i \ge V_i - k_2} \right], \tag{4.4}$$

with the convention that $\binom{n}{k} = 0$ for $k > n$. Clearly, (4.4) reduces to (4.2) if $k = 1$.

4.1.4 Bootstrap Error

As discussed in Section 2.6, bootstrap samples are drawn from an empirical distribution F_{XY}^*, which assigns discrete probability mass $\frac{1}{n}$ at each observed data point in S_n. In the discrete case, it is straightforward to see that F_{XY}^* is a discrete distribution specified by the parameters

$$\hat{c}_0 = \frac{N_0}{n},\ \hat{c}_1 = \frac{N_1}{n} \quad \text{and} \quad \hat{p}_i = \frac{U_i}{N_0},\ \hat{q}_i = \frac{V_i}{N_1}, \quad \text{for } i = 1, \ldots, b. \tag{4.5}$$

These are the maximum-likelihood (ML) estimators for the model parameters of the original discrete distribution, given in (1.58). A bootstrap sample $S_n^* = \{(X_1^*, Y_1^*), \ldots, (X_n^*, Y_n^*)\}$ from the empirical distribution corresponds to n draws with replacement from an urn model with n balls of $2b$ distinct colors, with U_1 balls of the first kind, U_2 balls of the second kind, and so on, up to V_b balls of the $(2b)$-th kind. The counts for each category in the bootstrap sample are given by the vector $\{U_1^*, \ldots, U_b^*, V_1^*, \ldots, V_b^*\}$, which is multinomially distributed with parameters

$$(n, \hat{c}_0\hat{p}_1, \ldots, \hat{c}_0\hat{p}_b, \hat{c}_1\hat{q}_1, \ldots, \hat{c}_1\hat{q}_b) = \left(n, \frac{u_1}{n}, \ldots, \frac{u_b}{n}, \frac{v_1}{n}, \ldots, \frac{v_b}{n}\right), \tag{4.6}$$

where u_i, v_i are the observed sample values of the random variables U_i and V_i, respectively, for $i = 1, \ldots, b$.

The zero bootstrap error estimation rule was introduced in Section 2.6. Given a sample S_n, independent bootstrap samples $S_n^{*,j} = \{(X_1^{*,j}, Y_1^{*,j}), \ldots, (X_n^{*,j}, Y_n^{*,j})\}$, for $j = 1, \dots, B$, are generated. In the discrete case, each bootstrap sample $S_n^{*,j}$ is determined by the vector of counts $\{U_1^{*,j}, \dots, U_b^{*,j}, V_1^{*,j}, \dots, V_b^{*,j}\}$. The zero bootstrap error estimator can be written in the discrete case as:

$$\hat{\varepsilon}_n^{\text{boot}} = \frac{\sum_{j=1}^{B} \sum_{i=1}^{n} (I_{U_{X_i}^{*,j} < V_{X_i}^{*,j}} I_{Y_i=0} + I_{U_{X_i}^{*,j} \geq V_{X_i}^{*,j}} I_{Y_i=1}) I_{(X_i,Y_i) \notin S_n^{*,j}}}{\sum_{j=1}^{B} \sum_{i=1}^{n} I_{(X_i,Y_i) \notin S_n^{*,j}}}. \tag{4.7}$$

The complete zero bootstrap estimator introduced in (2.48) is given by

$$\begin{aligned}
\hat{\varepsilon}_n^{\text{zero}} &= E\left[\hat{\varepsilon}_n^{\text{boot}} \mid S_n\right] \\
&= P\left(U_{X^*}^* < V_{X^*}^*, Y^* = 0 \mid S_n\right) + P\left(U_{X^*}^* \geq V_{X^*}^*, Y^* = 1 \mid S_n\right) \\
&= P\left(U_{X^*}^* < V_{X^*}^* \mid Y^* = 0, S_n\right) P\left(Y^* = 0 \mid S_n\right) \\
&\quad + P\left(U_{X^*}^* \geq V_{X^*}^* \mid Y^* = 1, S_n\right) P\left(Y^* = 1 \mid S_n\right),
\end{aligned} \tag{4.8}$$

where $(X^*, Y^*) \sim F_{XY}^*$ and is independent of S_n^*. Now, $P(Y^* = i \mid S_n) = \hat{c}_i$, for $i = 0, 1$, and we can develop (4.8) further:

$$\begin{aligned}
\hat{\varepsilon}_n^{\text{zero}} &= \hat{c}_0\, E\left[P\left(U_{X^*}^* < V_{X^*}^* \mid X^*, Y^* = 0, S_n\right)\right] \\
&\quad + \hat{c}_1\, E\left[P\left(U_{X^*}^* \geq V_{X^*}^* \mid X^*, Y^* = 1, S_n\right)\right] \\
&= \hat{c}_0 \sum_{i=1}^{b} P\left(U_{X^*}^* < V_{X^*}^* \mid X^* = i, Y^* = 0, S_n\right) \hat{p}_i \\
&\quad + \hat{c}_1 \sum_{i=1}^{b} P\left(U_{X^*}^* \geq V_{X^*}^* \mid X^* = i, Y^* = 1, S_n\right) \hat{q}_i \\
&= \sum_{i=1}^{b} \hat{c}_0 \hat{p}_i P\left(U_i^* < V_i^* \mid S_n\right) + \sum_{i=1}^{b} \hat{c}_1 \hat{q}_i P\left(U_i^* \geq V_i^* \mid S_n\right).
\end{aligned} \tag{4.9}$$

Using the fact, mentioned in Section 1.5, that the vector (U_i, V_i) is multinomial with parameters $(n, c_0 p_i, c_1 q_i, 1 - c_0 p_i - c_1 q_i)$, while the vector (U_i^*, V_i^*) is multinomial with parameters $(n, \hat{c}_0 \hat{p}_i, \hat{c}_1 \hat{q}_i, 1 - \hat{c}_0 \hat{p}_i - \hat{c}_1 \hat{q}_i)$, we see that $P(U_i^* < V_i^* \mid S_n)$ and $P(U_i^* \geq V_i^* \mid S_n)$ are obtained from $P(U_i < V_i)$ and $P(U_i \geq V_i)$, respectively, by substituting the MLEs of the distributional parameters. If we then compare the expression for $\hat{\varepsilon}_n^{\text{zero}}$ in (4.9) with the expression for $E[\varepsilon_n]$ in (1.60), we can see that the former can be obtained from the latter by substituting everywhere the parameter MLEs. Hence, we obtain the interesting conclusion that the complete zero bootstrap estimator $\hat{\varepsilon}_n^{\text{zero}}$ is a maximum-likelihood estimator of the expected error rate $E[\varepsilon_n]$. Since the zero bootstrap is generally positively biased, this is a biased MLE.

Finally, the .632 bootstrap error estimator can then be defined by applying (2.49),

$$\hat{\varepsilon}_n^{632} = 0.368\,\hat{\varepsilon}_n^{r} + 0.632\,\hat{\varepsilon}_n^{\text{boot}}\,, \tag{4.10}$$

and using (4.1) and (4.7). Alternatively, the complete zero bootstrap $\hat{\varepsilon}_n^{\text{zero}}$ in (4.8) can be substituted for $\hat{\varepsilon}_n^{\text{boot}}$.

4.2 SMALL-SAMPLE PERFORMANCE

The fact that the distribution of the vector of bin counts $(U_1, \dots, U_b, V_1, \dots, V_b)$ is multinomial, and thus easily computable, together with the simplicity of the expressions derived for the various error estimators in the previous section, allows the detailed analytical study of small-sample performance in terms of bias, deviation variance, RMS, and correlation coefficient between true and estimated errors (Braga-Neto and Dougherty, 2005, 2010).

4.2.1 Bias

The leave-one-out and cross-validation estimators display the general well-known bias properties discussed in Section 2.5. In this section, we focus on the bias of the resubstitution estimator.

Taking expectation on both sides of (4.1) results in

$$E\left[\hat{\varepsilon}_n^{r}\right] = \frac{1}{n}\sum_{i=1}^{b}[E[U_i I_{V_i>U_i}] + E[V_i I_{U_i\geq V_i}]]. \tag{4.11}$$

Using the expression for the expected true error rate obtained in (1.60) leads to the following expression for the bias:

$$\begin{aligned}\text{Bias}\left[\hat{\varepsilon}_n^{r}\right] &= E\left[\hat{\varepsilon}_n^{r}\right] - E[\varepsilon_n]\\ &= \sum_{i=1}^{b}\left\{E\left[\left(\frac{U_i}{n} - c_0 p_i\right) I_{V_i>U_i}\right] + E\left[\left(\frac{V_i}{n} - c_1 q_i\right) I_{U_i\geq V_i}\right]\right\},\end{aligned} \tag{4.12}$$

where

$$E\left[\left(\frac{U_i}{n} - c_0 p_i\right) I_{V_i>U_i}\right] =$$

$$\sum_{\substack{0\leq k+l\leq n\\ k<l}} \left(\frac{k}{n} - c_0 p_i\right) \frac{n!}{k!\,l!\,(n-k-l)!}(c_0 p_i)^k (c_1 q_i)^l (1-c_0 p_i-c_1 q_i)^{n-k-l}, \tag{4.13}$$

with a similar expression for $E[(\frac{V_i}{n} - c_1 q_i) I_{U_i \geq V_i}]$. We can interpret this as follows. The average contributions to the expected true error and expected resubstitution error are $c_0 p_i$ and $\frac{U_i}{n}$, respectively, if bin i is assigned label 1; they are $c_1 q_i$ and $\frac{V_i}{n}$, respectively, if bin i is assigned label 0. The bias of resubstitution is the average difference between these contributions.

Next we show that the resubstitution bias is always negative, except in trivial cases, for any sample size and distribution of the data, thereby providing an example where the often assumed optimistic bias of resubstitution is indeed guaranteed (resubstitution is not always optimistic for a general classification rule). In fact, we show that

$$E\left[\hat{\varepsilon}_n^r\right] \leq \varepsilon_{\text{bay}} \leq E[\varepsilon_n]. \tag{4.14}$$

Since $\varepsilon_{\text{bay}} = E[\varepsilon_n]$ only in trivial cases, this proves that the bias is indeed negative. The proof of the first inequality is as follows:

$$\begin{aligned} E\left[\hat{\varepsilon}_n^r\right] &= E\left[\frac{1}{n}\sum_{i=1}^{b} \min\{U_i, V_i\}\right] \\ &= \frac{1}{n}\sum_{i=1}^{b} E[\min\{U_i, V_i\}] \\ &\leq \frac{1}{n}\sum_{i=1}^{b} \min\{E[U_i], E[V_i]\} \\ &= \frac{1}{n}\sum_{i=1}^{b} \min\{nc_0 p_i, nc_1 q_i\} \\ &= \sum_{i=1}^{b} \min\{c_0 p_i, c_1 q_i\} = \varepsilon_{\text{bay}}, \end{aligned} \tag{4.15}$$

where we used Jensen's inequality and the fact that U_i and V_i are distributed as binomial random variables with parameters $(n, c_0 p_i)$ and $(n, c_1 q_i)$, respectively. Hence, the expected resubstitution estimate bounds from below even the Bayes error. This fact seems to have been demonstrated for the first time in Cochran and Hopkins (1961) (see also Hills (1966)). Observe, however, that (4.14) applies to the expected values. There are no guarantees that $\hat{\varepsilon}_n^r \leq \varepsilon_{\text{bay}}$ or even $\hat{\varepsilon}_n^r \leq \varepsilon_n$ for a given training data set. The optimistic bias of resubstitution tends to be larger when the number of bins b is large compared to the sample size; in other words, there is more overfitting of the classifier to the training data in such cases. On the other hand, resubstitution tends to have small variance. In cases where the bias is not too large, this makes resubstitution a very competitive option as an error estimator (e.g., see the results in Section 4.2.4).

In addition, it can be shown that, as the sample size increases, both the bias and variance of resubstitution vanish (see Section 4.3).

4.2.2 Variance

It follows from (4.1) that the variance of the resubstitution error estimator is given by

$$\begin{aligned}\text{Var}\left(\hat{\varepsilon}_n^r\right) &= \frac{1}{n^2}\sum_{i=1}^{b} \text{Var}(U_i I_{U_i<V_i}) + \text{Var}(V_i I_{U_i\geq V_i}) + 2\text{Cov}(U_i I_{U_i<V_i}, V_i I_{U_i\geq V_i}) \\ &+ \frac{2}{n^2}\sum_{i<j}\Big[\text{Cov}(U_i I_{U_i<V_i}, U_j I_{U_j<V_j}) + \text{Cov}(U_i I_{U_i<V_i}, V_j I_{U_j\geq V_j}) \\ &+ \text{Cov}(U_j I_{U_j<V_j}, V_i I_{U_i\geq V_i}) + \text{Cov}(V_i I_{U_i\geq V_i}, V_j I_{U_j\geq V_j})\Big]. \end{aligned} \tag{4.16}$$

Similarly, the variance of the leave-one-out error estimator follows from (4.2):

$$\begin{aligned}\text{Var}\left(\hat{\varepsilon}_n^l\right) &= \frac{1}{n^2}\sum_{i=1}^{b} \text{Var}(U_i I_{U_i\leq V_i}) + \text{Var}(V_i I_{U_i\geq V_i-1}) + 2\text{Cov}(U_i I_{U_i\leq V_i}, V_i I_{U_i\geq V_i-1}) \\ &+ \frac{2}{n^2}\sum_{i<j}\Big[\text{Cov}(U_i I_{U_i\leq V_i}, U_j I_{U_j\leq V_j}) + \text{Cov}(U_i I_{U_i\leq V_i}, V_j I_{U_j\geq V_j-1}) \\ &+ \text{Cov}(U_j I_{U_j\leq V_j}, V_i I_{U_i\geq V_i-1}) + \text{Cov}(V_i I_{U_i\geq V_i-1}, V_j I_{U_j\geq V_j-1})\Big]. \end{aligned} \tag{4.17}$$

Exact expressions for the variances and covariances appearing in (4.16) and (4.17) can be obtained by using the joint multinomial distribution of the bin counts U_i, V_i. For (4.16),

$$\begin{aligned}\text{Var}(U_i I_{U_i<V_i}) &= E[U_i^2 I_{U_i<V_i}] - \left(E[U_i I_{U_i<V_i}]\right)^2 \\ &= \sum_{k<l} k^2 P(U_i = k, V_i = l) - \left(\sum_{k<l} k\, P(U_i = k, V_i = l)\right)^2, \end{aligned} \tag{4.18}$$

$$\begin{aligned}\text{Var}(V_i I_{U_i\geq V_i}) &= E\left[V_i^2 I_{U_i\geq V_i}\right] - (E[V_i I_{U_i\geq V_i}])^2 \\ &= \sum_{k\geq l} l^2 P(U_i = k, V_i = l) - \left(\sum_{k\geq l} l\, P(U_i = k, V_i = l)\right)^2, \end{aligned} \tag{4.19}$$

$$\begin{aligned}\text{Cov}(U_i I_{U_i<V_i}, V_i I_{U_i\geq V_i}) &= E[U_i V_i I_{U_i<V_i} I_{U_i\geq V_i}] - E[U_i I_{U_i<V_i}]E[V_i I_{U_i\geq V_i}] \\ &= -\left(\sum_{k<l} k\, P(U_i = k, V_i = l)\right)\left(\sum_{k\geq l} l\, P(U_i = k, V_i = l)\right), \end{aligned} \tag{4.20}$$

$$\begin{aligned}\text{Cov}(U_i I_{U_i<V_i}, U_j I_{U_j<V_j}) &= E[U_i U_j I_{U_i<V_i} I_{U_j<V_j}] - E[U_i I_{U_i<V_i}]E[U_j I_{U_j<V_j}] \\ &= \sum_{\substack{k<l \\ r<s}} krP(U_i = k, V_i = l, U_j = r, V_j = s) \\ &\quad - \left(\sum_{k<l} kP(U_i = k, V_i = l)\right)\left(\sum_{k<l} kP(U_j = k, V_j = l)\right),\end{aligned} \tag{4.21}$$

while the covariances $\text{Cov}(U_i I_{U_i<V_i}, V_j I_{U_j \geq V_j})$, $\text{Cov}(U_j I_{U_j<V_j}, V_i I_{U_i \geq V_i})$, and $\text{Cov}(V_i I_{U_i \geq V_i}, V_j I_{U_j \geq V_j})$ are found analogously as in (4.21).

Similarly, for (4.17),

$$\text{Var}(U_i I_{U_i \leq V_i}) = \sum_{k \leq l} k^2 P(U_i = k, V_i = l) - \left(\sum_{k \leq l} k\, P(U_i = k, V_i = l)\right)^2, \tag{4.22}$$

$$\text{Var}(V_i I_{U_i \geq V_i - 1}) = \sum_{k \geq l-1} l^2 P(U_i = k, V_i = l) - \left(\sum_{k \geq l-1} l\, P(U_i = k, V_i = l)\right)^2, \tag{4.23}$$

$$\begin{aligned}\text{Cov}(U_i I_{U_i \leq V_i}, V_i I_{U_i \geq V_i - 1}) &= \sum_{l-1 \leq k \leq l} kl\, P(U_i = k, V_i = l) \\ &\quad - \left(\sum_{k \leq l} k\, P(U_i = k, V_i = l)\right)\left(\sum_{k \geq l-1} l\, P(U_i = k, V_i = l)\right),\end{aligned} \tag{4.24}$$

$$\begin{aligned}\text{Cov}(U_i I_{U_i \leq V_i}, U_j I_{U_j \leq V_j}) &= \sum_{\substack{k \leq l \\ r \leq s}} krP(U_i = k, V_i = l, U_j = r, V_j = s) \\ &\quad - \left(\sum_{k \leq l} kP(U_i = k, V_i = l)\right)\left(\sum_{k \leq l} kP(U_j = k, V_j = l)\right),\end{aligned} \tag{4.25}$$

while the covariances $\text{Cov}(U_i I_{U_i \leq V_i}, V_j I_{U_j \geq V_j - 1})$, $\text{Cov}(U_j I_{U_j \leq V_j}, V_i I_{U_i \geq V_i - 1})$, and $\text{Cov}(V_i I_{U_i \geq V_i - 1}, V_j I_{U_j \geq V_j - 1})$ are found analogously as in (4.25).

With a little more effort, it is possible to follow the same approach and obtain expressions for exact calculation of the variance $\mathrm{Var}(\hat{\varepsilon}_n^{l(k)})$ of the leave-k-out estimator in (4.4).

4.2.3 Deviation Variance, RMS, and Correlation

From (2.5), (2.6), and (2.9), it follows that to find the deviation variance, RMS, and correlation coefficient, respectively, one needs the first moment and variance of the true and estimated errors, which have already been discussed, and the covariance between true and estimated errors. We give expressions for the latter below.

For resubstitution,

$$\begin{aligned}\mathrm{Cov}\left(\varepsilon_n, \hat{\varepsilon}_n^r\right) = \frac{1}{n}\sum_{i,j} & c_0 p_i(\mathrm{Cov}(I_{U_i<V_i}, U_j I_{U_j<V_j}) + \mathrm{Cov}(I_{U_i<V_i}, V_j I_{U_j \geq V_j})) \\ & + c_1 q_i(\mathrm{Cov}(I_{U_i \geq V_i}, U_j I_{U_j<V_j}) + \mathrm{Cov}(I_{U_i \geq V_i}, V_j I_{U_j \geq V_j})),\end{aligned} \tag{4.26}$$

while for leave-one-out, the covariance is given by

$$\begin{aligned}\mathrm{Cov}\left(\varepsilon_n, \hat{\varepsilon}_n^l\right) = \frac{1}{n}\sum_{i,j} & c_0 p_i(\mathrm{Cov}(I_{U_i<V_i}, U_j I_{U_j \leq V_j}) + \mathrm{Cov}(I_{U_i<V_i}, V_j I_{U_j \geq V_j - 1})) \\ & + c_1 q_i(\mathrm{Cov}(I_{U_i \geq V_i}, U_j I_{U_j \leq V_j}) + \mathrm{Cov}(I_{U_i \geq V_i}, V_j I_{U_j \geq V_j - 1})).\end{aligned} \tag{4.27}$$

The covariances involved in the previous equations can be calculated as follows. For (4.26),

$$\begin{aligned}\mathrm{Cov}(I_{U_i<V_i}, U_j I_{U_j<V_j}) &= E[U_j I_{U_i<V_i} I_{U_j<V_j}] - E[I_{U_i<V_i}] E[U_j I_{U_j<V_j}] \\ &= \sum_{\substack{k<l \\ r<s}} r P(U_i = k, V_i = l, U_j = r, V_j = s) \\ &\quad - P(U_i < V_i)\left(\sum_{k<l} k P(U_j = k, V_j = l)\right).\end{aligned} \tag{4.28}$$

When $i = j$, this reduces to

$$\begin{aligned}\mathrm{Cov}(I_{U_i<V_i}, U_i I_{U_i<V_i}) &= E[U_i I_{U_i<V_i}] - E[I_{U_i<V_i}] E[U_i I_{U_i<V_i}] \\ &= [1 - P(U_i < V_i)] E[U_i I_{U_i<V_i}] \\ &= P(U_i \geq V_i)\left(\sum_{k<l} k P(U_i = k, V_i = l)\right),\end{aligned} \tag{4.29}$$

while the covariances $\text{Cov}(I_{U_i<V_i}, V_j I_{U_j \geq V_j})$, $\text{Cov}(I_{U_i \geq V_i}, U_j I_{U_j<V_j})$, and $\text{Cov}(I_{U_i \geq V_i}, V_j I_{U_j \geq V_j})$ are found analogously as in (4.28) and (4.29). Similarly, for (4.27),

$$\begin{aligned}\text{Cov}(I_{U_i<V_i}, U_j I_{U_j \leq V_j}) &= E[U_j I_{U_i<V_i} I_{U_j \leq V_j}] - E[I_{U_i<V_i}]E[U_j I_{U_j \leq V_j}] \\ &= \sum_{\substack{k<l \\ r \leq s}} r P(U_i = k, V_i = l, U_j = r, V_j = s) \\ &\quad - P(U_i < V_i)\left(\sum_{k \leq l} k P(U_j = k, V_j = l)\right). \end{aligned} \tag{4.30}$$

When $i = j$, this reduces to

$$\begin{aligned}\text{Cov}(I_{U_i<V_i}, U_i I_{U_i \leq V_i}) &= E[U_i I_{U_i<V_i}] - E[I_{U_i<V_i}]E[U_i I_{U_i \leq V_i}] \\ &= \left(\sum_{k<l} k P(U_i = k, V_i = l)\right) \\ &\quad - P(U_i < V_i)\left(\sum_{k \leq l} k P(U_i = k, V_i = k)\right), \end{aligned} \tag{4.31}$$

while the covariances $\text{Cov}(I_{U_i<V_i}, V_j I_{U_j \geq V_j - 1})$, $\text{Cov}(I_{U_i \geq V_i}, U_j I_{U_j \leq V_j})$, and $\text{Cov}(I_{U_i \geq V_i}, V_j I_{U_j \geq V_j - 1})$ are found analogously as in (4.30) and (4.31).

Note that, for computational efficiency, one can pre-compute and store many of the terms in the formulas above, such as $P(U_i < V_i)$, $E[U_i I_{U_i<V_i}]$, $E[V_i I_{U_i<V_i}]$, and so forth, and reuse them multiple times in the calculations.

4.2.4 Numerical Example

As an illustration, Figure 4.1 displays the exact bias, deviation standard deviation, and RMS of resubstitution and leave-one-out, plotted as functions of the number of bins. For comparison, also plotted are Monte-Carlo estimates (employing $M = 20000$ Monte-Carlo samples) of 10 repetitions of 4-fold cross-validation (cv) — this is not complete cross-validation, but the ordinary randomized version — and the .632 bootstrap (b632), with $B = 100$ bootstrap samples. A simple power-law model (a "Zipf" model) is used for the probabilities: $p_i = Ki^{\alpha}$ and $q_i = p_{b-i+1}$, for $i = 1, \ldots, b$, where $\alpha > 0$ and K is a normalizing constant to get the probabilities to sum to 1. In this example, $n_0 = n_1 = 10$, $c_0 = c_1 = 0.5$, and the parameter α is adjusted to yield $E[\varepsilon_n] \approx 0.20$, which corresponds to intermediate classification difficulty under a small sample size. One can see some jitter in the curves for the standard deviation and RMS of cross-validation and bootstrap; that is due to the Monte-Carlo approximation. By contrast, no such jitter is present in the curves for resubstitution and leave-one-out, since those results are exact, following the expressions obtained in the previous subsections. From the plots one can see that resubstitution is the most

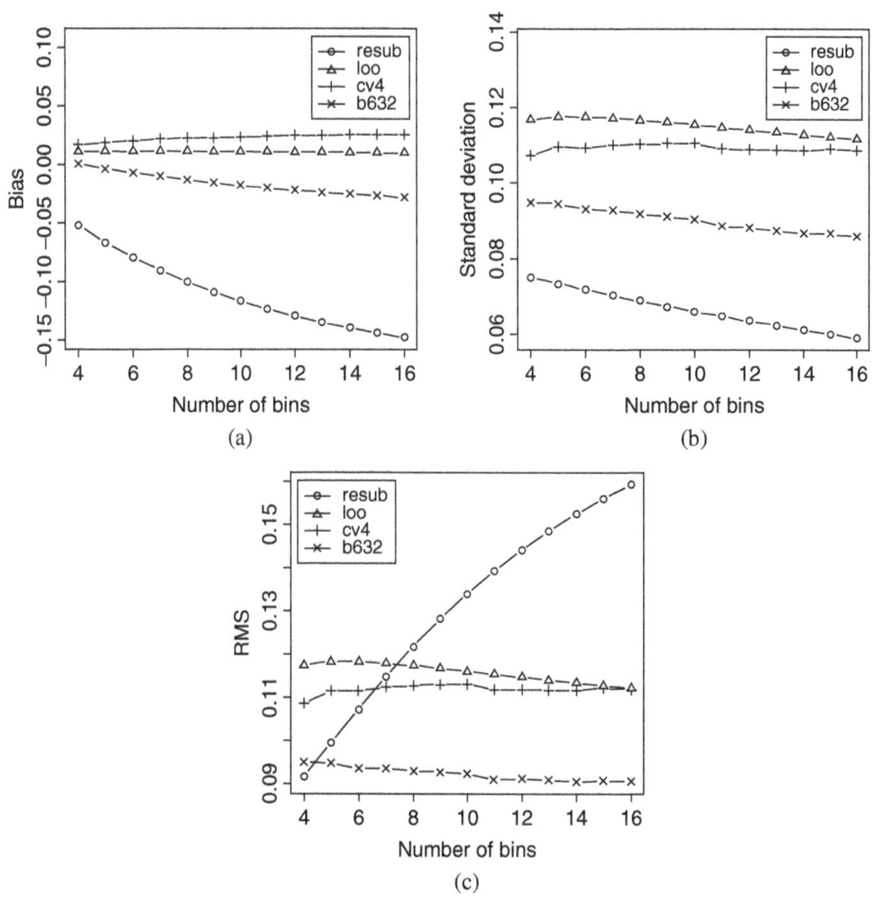

FIGURE 4.1 Performance of error estimators: (a) bias, (b) deviation standard deviation, and (c) RMS, plotted versus number of bins, for $n = 20$ and $E[\varepsilon_n] = 0.2$. The curves for resubstitution and leave-one-out are exact; the curves for the other error estimators are approximations based on Monte-Carlo computation.

optimistically biased estimator, with bias that increases with complexity, but it is also much less variable than all other estimators, including the bootstrap. Both leave-one-out and 4-fold cross-validation estimators are pessimistically biased, with the bias being small in both cases, and a little bigger in the case of cross-validation. Those two estimators are also the most variable. The bootstrap estimator is affected by the bias of resubstitution when complexity is high, since it incorporates the resubstitution estimate in its computation, but it is clearly superior to the cross-validation estimators in RMS. Perhaps the most remarkable observation is that, for low complexity classifiers ($b < 8$), the simple resubstitution estimator becomes more accurate, according to RMS, than both cross-validation error estimators and slightly more accurate than even the .632 bootstrap error estimator for $b = 4$. This

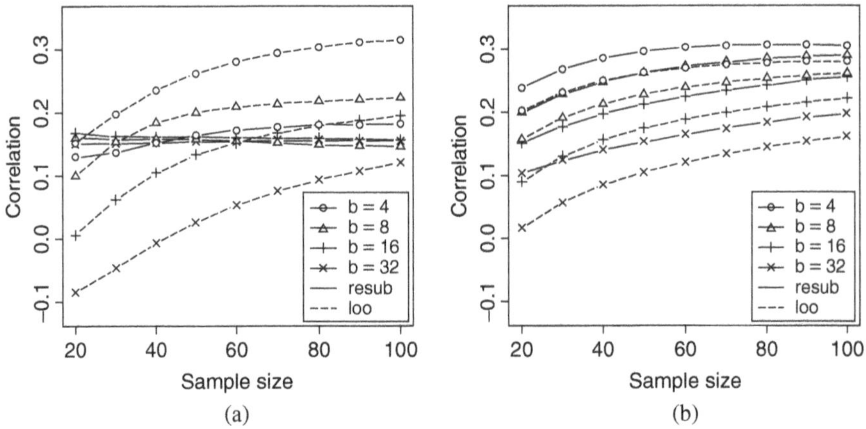

FIGURE 4.2 Exact correlation between the true error and the resubstitution and leave-one-out error estimators for feature–label distributions of distinct difficulty, as determined by the Bayes error. Left panel: Bayes error = 0.10. Right panel: Bayes error = 0.40.

is despite the fact that resubstitution is typically much faster to compute than those error estimators (in some cases considered in Braga-Neto and Dougherty (2004b), hundreds of times faster). Therefore, under small sample sizes, resubstitution is the estimator of choice, provided that there is a small number of bins compared to sample size.

Figure 4.2 displays the exact correlation between the true error and the resubstitution and leave-one-out error estimators, plotted versus sample size, for different bin sizes. In this example, we assume the Zipf parametric model, with $c_0 = c_1 = 0.5$ and parameter α adjusted to yield two cases: easy (Bayes error = 0.1) and difficult (Bayes error = 0.4) classification. We observe that the correlation is generally low (below 0.3). We also observe that at small sample sizes correlation for resubstitution is larger than for leave-one-out and, with a larger difficulty of classification, this is true even at moderate sample sizes. Correlation generally decreases with increasing bin size; in one striking case, the correlation for leave-one-out becomes negative at the acute small-sample situation of $n = 20$ and $b = 32$ (recall the example in Section 2.5 where leave-one-out also displays negative correlation with the true error). This poor correlation for leave-one-out mirrors the poor deviation variance of this error estimator, which is known to be large under complex models and small sample sizes (Braga-Neto and Dougherty, 2004b; Devroye et al., 1996; Hanczar et al., 2007). Note that this is in accord with (2.10), which states that the deviation variance is in inverse relation with the correlation coefficient.

4.2.5 Complete Enumeration Approach

All performance metrics of interest for the true error ε_n and any given error estimator $\hat{\varepsilon}_n$ can be derived from the joint sampling distribution of the pair of random variables

$(\varepsilon_n, \hat{\varepsilon}_n)$. In addition to bias, deviation variance, RMS, tail probabilities, and correlation coefficient, this includes conditional metrics, such as the conditional expected true error given the estimated error and confidence intervals on the true error conditioned on the estimated error.

The latter conditional measures are difficult to study via the analytical approach outlined in Subsections 4.2.1–4.2.3, due to the complexity of the expressions involved. However, owing to the finiteness of the discrete problem, exact computation of these measures is still possible, since the joint sampling distribution of true and estimated errors in the discrete case can be computed exactly by means of a *complete enumeration* approach. Basically, complete enumeration relies on intensive computational power to list all possible configurations of the discrete data and their probabilities, and from this exact statistical properties of the metrics of interest are obtained. The availability of efficient algorithms to enumerate all possible cases on fast computers has made possible the use of complete enumeration in an increasingly wider variety of settings. In particular, such methods have been extensively used in categorical data analysis (Agresti, 1992, 2002; Hirji et al., 1987; Klotz, 1966; Verbeek, 1985). For example, complete enumeration approaches have been useful in the computation of exact distributions and critical regions for tests based on contingency tables, as in the case of the well-known Fisher exact test and the chi-square approximate test (Agresti, 1992; Verbeek, 1985).

In the case of discrete classification, recall that the random sample is specified by the vector of bin counts $\mathbf{W}_n = (U_1, \ldots, U_b, V_1, \ldots, V_b)$ defined previously, so that we can write $\varepsilon_n = \varepsilon_n(\mathbf{W}_n)$ and $\hat{\varepsilon}_n = \hat{\varepsilon}_n(\mathbf{W}_n)$. The random vector $\mathbf{W}_n$ has a multinomial distribution, as discussed previously. In addition, notice that $\mathbf{W}_n$ is discrete, and thus the random variables ε_n and $\hat{\varepsilon}_n$ are also discrete, and so is the configuration space D_n of all possible distinct sample values $\mathbf{w}_n$ that can be taken on by $\mathbf{W}_n$. The joint probability distribution of $(\varepsilon_n, \hat{\varepsilon}_n)$ can be written, therefore, as a discrete bidimensional probability mass function

$$P(\varepsilon_n = k, \hat{\varepsilon}_n = l) = \sum_{\mathbf{w}_n \in D_n} I_{\{\varepsilon_n(\mathbf{w}_n)=k,\, \hat{\varepsilon}_n(\mathbf{w}_n)=l\}} P(\mathbf{W}_n = \mathbf{w}_n), \tag{4.32}$$

where $P(\mathbf{W}_n = \mathbf{w}_n)$ is a multinomial probability. Even though the configuration space D_n is finite, it quickly becomes huge with increasing sample size n and bin size b. In Braga-Neto and Dougherty (2005) an algorithm is given to traverse D_n efficiently, which leads to reasonable computational times to evaluate the joint sampling distribution when n and b are not too large — an even more efficient algorithm for generating all the configurations in D_n is the NEXCOM routine given in Nijenhuis and Wilf (1978, Chap. 5). As an illustration, Figure 4.3 displays graphs of $P(\varepsilon_n = k, \hat{\varepsilon}_n = l)$ for the resubstitution and leave-one-out error estimators for a Zipf parametric model, with $c_0 = c_1 = 0.5$, $n = 20$, $b = 8$, and parameter α adjusted to yield intermediate classification difficulty (Bayes error $= 0.2$). One can observe that the joint distribution for resubstitution is more compact than that for leave-one-out, which explains its smaller variance and partly accounts for its larger correlation with the true error in small-sample settings.

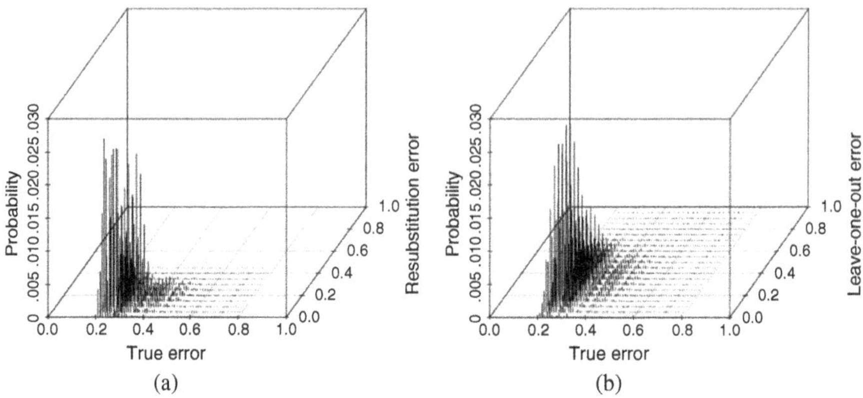

FIGURE 4.3 Exact joint distribution between the true error and the resubstitution and leave-one-out error estimators for $n = 20$ and $b = 8$, and a Zipf probability model of intermediate difficulty (Bayes error = 0.2). Reproduced from Braga-Neto and Dougherty (2010) with permission from Elsevier.

The complete enumeration approach can also be applied to compute the conditional sampling distribution $P(\varepsilon_n = k \mid \hat{\varepsilon}_n = l)$. This was done in Xu et al. (2006) in order to find exact conditional metrics of performance for resubstitution and leave-one-out error estimators. Those included the conditional expectation $E[\varepsilon_n|\hat{\varepsilon}_n]$ and conditional variance $\text{Var}(\varepsilon_n|\hat{\varepsilon}_n)$ of the true error given the estimated error, as well as the $100(1-\alpha)\%$ upper confidence bound γ_α, such that the true error is contained in a confidence interval $[0, \gamma_\alpha]$ with probability $1-\alpha$, that is, $P(\varepsilon_n < \gamma_\alpha \mid \hat{\varepsilon}_n) = 1-\alpha$. This is illustrated in Figure 4.4, where the aforementioned conditional metrics of performance for resubstitution and leave-one-out are plotted versus conditioning estimated errors for different bin sizes. In this example, we employ the Zipf parametric model, with $n_0 = n_1 = 10$, and parameter α adjusted to yield $E[\varepsilon_n] \approx 0.25$, which corresponds to intermediate classification difficulty under a small sample size. Two bin sizes are considered, $b = 4, 8$, which corresponds to truth tables with 2 and 3 variables, respectively. The conditional expectation increases with the estimated error. In addition, the conditional expected true error is larger than the estimated error for small estimated errors and smaller than the estimated error for large estimated errors. Note the flatness of the leave-one-out curves. The 95% upper confidence bounds are nondecreasing with respect to increasing estimated error, as expected. The flat spots observed in the bounds result from the discreteness of the estimation rule (a phenomenon that is more pronounced when the number of bins is smaller).

4.3 LARGE-SAMPLE PERFORMANCE

Large-sample analysis of performance has to do with the behavior of the various error estimators as sample size increases without bound, that is, as $n \to \infty$. From a

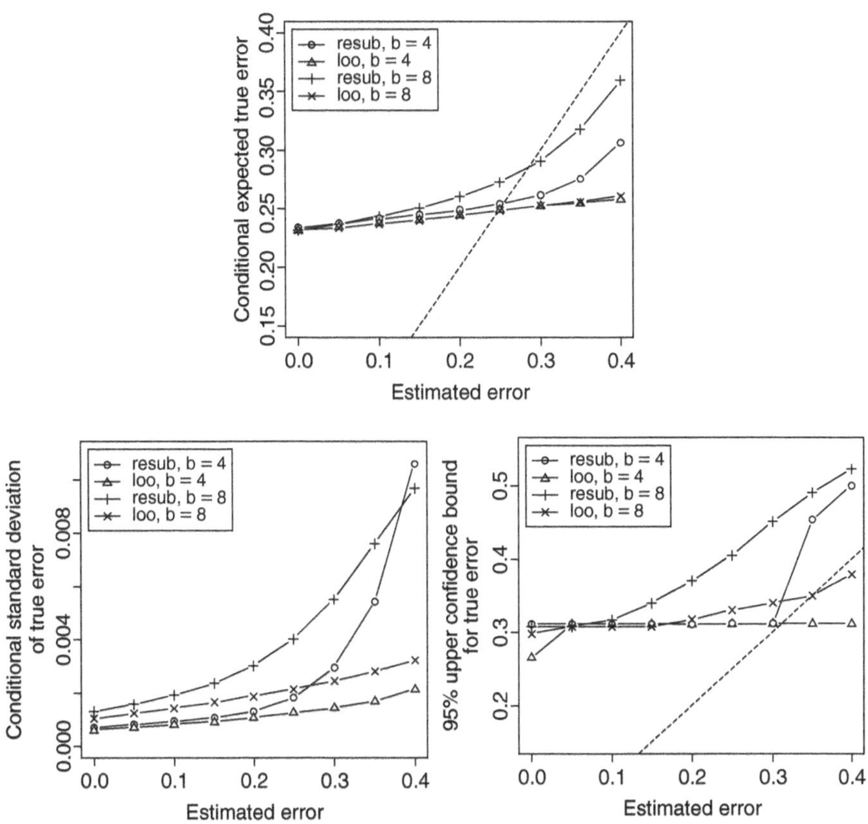

FIGURE 4.4 Exact conditional metrics of performance for resubstitution and leave-one-out error estimators. The dashed line indicates the $y = x$ line.

practical perspective, one would expect performance, both of the error estimators and of the classification rule itself, to improve with larger samples. It turns out that, not only is this true for the discrete histogram rule, but also it is possible in several cases to obtain fast (exponential) rates of convergence. Critical results in this area are due to Cochran and Hopkins (1961), Glick (1972, 1973), and Devroye et al. (1996). We will review briefly these results in this section.

A straightforward application of the Strong Law of Large Numbers (SLLN) (Feller, 1968) leads to $U_i/n \to c_0 p_i$ and $V_i/n \to c_1 q_i$ as $n \to \infty$, with probability 1. From this, it follows immediately that

$$\lim_{n\to\infty} \psi_n(i) = \lim_{n\to\infty} I_{U_i<V_i} = I_{c_0 p_i < c_1 q_i} = \psi_{\text{bay}}(i) \text{ with probability 1;} \qquad (4.33)$$

that is, the discrete histogram classifier designed from sample data converges to the optimal classifier over each bin, with probability 1. This is a distribution-free result, as the SLLN itself is distribution-free. In other words, the discrete histogram rule is universally strongly consistent.

The exact same argument shows that

$$\lim_{n\to\infty} \varepsilon_n = \lim_{n\to\infty} \hat{\varepsilon}_n^r = \varepsilon_{\text{bay}} \text{ with probability } 1, \tag{4.34}$$

so that the classification error, and also the apparent error, converge to the Bayes error as sample size increases. From (4.34) it also follows that

$$\lim_{n\to\infty} E[\varepsilon_n] = \lim_{n\to\infty} E\left[\hat{\varepsilon}_n^r\right] = \varepsilon_{\text{bay}} \,. \tag{4.35}$$

In particular, $\lim_{n\to\infty} E[\hat{\varepsilon}_n^r - \varepsilon_n] = 0$; that is, the resubstitution estimator is asymptotically unbiased. Recalling (4.14), one always has $E[\hat{\varepsilon}_n^r] \leq \varepsilon_{\text{bay}} \leq E[\varepsilon_n]$. Together with (4.35), this implies that $E[\varepsilon_n]$ converges to ε_{bay} from above, while $E[\hat{\varepsilon}_n^r]$ converges to ε_{bay} from below, as $n \to \infty$. These results also follow as a special case of the consistency results presented in Section 2.4, given that the discrete histogram rule has finite VC dimension $V_C = b$ (c.f. Section B.3).

The previous results are all distribution-free. The question arises as to the speed with which limiting performance metrics, such as the bias, are attained, as distribution-free results, such as the SLLN, can yield notoriously slow rates of convergence. As was the case for the true error rate (c.f. Section 1.5), exponential rates of convergence can be obtained, if one is willing to drop the distribution-free requirement, as we see next.

Provided that ties in bin counts are assigned a class randomly (with equal probability), the following exponential bound on the convergence of $E[\hat{\varepsilon}_n^r]$ to ε_{bay} can be obtained (Glick, 1973, Theorem B):

$$\varepsilon_{\text{bay}} - E\left[\hat{\varepsilon}_n^r\right] \leq \frac{1}{2} bn^{-1/2} e^{-cn} , \tag{4.36}$$

provided that there is no cell over which $c_0 p_i = c_1 q_i$ — this makes the result distribution-dependent. Here, the constant $c > 0$ is the same as in (1.66)–(1.67). The presence of a cell where $c_0 p_i = c_1 q_i$ invalidates the bound in (4.36) and slows down convergence; in fact, it is shown in Glick (1973) that, in such a case, $\varepsilon_{\text{bay}} - E[\hat{\varepsilon}_n^r]$ has both upper and lower bounds that are $O(n^{-1/2})$, so that convergence cannot be exponential. Finally, observe that the bounds in (1.66) and (4.36) can be combined to bound the bias of resubstitution $E[\hat{\varepsilon}_n^r - \varepsilon_n]$. We can conclude, for example, that if there are no cells over which $c_0 p_i = c_1 q_i$, convergence of the bias to zero is exponentially fast.

Distribution-free bounds on the bias, as well as the variance and RMS, can be obtained, but they do not yield exponential rates of convergence. Some of these results are reviewed next. We return for the remainder of the chapter to the convention of breaking ties deterministically in the direction of class 0. For the resubstitution error estimator, one can apply (2.35) directly to obtain

$$\left|\text{Bias}\left(\hat{\varepsilon}_n^r\right)\right| \leq 8\sqrt{\frac{b\log(n+1)+4}{2n}}, \tag{4.37}$$

while the following bounds are proved in Devroye et al. (1996, Theorem 23.3):

$$\text{Var}\left[\hat{\varepsilon}_n^r\right] \leq \frac{1}{n} \tag{4.38}$$

and

$$\text{RMS}\left(\hat{\varepsilon}_n^r\right) \leq \sqrt{\frac{6b}{n}}. \tag{4.39}$$

Note that bias, variance, and RMS converge to zero as sample size increases, for fixed b.

For the leave-one-out error estimator, one has the following bound (Devroye et al., 1996, Theorem 24.7):

$$\text{RMS}\left(\hat{\varepsilon}_n^l\right) \leq \sqrt{\frac{1+6/e}{n}+\frac{6}{\sqrt{\pi(n-1)}}}. \tag{4.40}$$

This guarantees, in particular, convergence to zero as sample size increases.

An important factor in the comparison of the resubstitution and leave-one-out error estimators for discrete histogram classification resides in the different speeds of convergence of the RMS to zero. Convergence of the RMS bound in (4.39) to zero for the resubstitution estimator is $O(n^{-1/2})$ (for fixed b), whereas convergence of the RMS bound in (4.40) to zero for the leave-one-out estimator is $O(n^{-1/4})$. Now, it can be shown that there is a feature–label distribution for which (Devroye et al., 1996, p. 419)

$$\text{RMS}\left(\hat{\varepsilon}_n^l\right) \geq \frac{1}{2\sqrt[4]{2\pi n}}e^{-1/24}. \tag{4.41}$$

Therefore, in the worst case, the RMS of leave-one-out to zero is guaranteed to decrease as $O(n^{-1/4})$, and therefore is certain to converge to zero slower than the RMS of resubstitution. Note that the bad RMS of leave-one-out is due almost entirely to its large variance, typical of the cross-validation approach, since this estimator is essentially unbiased.

EXERCISES

4.1 Consider a discrete classification problem with $b = 2$, and the following sample data: $S_n = \{(1,0),(1,1),(1,1),(2,0),(2,0),(2,1)\}$. Compute manually the resubstitution, leave-one-out, leave-two-out, and leave-three-out error estimates for the discrete histogram classifier.

4.2 Repeat Problem 4.1 for the data in Figure 1.3. It will be necessary to use a computer to compute the leave-two-out and leave-three-out error estimates.

4.3 Derive an expression for the bias of the leave-one-out error estimator for the discrete histogram rule.

4.4 Generalize Problem 4.3 to the leave-k-out error estimator for the discrete histogram rule.

4.5 Derive expressions for the variances and covariances appearing in (4.16) and (4.17).

4.6 Derive expressions for the covariances referred to following (4.21), (4.25), (4.29), and (4.31).

4.7 Derive expressions for the exact calculation of the variance of the leave-k-out error estimator $\text{Var}(\hat{\varepsilon}_n^{l(k)})$ and its covariance with the true error $\text{Cov}(\varepsilon_n, \hat{\varepsilon}_n^{l(k)})$.

4.8 Compute and plot the exact bias, deviation standard deviation, and RMS of resubstitution and leave-one-out as a function of sample size $n = 10, 20, \ldots, 100$ for the following distributional parameters: $c_0 = c_1 = 0.5, b = 4, p_1 = p_2 = 0.4, p_3 = p_4 = 0.1$, and $q_i = p_{5-i}$, for $i = 1,2,3,4$.

4.9 Use the complete enumeration method to compute and plot the joint probability between the true error and the resubstitution and leave-one-out error estimators for $n = 10, 15, 20$ and the distributional parameters in Problem 4.8.

CHAPTER 5

DISTRIBUTION THEORY

This chapter provides results on sampling moments and distributions of error rates for classification rules specified by general discriminants, including the true error rate and the resubstitution, leave-one-out, cross-validation, and bootstrap error estimators. These general results provide the foundation for the material in Chapters 6 and 7, where concrete examples of discriminants will be considered, namely, linear and quadratic discriminants over Gaussian populations. We will also tackle in this chapter the issue of separate sampling. We waited until the current chapter to take up this issue in order to improve the clarity of exposition in the previous chapters. Separate sampling will be fully addressed in this chapter, side by side with the previous case of mixture sampling, in a unified framework.

5.1 MIXTURE SAMPLING VERSUS SEPARATE SAMPLING

To this point we have assumed that a classifier is to be designed from a random i.i.d. sample $S_n = \{(\mathbf{X}_1, Y_1), \dots, (\mathbf{X}_n, Y_n)\}$ from the mixture of the two populations Π_0 and Π_1, that is, from feature–label distribution $F_{\mathbf{X}Y}$. It is not unusual for this assumption to be made throughout a text on pattern recognition. This *mixture-sampling* assumption is so pervasive that often it is not even explicitly mentioned or, if it is, its consequences are matter-of-factly taken for granted. In this vein, Duda et al. state, "In typical supervised pattern classification problems, the estimation of the prior probabilities presents no serious difficulties" (Duda and Hart, 2001). This statement may be true under mixture sampling, because the prior probabilities c_0 and c_1 can be estimated

Error Estimation for Pattern Recognition, First Edition. Ulisses M. Braga-Neto and Edward R. Dougherty.

by $\hat{c}_0 = \frac{N_0}{n}$ and $\hat{c}_1 = \frac{N_1}{n}$, respectively, with $\hat{c}_0 \to c_0$ and $\hat{c}_1 \to c_1$ with probability 1 as $n \to \infty$, by virtue of the Law of Large Numbers (c.f. Theorem A.2). Therefore, provided that the sample size is large enough, $\hat{c}_0$ and $\hat{c}_1$ provide decent estimates of the prior probabilities under mixture sampling.

However, suppose sampling is not from the mixture of populations, but rather from each population separately, such that a nonrandom number n_0 of points are drawn from Π_0, while a nonrandom number n_1 of points are drawn from Π_1, where n_0 and n_1 are chosen prior to sampling. In this *separate-sampling* case, the labels $Y_1, \ldots, Y_n$ are not i.i.d. Instead, there is a deterministic number n_0 of labels $Y_i = 0$ and a deterministic number n_1 of labels $Y_i = 1$. One can interpret the data as two *separate samples*, where each sample consists of i.i.d. feature vectors $\mathbf{X}_i$ drawn from its respective population. In this case, we have no sensible estimate of c_0 and c_1. Sample data, whether obtained by mixture or separate sampling, will be denoted by S. When it is important to make the distinction between the two sampling designs, subscripts "n" or "n_0, n_1" will be appended: S_n will denote mixture-sampling data (which agrees with the notation in previous chapters), while S_{n_0,n_1} will denote separate-sampling data. The same convention will apply to other symbols defined in the previous chapters.

The sampling issue creates problems for classifier design and error estimation. In the case of sample-based discriminant rules, such as linear discriminant analysis (LDA),

$$\Psi_L(S)(\mathbf{x}) = \begin{cases} 1, & \left(\mathbf{x} - \frac{\bar{\mathbf{X}}_0+\bar{\mathbf{X}}_1}{2}\right)^T \hat{\Sigma}^{-1}\left(\bar{\mathbf{X}}_0 - \bar{\mathbf{X}}_1\right) \leq \log \frac{\hat{c}_1}{\hat{c}_0}, \\ 0, & \text{otherwise}, \end{cases} \tag{5.1}$$

for $\mathbf{x} \in R^d$, the estimates $\hat{c}_0$ and $\hat{c}_1$ are required explicitly and the problem is obvious. The difficulty is less transparent with model-free classification rules, such as support vector machines (SVM), which do not explicitly require estimates of c_0 and c_1. Although mixture sampling is tacitly assumed in most contemporary pattern recognition studies, in fields such as biomedical science, separate sampling is prevalent. In biomedicine, it is often the case that one of the populations is small (e.g., a rare disease) so that sampling from the mixture of populations is difficult and would not obtain enough members of the small population. Therefore, *retrospective* studies employ a fixed number of members of each population, by assuming that the outcomes (labels) are known in advance (Zolman, 1993), which constitutes separate sampling.

Failure to account for distinct sampling mechanisms will manifest itself when determining the average performance of classifiers and error estimators; for example, it affects fundamental quantities such as expected classification error, bias, variance, and RMS of error estimators. An example from Esfahani and Dougherty (2014a) illustrates the discrepancy in expected LDA classification error between the two sampling mechanisms. Let $r = \hat{c}_0$ denote the (deterministic) proportion n_0/n of sample points from population Π_0 under separate sampling, and let $E[\varepsilon_n]$ and $E[\varepsilon_n|r]$ denote the expected classification error of the LDA classification rule under

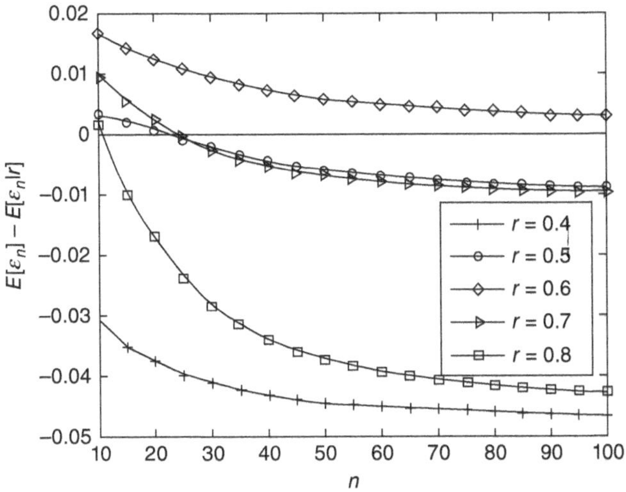

FIGURE 5.1 The difference $E[\varepsilon_n] - E[\varepsilon_n \mid r]$ as a function of sample size n, for an LDA classification rule under a homoskedastic Gaussian model with $c_0 = 0.6$ and different values of r. Reproduced from Esfahani and Dougherty (2014a) with permission from Oxford University Press.

mixture and separate sampling, respectively (later on, we will adopt a different notation for separate sampling, but for the time being this will do). Figure 5.1 displays the difference $E[\varepsilon_n] - E[\varepsilon_n|r]$, computed using Monte-Carlo sampling, for the LDA classification rule in (5.1) and different values of r as a function of sample size n, for a homoskedastic Gaussian model with $c_0 = 0.6$. The distributional parameters are as follows: $\boldsymbol{\mu}_0 = (0.3, 0.3)$, $\boldsymbol{\mu}_1 = (0.8, 0.8)$, and $\Sigma_0 = \Sigma_1 = 0.4I_d$, leading to the Bayes error $\varepsilon_{\text{bay}} = 0.27$. Note that the threshold parameter in (5.1) is fixed and equal to $\log(1-r)/r$ in the separate-sampling case, whereas it is random and equal to $\log N_1/N_0$ in the mixture-sampling case. We observe in Figure 5.1 that, if $r = c_0 = 0.6$, then $E[\varepsilon_n|r] < E[\varepsilon_n]$ for all n, so that it is actually advantageous to use separate sampling. However, this is only true if r is equal to the true prior probability c_0. For other values of r, $E[\varepsilon_n|r]$ may be less than $E[\varepsilon_n]$ only for sufficiently small n (when the estimates $\hat{c}_0$ and $\hat{c}_1$ become unreliable even under mixture sampling). For the case $r = 0.4$, however, $E[\varepsilon_n|r] > E[\varepsilon_n]$ for all n by a substantial margin.

We conclude that the use of fixed sample sizes with the incorrect proportions can be very detrimental to the performance of the classification rule, especially at larger sample sizes. On the other hand, if c_0 is accurately known, then, whatever the sample size, it is best to employ separate sampling with $n_0 \approx c_0 n$. If c_0 is approximately known through an estimate c_0', then, for small n, it may be best, or at least acceptable, to do separate sampling with $n_0 \approx c_0' n$, and the results will likely still be acceptable for large n, though not as good as with mixture sampling. Absent knowledge of c_0, mixture sampling is necessary because the penalty for separate sampling with an incorrect sample proportion can be very large. While these points were derived for LDA in Figure 5.1, they apply quite generally.

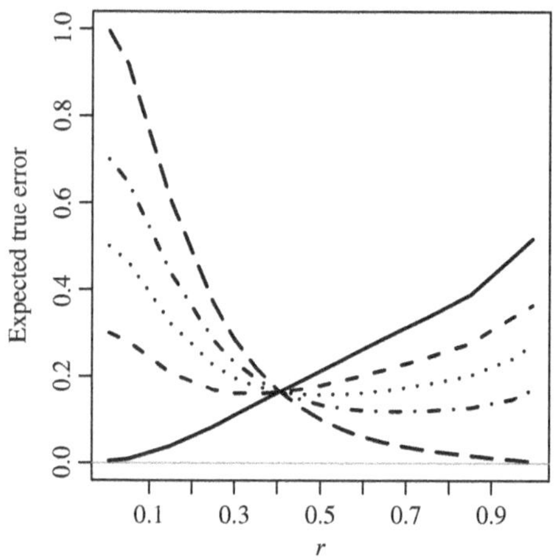

FIGURE 5.2 Effect of separate sampling on the expected true error as a function of r, for an LDA classification rule under a heteroskedastic Gaussian model with different values of c_0. Plot key: $c_0 = 0.001$ (solid), $c_0 = 0.3$ (dashed), $c_0 = 0.5$ (dotted), $c_0 = 0.7$ (dash-dot), $c_0 = 0.999$ (long dash).

Figure 5.2 shows the effect of separate sampling on the expected true error as a function of r for different values of c_0. It assumes LDA and a model composed of multivariate Gaussian distributions with a covariance matrix that has σ^2 on the diagonal and ρ_y off the diagonal with y indicating the class label. We let $\sigma^2 = 0.6$, $\rho_0 = 0.8$, and $\rho_1 = 0.4$. Feature-set size is 3. For a given sample size $n = 10{,}000$ and sampling ratio $r = \frac{n_0}{n}$, $E[\varepsilon_n|r]$ is plotted for different class prior probabilities c_0. We observe that expected error is close to minimal when $r = c_0$ and that it can greatly increase when $r \neq c_0$. An obvious characteristic of the curves in Figure 5.2 is that they appear to cross at a single value of r. Keeping in mind that the figure shows continuous curves but r is a discrete variable, this phenomenon is explored in depth in Esfahani and Dougherty (2014a), where similar graphs for different classification rules are provided.

Let us now consider the effect of separate sampling on the bias of cross-validation error estimation — this problem is studied at length in Braga-Neto et al. (2014). Figure 5.3 treats the same scenario as Figure 5.2, except that now the curves give the bias of 5-fold cross-validation (see Braga-Neto et al. (2014) for additional experimental results). As opposed to small, slightly optimistic bias, which is the case with mixture sampling, we observe in Figure 5.3 that for a separate-sampling ratio r not close to c_0 there can be large bias, optimistic or pessimistic. The following trends are typical: (1) with c_0 near 0.5 there is significant optimistic bias for large $|r - c_0|$; (2) with small or large c_0, there is optimistic bias for large $|r - c_0|$ and pessimistic bias for small $|r - c_0|$ so long as $|r - c_0|$ is not very close to 0. In the sequel, we shall provide analytic insight into this bias behavior. Whereas the deviation variance of classical cross-validation can be mitigated by large samples, the bias issue remains just as bad for large samples.

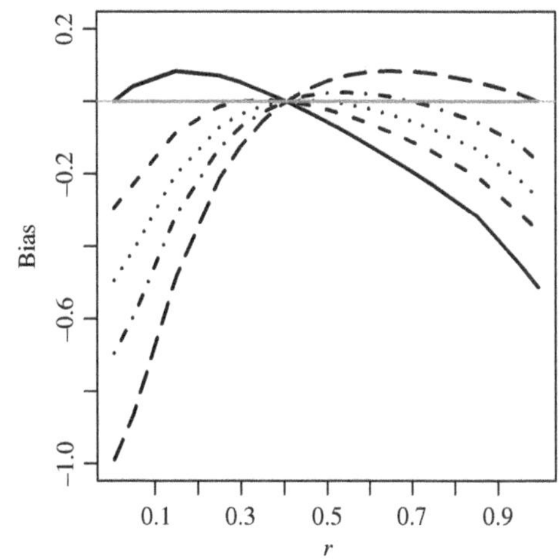

FIGURE 5.3 Effect of separate sampling on the bias of the 5-fold cross-validation error estimator as a function of r, for an LDA classification rule and a heteroskedastic Gaussian model with different values of c_0. Plot key: $c_0 = 0.001$ (solid), $c_0 = 0.3$ (dashed), $c_0 = 0.5$ (dotted), $c_0 = 0.7$ (dash-dot), $c_0 = 0.999$ (long dash).

A serious consequence of this behavior can be ascertained by looking at Figures 5.2 and 5.3 in conjunction. Whereas the expected true error of the designed classifier grows large when r greatly deviates from c_0, a large optimistic cross-validation bias when r is far from c_0 can obscure the large error and leave one with the illusion of good performance – and this illusion is not mitigated by large samples.

5.2 SAMPLE-BASED DISCRIMINANTS REVISITED

We must now revisit sample-based discriminants with an eye toward differentiating between mixture and separate sampling. Given a sample S, a sample-based discriminant was defined in Section 1.4 as a real function $W(S, \mathbf{x})$. A discriminant defines a classification rule Ψ via

$$\Psi(S)(\mathbf{x}) = \begin{cases} 1, & W(S, \mathbf{x}) \leq 0, \\ 0, & \text{otherwise}, \end{cases} \tag{5.2}$$

for $\mathbf{x} \in R^d$, where we assumed the discriminant threshold to be 0, without loss of generality. When there is no danger of confusion, we will denote a discriminant by $W(\mathbf{x})$.

We introduce next the concept of class label flipping, which will be useful in this chapter as well as in Chapter 7. If $S = \{(\mathbf{X}_1, Y_1), \ldots, (\mathbf{X}_n, Y_n)\}$ is a sample from the feature–label distribution $F_{\mathbf{X}Y}$, flipping all class labels produces the sample

$\overline{S} = \{(\mathbf{X}_1, 1 - Y_1), \ldots, (\mathbf{X}_n, 1 - Y_n)\}$, which is a sample from the feature–label distribution $F_{\mathbf{X}\overline{Y}}$, where $\overline{Y} = 1 - Y$. Note that $F_{\mathbf{X}\overline{Y}}$ is obtained from $F_{\mathbf{X}Y}$ by switching the populations — in a parametric setting, this means flipping the class labels of all distributional parameters in an expression or statistical representation (this may require some care). In the mixture-sampling case, this is all there is to be said. If S_n is a sample of size n from $F_{\mathbf{X}Y}$ then $\overline{S_n}$ is a sample of size n from $F_{\mathbf{X}\overline{Y}}$. However, in the separate-sampling case, one must also deal with the switch of the separate sample sizes. Namely, if S_{n_0,n_1} is a sample of sizes n_0 and n_1 from $F_{\mathbf{X}Y}$, then $\overline{S_{n_0,n_1}}$ is a sample of sizes n_1 and n_0 from $F_{\mathbf{X}\overline{Y}}$.

The following common-sensical properties of discriminants will be assumed without further notice: (i) the order of the sample points within a sample does not matter; (ii) flipping all class labels in the sample results in the complement of the original classifier; that is, $W(S, \mathbf{x}) \leq 0$ if and only if $W(\overline{S}, \mathbf{x}) > 0$. For some discriminants, a stronger version of (ii) holds: flipping all class labels in the sample results in the negative of the discriminant, in which case we say that the discriminant is *anti-symmetric*:

$$W(S, \mathbf{x}) = -W(\overline{S}, \mathbf{x}), \quad \text{for all } \mathbf{x} \in R^d. \tag{5.3}$$

Anti-symmetry is a key property that is useful in the distributional study of discriminants, as will be seen in the sequel. The sample-based quadratic discriminant analysis (QDA) and LDA discriminants are clearly both anti-symmetric. Kernel discriminants, defined in Section 1.4.3, are also anti-symmetric, provided that the sample-based kernel bandwidth h is invariant to the label switch (a condition that is typically satisfied).

5.3 TRUE ERROR

The true error rate of the classifier defined by (5.2) is given by

$$\begin{aligned}\varepsilon &= P(W(\mathbf{X}) \leq 0, Y = 0 \mid S) + P(W(\mathbf{X}) > 0, Y = 1 \mid S) \\ &= P(W(\mathbf{X}) \leq 0 \mid Y = 0, S)P(Y = 0) \\ &\quad + P(W(\mathbf{X}) > 0 \mid Y = 1, S)P(Y = 1) \\ &= c_0\varepsilon^0 + c_1\varepsilon^1, \end{aligned} \tag{5.4}$$

where $(\mathbf{X}, Y)$ is an independent test point and

$$\begin{aligned}\varepsilon^0 &= P(W(\mathbf{X}) \leq 0 \mid Y = 0, S), \\ \varepsilon^1 &= P(W(\mathbf{X}) > 0 \mid Y = 1, S)\end{aligned} \tag{5.5}$$

are the *population-specific error rates*, that is, the probabilities of misclassifying members of each class separately.

Notice that, for a given fixed sample S, all previous error rates are obviously unchanged regardless of whether the data are obtained under mixture- or separate-sampling schemes. However, the sampling moments and sampling distributions of these error rates will be distinct under the two sampling schemes. When it is desirable to make the distinction, subindices "n" for mixture sampling or "n_0, n_1" for separate sampling will be appended.

5.4 ERROR ESTIMATORS

5.4.1 Resubstitution Error

The classical resubstitution estimator of the true error rate was defined for the mixture-sampling case in (2.30). Here we approach the issue from a slightly different angle.

Given the sample data, the resubstitution estimates of the population-specific true error rates are

$$\begin{aligned}\hat{\varepsilon}^{r,0} &= \frac{1}{n_0}\sum_{i=1}^{n} I_{W(\mathbf{X}_i)\leq 0} I_{Y_i=0},\\ \hat{\varepsilon}^{r,1} &= \frac{1}{n_1}\sum_{i=1}^{n} I_{W(\mathbf{X}_i)> 0} I_{Y_i=1}.\end{aligned} \tag{5.6}$$

As in the case of the true error rates, subindices "n" or "n_0, n_1" can be appended if it is desirable to distinguish between mixture and separate sampling. Notice that in the mixture-sampling case, n_0 and n_1 are random variables. In this chapter, we will write n_0 and n_1 in both the mixture- and separate-sampling cases when there is no danger of confusion and it allows us to write a single equation in both cases, as was done in (5.6).

We distinguish two cases. If the prior probabilities c_0 and c_1 are known, then a resubstitution estimate of the overall error rate is given simply by

$$\hat{\varepsilon}^{r,01} = c_0\hat{\varepsilon}^{r,0} + c_1\hat{\varepsilon}^{r,1}. \tag{5.7}$$

This estimator is appropriate regardless of sampling scheme. Now, if the prior probabilities are not known, then the situation changes. In the mixture-sampling case, the prior probabilities can be estimated by n_0/n and n_1/n, respectively, so it is natural to use the estimator

$$\hat{\varepsilon}^{r}_n = \frac{n_0}{n}\hat{\varepsilon}^{r,0}_n + \frac{n_1}{n}\hat{\varepsilon}^{r,1}_n = \frac{1}{n}\sum_{i=1}^{n}(I_{W(\mathbf{X}_i)\leq 0} I_{Y_i=0} + I_{W(\mathbf{X}_i)> 0} I_{Y_i=1}). \tag{5.8}$$

This is equivalent to the resubstitution estimator introduced in Chapter 2. Now, in the separate-sampling case, there is no information about c_0 and c_1 in the sample, so there is no resubstitution estimator of the overall error rate. It is important to emphasize this point. The estimator in (5.8) is inappropriate for the separate-sampling case, generally displaying very poor performance in this case. Unfortunately, the use of (5.8) as an estimator of the overall error rate in the separate-sampling case is widespread in practice.

5.4.2 Leave-One-Out Error

The development here is similar to the one in the previous section. Let $S^{(i)}$ denote the sample data S with sample point $(\mathbf{X}_i, Y_i)$ deleted, for $i = 1, \ldots, n$. Define the *deleted sample-based discriminant* $W^{(i)}(S, \mathbf{x}) = W(S^{(i)}, \mathbf{x})$, for $i = 1, \ldots, n$. Given the sample data, the leave-one-out estimates of the population-specific true error rates are

$$\begin{aligned}\hat{\varepsilon}^{l,0} &= \frac{1}{n_0} \sum_{i=1}^{n} I_{W^{(i)}(\mathbf{X}_i) \le 0} I_{Y_i=0},\\ \hat{\varepsilon}^{l,1} &= \frac{1}{n_1} \sum_{i=1}^{n} I_{W^{(i)}(\mathbf{X}_i) > 0} I_{Y_i=1}.\end{aligned} \tag{5.9}$$

As before, subindices "n" or "n_0, n_1" can be appended if it is desirable to distinguish between mixture- and separate-sampling schemes, respectively.

As before, we distinguish two cases. If the prior probabilities c_0 and c_1 are known, then the leave-one-out estimate of the overall error rate is given by

$$\hat{\varepsilon}^{l,01} = c_0 \hat{\varepsilon}^{l,0} + c_1 \hat{\varepsilon}^{l,1}. \tag{5.10}$$

This estimator is appropriate under either sampling scheme. If the prior probabilities are not known, then the situation changes. In the mixture-sampling case, the prior probabilities can be estimated by n_0/n and n_1/n, respectively, so it is natural to use the estimator

$$\hat{\varepsilon}_n^{l} = \frac{n_0}{n} \hat{\varepsilon}_n^{l,0} + \frac{n_1}{n} \hat{\varepsilon}_n^{l,1} = \frac{1}{n} \sum_{i=1}^{n} (I_{W^{(i)}(\mathbf{X}_i) \le 0} I_{Y_i=0} + I_{W^{(i)}(\mathbf{X}_i) > 0} I_{Y_i=1}). \tag{5.11}$$

This is precisely the leave-one-out estimator that was introduced in Chapter 2. However, in the separate-sampling case, there is no information about c_0 and c_1 in the sample, so there is no leave-one-out estimator of the overall error rate. The estimator in (5.11), if applied in the separate-sampling case, can be severely biased. Unfortunately, as in the case of resubstitution, the use of (5.11) as an estimator of the overall error rate in the separate-sampling case is widespread in practice.

5.4.3 Cross-Validation Error

The classical cross-validation estimator of the overall error rate, introduced in Chapter 2, is suitable to the mixture-sampling case. It averages the results of deleting a fold from the data, designing the classifier on the remaining data points, and testing it on the data points in the removed fold (in Chapter 2, the folds were assumed to be disjoint, but we will lift this requirement here).

Let the size of all folds be equal to k, where it is assumed, for convenience, that k divides n. For a fold $U \subset \{1, \ldots, n\}$, let $S_n^{(U)}$ denote the sample S_n with the points with indices in U deleted, and define $W^{(U)}(S_n, \mathbf{x}) = W(S_n^{(U)}, \mathbf{x})$. Let Υ be a collection of folds and let $|\Upsilon| = L$. Then the k-fold cross-validation estimator can be written as

$$\hat{\varepsilon}_n^{\,\mathrm{cv}(k)} = \frac{1}{kL} \sum_{U \in \Upsilon} \sum_{i \in U} (I_{W_n^{(U)}(\mathbf{X}_i) \leq 0} I_{Y_i=0} + I_{W_n^{(U)}(\mathbf{X}_i) > 0} I_{Y_i=1}). \tag{5.12}$$

If all possible distinct folds are used, then $L = \binom{n}{k}$ and the resulting estimator is the complete k-fold cross-validation estimator, defined in Section 2.5, which is a nonrandomized error estimator. In practice, for computational reasons, $L < \binom{n}{k}$ and Υ contains L random folds, which leads to the standard randomized k-fold cross-validation estimator discussed in Chapter 2.

The cross-validation estimator has been adapted to separate sampling in Braga-Neto et al. (2014). Let $S^0_{n_0,n_1}$ and $S^1_{n_0,n_1}$ be the subsamples containing points coming from populations Π_0 and Π_1, respectively. There are n_0 and n_1 points in $S^0_{n_0,n_1}$ and $S^1_{n_0,n_1}$, respectively, with $S^0_{n_0,n_1} \cup S^1_{n_0,n_1} = S_{n_0,n_1}$. The idea is to resample these subsamples separately and simultaneously. Let I_0 and I_1 be the sets of indices of points in $S^0_{n_0,n_1}$ and $S^1_{n_0,n_1}$, respectively. Notice that $|I_0| = n_0$ and $|I_1| = n_1$, and $I_0 \cup I_1 = \{1, \dots, n\}$. Let $U \subset I_0$ and $V \subset I_1$ be folds of lengths k_0 and k_1, respectively, where it is assumed that k_0 divides n_0 and k_1 divides n_1. Let Υ and Ω be collections of folds from $S^0_{n_0,n_1}$ and $S^1_{n_0,n_1}$, respectively, where $|\Upsilon| = L$ and $|\Omega| = M$. For $U \in \Upsilon$ and $V \in \Omega$, define the deleted sample-based discriminant

$$W^{(U,V)}(S, \mathbf{x}) = W(S^{(U \cup V)}, \mathbf{x}). \tag{5.13}$$

Separate-sampling cross-validation estimators of the population-specific error rates are given by

$$\begin{aligned} \hat{\varepsilon}_{n_0,n_1}^{\,\mathrm{cv}(k_0,k_1),0} &= \frac{1}{k_0 LM} \sum_{\substack{U \in \Upsilon \\ V \in \Omega}} \sum_{i \in U} I_{W^{(U,V)}(\mathbf{X}_i) \leq 0}, \\ \hat{\varepsilon}_{n_0,n_1}^{\,\mathrm{cv}(k_0,k_1),1} &= \frac{1}{k_1 LM} \sum_{\substack{U \in \Upsilon \\ V \in \Omega}} \sum_{i \in V} I_{W^{(U,V)}(\mathbf{X}_i) > 0}. \end{aligned} \tag{5.14}$$

These are estimators of the population-specific true error rates $\varepsilon^0_{n_0,n_1}$ and $\varepsilon^1_{n_0,n_1}$, respectively. If the prior probabilities c_0 and c_1 are known, then a cross-validation estimator of the overall error rate is given by

$$\hat{\varepsilon}_{n_0,n_1}^{\,\mathrm{cv}(k_0,k_1),01} = c_0 \hat{\varepsilon}_{n_0,n_1}^{\,\mathrm{cv}(k_0,k_1),0} + c_1 \hat{\varepsilon}_{n_0,n_1}^{\,\mathrm{cv}(k_0,k_1),1}. \tag{5.15}$$

In the absence of knowledge about the mixture probabilities, no such cross-validation estimator of the overall error rate would be appropriate in the separate-sampling case.

5.4.4 Bootstrap Error

We will consider here and in the next two chapters, the complete zero bootstrap, introduced in Section 2.6. There are two cases: for mixture sampling, bootstrap sampling is performed over the entire data S_n, where as for separate sampling, bootstrap sampling is applied to each population sample separately.

Let us consider first the mixture-sampling case. Let $\mathbf{C} = (C_1, \dots, C_n)$ be a multinomial vector of length n, that is, a random vector consisting of n nonnegative random integers that add up to n. A bootstrap sample $S_n^{\mathbf{C}}$ (here we avoid the previous notation S_n^* for a bootstrap sample, as we want to make the dependence on $\mathbf{C}$ explicit) from S_n corresponds to taking C_i repetitions of the ith data point of S_n, for $i = 1, \dots, n$. Given S_n, there is thus a one-to-one relation between a bootstrap sample $S_n^{\mathbf{C}}$ and a multinomial vector $\mathbf{C}$ of length n. The number of distinct bootstrap samples (distinct values of $\mathbf{C}$) is $\binom{2n-1}{n}$. The bootstrap uniform sample-with-replacement strategy implies that $\mathbf{C} \sim \text{Multinomial}(n, \frac{1}{n}, \dots, \frac{1}{n})$, that is,

$$P(\mathbf{C}) = P(C_1 = c_1, \dots, C_n = c_n) = \frac{1}{n^n} \frac{n!}{c_1! \cdots c_n!}, \quad c_1 + \cdots + c_n = n. \tag{5.16}$$

Define $\hat{\varepsilon}_n^{\mathbf{C}}$ as the error rate committed by the bootstrap classifier on the data left out of the bootstrap sample:

$$\hat{\varepsilon}_n^{\mathbf{C}} = \frac{1}{n(\mathbf{C})} \sum_{i=1}^{n} (I_{W(S_n^{\mathbf{C}}, \mathbf{X}_i) \le 0} I_{Y_i=0} + I_{W(S_n^{\mathbf{C}}, \mathbf{X}_i) > 0} I_{Y_i=1}) I_{C_i=0}, \tag{5.17}$$

where $n(\mathbf{C}) = \sum_{i=1}^{n} I_{C_i=0}$ is the number of zeros in $\mathbf{C}$. The complete zero bootstrap error estimator in (2.48) is the expected value of $\hat{\varepsilon}_n^{\mathbf{C}}$ over the bootstrap sampling mechanism, which is simply the distribution of $\mathbf{C}$:

$$\hat{\varepsilon}_n^{\text{zero}} = E[\hat{\varepsilon}_n^{\mathbf{C}} \mid S_n] = \sum_{\mathbf{C}} \hat{\varepsilon}_n^{\mathbf{C}} P(\mathbf{C}), \tag{5.18}$$

where $P(\mathbf{C})$ is given by (5.16) and the sum is taken over all possible values of $\mathbf{C}$ (an efficient procedure for listing all multinomial vectors is provided by the NEXCOM routine given in Nijenhuis and Wilf (1978, Chapter 5)).

As with cross-validation, bootstrap error estimation can be adapted for separate sampling. We employ here the same notation used previously. The idea is to take bootstrap samples from the individual population samples S_{n_0,n_1}^0 and S_{n_0,n_1}^1 separately and simultaneously. To facilitate notation, let the sample indices be sorted such that $Y_1 = \cdots = Y_{n_0} = 0$ and $Y_{n_0+1} = \cdots = Y_{n_0+n_1} = 1$, and consider multinomial random vectors $\mathbf{C}_0 = (C_{01}, \dots, C_{0n_0})$ and $\mathbf{C}_1 = (C_{11}, \dots, C_{1n_1})$ of length n_0 and n_1, respectively. Let $S_{n_0,n_1}^{0,\mathbf{C}_0}$ and $S_{n_0,n_1}^{1,\mathbf{C}_1}$ be bootstrap samples from S_{n_0,n_1}^0 and S_{n_0,n_1}^1, according to $\mathbf{C}_0$ and $\mathbf{C}_1$, respectively, and let $S_{n_0,n_1}^{\mathbf{C}_0,\mathbf{C}_1} = S_{n_0,n_1}^{0,\mathbf{C}_0} \cup S_{n_0,n_1}^{1,\mathbf{C}_1}$. We define

$$\begin{aligned}\hat{\varepsilon}_{n_0,n_1}^{\mathbf{C}_0,\mathbf{C}_1,0} &= \frac{1}{n(\mathbf{C}_0)} \sum_{i=1}^{n_0} I_{W\left(S_{n_0,n_1}^{\mathbf{C}_0,\mathbf{C}_1},\mathbf{X}_i\right)\leq 0} I_{C_{0i}=0},\\ \hat{\varepsilon}_{n_0,n_1}^{\mathbf{C}_0,\mathbf{C}_1,1} &= \frac{1}{n(\mathbf{C}_1)} \sum_{i=1}^{n_1} I_{W\left(S_{n_0,n_1}^{\mathbf{C}_0,\mathbf{C}_1},\mathbf{X}_{n_0+i}\right)> 0} I_{C_{1i}=0},\end{aligned} \tag{5.19}$$

where $n(\mathbf{C}_0) = \sum_{i=1}^{n_0} I_{C_{0i}=0}$ and $n(\mathbf{C}_1) = \sum_{i=1}^{n_1} I_{C_{1i}=0}$ are the number of zeros in $\mathbf{C}_0$ and $\mathbf{C}_1$, respectively. Separate-sampling complete zero bootstrap estimators of the population-specific error rates are given by the expected value of these estimators over the bootstrap sampling mechanism:

$$\hat{\varepsilon}_{n_0,n_1}^{\,\text{zero},i} = E\left[\hat{\varepsilon}_{n_0,n_1}^{\mathbf{C}_0,\mathbf{C}_1,i} \mid S_{n_0,n_1}\right] = \sum_{\mathbf{C}_0,\mathbf{C}_1} \hat{\varepsilon}_{n_0,n_1}^{\mathbf{C}_0,\mathbf{C}_1,i} P(\mathbf{C}_0)P(\mathbf{C}_1), \quad i = 0, 1, \tag{5.20}$$

where $P(\mathbf{C}_0)$ and $P(\mathbf{C}_1)$ are defined as in (5.16), with the appropriate sample sizes. These are estimators of the population-specific true error rates ε_{n_0,n_1}^0 and ε_{n_0,n_1}^1, respectively. If the prior probabilities c_0 and c_1 are known, then a complete zero bootstrap estimator of the overall error rate is given by

$$\hat{\varepsilon}_{n_0,n_1}^{\,\text{zero},01} = c_0 \hat{\varepsilon}_{n_0,n_1}^{\,\text{zero},0} + c_1 \hat{\varepsilon}_{n_0,n_1}^{\,\text{zero},1}. \tag{5.21}$$

In the absence of knowledge about the mixture probabilities, no such bootstrap estimator of the overall error rate would be appropriate in the separate sampling case.

Convex combinations of the zero bootstrap and resubstitution, including the .632 bootstrap error estimator, can be made as before, as long as estimators of different kinds are not mixed.

5.5 EXPECTED ERROR RATES

In this section we analyze the expected error rates associated with sample-based discriminants in the mixture- and separate-sampling cases. We consider both the true error rates as well as the error estimators defined in the previous section. It will be seen that the expected error rates are expressed in terms of the sampling distributions of discriminants.

5.5.1 True Error

It follows by taking expectations on both sides of (5.4) and (5.5) that the expected true error rate is given by

$$E[\varepsilon] = c_0 E[\varepsilon^0] + c_1 E[\varepsilon^1], \tag{5.22}$$

where

$$E[\varepsilon^0] = P(W(\mathbf{X}) \leq 0 \mid Y = 0),$$
$$E[\varepsilon^1] = P(W(\mathbf{X}) > 0 \mid Y = 1) \tag{5.23}$$

are the population-specific expected error rates. Therefore, the expected true error rates can be obtained given knowledge of the sampling distribution of W (conditional on the population).

The previous expressions are valid for both the mixture- and separate-sampling schemes. The difference between them arises in the probability measure employed. In the mixture-sampling case, the expectation is over the mixture-sampling mechanism, yielding the "average" error rate over "all possible" random samples of total size n. In the separate-sampling case, the averaging occurs only over a subset of samples containing n_0 points from class 0 and n_1 points from class 1 (n_1 is redundant once n and n_0 are specified). Therefore, the expected error rates under the two sampling schemes are related by

$$E\left[\varepsilon^0_{n_0,n_1}\right] = E\left[\varepsilon^0_n \mid N_0 = n_0\right],$$
$$E\left[\varepsilon^1_{n_0,n_1}\right] = E\left[\varepsilon^1_n \mid N_0 = n_0\right],$$
$$E\left[\varepsilon_{n_0,n_1}\right] = E\left[\varepsilon_n \mid N_0 = n_0\right]. \tag{5.24}$$

As a consequence, the expected error rates in the mixture-sampling case can be easily obtained from the corresponding expected error rates in the separate- sampling case:

$$\begin{aligned}
E\left[\varepsilon^0_n\right] &= \sum_{n_0=0}^{n} E\left[\varepsilon^0_n \mid N_0 = n_0\right] P(N_0 = n_0) \\
&= \sum_{n_0=0}^{n} E\left[\varepsilon^0_{n_0,n_1}\right] P(N_0 = n_0) \\
&= \sum_{n_0=0}^{n} \binom{n}{n_0} c_0^{n_0} c_1^{n_1} E\left[\varepsilon^0_{n_0,n_1}\right], \\
E\left[\varepsilon^1_n\right] &= \sum_{n_0=0}^{n} \binom{n}{n_0} c_0^{n_0} c_1^{n_1} E\left[\varepsilon^1_{n_0,n_1}\right], \\
E[\varepsilon_n] &= \sum_{n_0=0}^{n} \binom{n}{n_0} c_0^{n_0} c_1^{n_1} E[\varepsilon_{n_0,n_1}] \\
&= \sum_{n_0=0}^{n} \binom{n}{n_0} c_0^{n_0} c_1^{n_1} \left(c_0 E\left[\varepsilon^0_{n_0,n_1}\right] + c_1 E\left[\varepsilon^1_{n_0,n_1}\right]\right).
\end{aligned} \tag{5.25}$$

The previous relationships imply the fundamental principle that all results concerning the expected true error rates for both mixture- and separate-sampling cases can be obtained from results for the separate sampling population-specific expected true error rates. This principle also applies to the resubstitution and leave-one-out error rates, as we shall see. This means that the distributional study of these error rates can concentrate on the separate-sampling population-specific expected error rates. This is the approach that will be taken in Chapters 6 and 7, when these error rates are studied for LDA and QDA discriminants on Gaussian populations.

Let S_{n_0,n_1} and S_{n_1,n_0} be samples from the feature–label distribution $F_{\mathbf{X}Y}$. We define the following *exchangeability* property for the separate-sampling distribution of the discriminant applied on a test point:

$$W(S_{n_0,n_1}, \mathbf{X}) \,|\, Y = 0 \sim W(\overline{S_{n_1,n_0}}, \mathbf{X}) \,|\, Y = 1. \tag{5.26}$$

Recall from Section 5.2 that $\overline{S_{n_1,n_0}}$ is obtained by flipping all class labels in S_{n_1,n_0}. In addition, $\overline{S_{n_1,n_0}}$ is a sample of separate sample sizes n_0 and n_1 from $F_{\mathbf{X}\overline{Y}}$, which is the feature–label distribution obtained from $F_{\mathbf{X}Y}$ by switching the populations. Examples of relation (5.26) will be given in Chapter 7.

Now, if the discriminant is anti-symmetric, c.f. (5.3), then $W(S_{n_0,n_1}, \mathbf{X}) = -W(\overline{S_{n_0,n_1}}, \mathbf{X})$. If in addition (5.26) holds, then one can write

$$W(S_{n_0,n_1}, \mathbf{X}) \,|\, Y = 0 \sim -W(\overline{S_{n_0,n_1}}, \mathbf{X}) \,|\, Y = 0 \sim -W(S_{n_1,n_0}, \mathbf{X}) \,|\, Y = 1, \tag{5.27}$$

and it follows from (5.23) that

$$E\left[\varepsilon^0_{n_0,n_1}\right] = E\left[\varepsilon^1_{n_1,n_0}\right] \text{ and } E\left[\varepsilon^1_{n_0,n_1}\right] = E\left[\varepsilon^0_{n_1,n_0}\right]. \tag{5.28}$$

If (5.26) holds, the discriminant is anti-symmetric, and the populations are equally likely ($c_0 = 0.5$), then it follows from (5.22) and (5.28) that the overall error rates are interchangeable in the sense that $E[\varepsilon_{n_0,n_1}] = E[\varepsilon_{n_1,n_0}]$. In addition, it follows from (5.25) that $E[\varepsilon_n] = E[\varepsilon^0_n] = E[\varepsilon^1_n]$ in the mixture- sampling case. This is the important case where the two population-specific error rates are perfectly balanced.

A similar result in the separate-sampling case can be obtained if the design is *balanced*, that is, $n_0 = n_1 = n/2$. Then, if (5.26) holds and the discriminant is anti-symmetric, we have that

$$E[\varepsilon_{n/2,n/2}] = E\left[\varepsilon^0_{n/2,n/2}\right] = E\left[\varepsilon^1_{n/2,n/2}\right], \tag{5.29}$$

regardless of the value of the prior probabilities c_0 and c_1. The advantage of this result is that it holds irrespective of c_0 and c_1. Hence, in this case, it suffices to study one of the population-specific expected error rates, irrespective of the prior probabilities.

5.5.2 Resubstitution Error

It follows by taking expectations on both sides of (5.6) that the expected resubstitution population-specific error rates are given simply by

$$
\begin{aligned}
E[\hat{\varepsilon}^{r,0}] &= P(W(\mathbf{X}_1) \le 0 \mid Y_1 = 0),\\
E[\hat{\varepsilon}^{r,1}] &= P(W(\mathbf{X}_1) > 0 \mid Y_1 = 1).
\end{aligned} \tag{5.30}
$$

The previous expressions are valid for both the mixture- and separate-sampling schemes. Once again, the expected error rates under the two sampling mechanisms are related by

$$
\begin{aligned}
E\left[\hat{\varepsilon}^{r,0}_{n_0,n_1}\right] &= E\left[\hat{\varepsilon}^{r,0}_{n} \mid N_0 = n_0\right],\\
E\left[\hat{\varepsilon}^{r,1}_{n_0,n_1}\right] &= E\left[\hat{\varepsilon}^{r,1}_{n} \mid N_0 = n_0\right].
\end{aligned} \tag{5.31}
$$

As a consequence, the expected error rates in the mixture-sampling case can be easily obtained from the corresponding expected error rates in the separate-sampling case:

$$
\begin{aligned}
E\left[\hat{\varepsilon}^{r,0}_{n}\right] &= \sum_{n_0=0}^{n} E\left[\hat{\varepsilon}^{r,0}_{n} \mid N_0 = n_0\right] P(N_0 = n_0)\\
&= \sum_{n_0=0}^{n} E\left[\hat{\varepsilon}^{r,0}_{n_0,n_1}\right] P(N_0 = n_0)\\
&= \sum_{n_0=0}^{n} \binom{n}{n_0} c_0^{n_0} c_1^{n_1} E\left[\hat{\varepsilon}^{r,0}_{n_0,n_1}\right],\\
E\left[\hat{\varepsilon}^{r,1}_{n}\right] &= \sum_{n_0=0}^{n} \binom{n}{n_0} c_0^{n_0} c_1^{n_1} E\left[\hat{\varepsilon}^{r,1}_{n_0,n_1}\right].
\end{aligned} \tag{5.32}
$$

If the prior probabilities c_0 and c_1 are known, then the estimator in (5.7) can be used. Taking expectations on both sides of that equation yields

$$
E[\hat{\varepsilon}^{r,01}] = c_0 E[\hat{\varepsilon}^{r,0}] + c_1 E[\hat{\varepsilon}^{r,1}]. \tag{5.33}
$$

As before, the mixture- and separate-sampling cases are related by

$$
E\left[\hat{\varepsilon}^{r,01}_{n_0,n_1}\right] = E\left[\hat{\varepsilon}^{r,01}_{n} \mid N_0 = n_0\right] \tag{5.34}
$$

and

$$
\begin{aligned}
E\left[\hat{\varepsilon}^{r,01}_{n}\right] &= \sum_{n_0=0}^{n} \binom{n}{n_0} c_0^{n_0} c_1^{n_1} E\left[\hat{\varepsilon}^{r,01}_{n_0,n_1}\right]\\
&= \sum_{n_0=0}^{n} \binom{n}{n_0} c_0^{n_0} c_1^{n_1} \left(c_0 E\left[\hat{\varepsilon}^{r,0}_{n_0,n_1}\right] + c_1 E\left[\hat{\varepsilon}^{r,1}_{n_0,n_1}\right]\right).
\end{aligned} \tag{5.35}
$$

In addition, it is easy to show that for the standard resubstitution error estimator in (5.8), we have an expression similar to (5.33):

$$E\left[\hat{\varepsilon}_n^r\right] = c_0 E\left[\hat{\varepsilon}_n^{r,0}\right] + c_1 E\left[\hat{\varepsilon}_n^{r,1}\right]. \tag{5.36}$$

The previous relationships imply that results concerning the overall expected resubstitution error rate, both in the separate- and mixture-sampling cases, can be obtained from results for the separate-sampling population-specific expected resubstitution error rates.

Let $S_{n_0,n_1} = \{(\mathbf{X}_1, Y_1), \ldots, (\mathbf{X}_n, Y_n)\}$ and $S_{n_1,n_0} = \{(\mathbf{X}'_1, Y'_1), \ldots, (\mathbf{X}'_n, Y'_n)\}$ be samples from the feature–label distribution $F_{\mathbf{X}Y}$. We define the following exchangeability property for the separate-sampling distribution of the discriminant applied on a training point:

$$W(S_{n_0,n_1}, \mathbf{X}_1) \mid Y_1 = 0 \sim W(\overline{S_{n_1,n_0}}, \mathbf{X}'_1) \mid Y'_1 = 1. \tag{5.37}$$

Examples of relation (5.37) will be given in Chapter 7.

If the discriminant is anti-symmetric, c.f. (5.3), then $W(S_{n_0,n_1}, \mathbf{X}_1) = -W(\overline{S_{n_0,n_1}}, \mathbf{X}_1)$. If in addition (5.37) holds, then one can write

$$\begin{aligned} W(S_{n_0,n_1}, \mathbf{X}_1) \mid Y_1 = 0 &\sim -W(\overline{S_{n_0,n_1}}, \mathbf{X}_1) \mid Y_1 = 0 \\ &\sim -W(S_{n_1,n_0}, \mathbf{X}'_1) \mid Y'_1 = 1, \end{aligned} \tag{5.38}$$

and it follows from (5.30) that

$$E\left[\hat{\varepsilon}_{n_0,n_1}^{r,0}\right] = E\left[\hat{\varepsilon}_{n_1,n_0}^{r,1}\right] \text{ and } E\left[\hat{\varepsilon}_{n_0,n_1}^{r,1}\right] = E\left[\hat{\varepsilon}_{n_1,n_0}^{r,0}\right]. \tag{5.39}$$

If (5.37) holds, the discriminant is anti-symmetric, the populations are equally likely ($c_0 = 0.5$), and the estimator in (5.6) is used, then $E[\hat{\varepsilon}_{n_0,n_1}^{r,01}] = E[\hat{\varepsilon}_{n_1,n_0}^{r,01}]$. In addition, it follows from (5.32) that $E[\hat{\varepsilon}_n^{r,0}] = E[\hat{\varepsilon}_n^{r,1}]$ in the mixture-sampling case. It then follows from (5.33) and (5.36) that $E[\hat{\varepsilon}_n^{r,01}] = E[\hat{\varepsilon}_n^r] = E[\hat{\varepsilon}_n^{r,0}] = E[\hat{\varepsilon}_n^{r,1}]$.

A similar result in the separate-sampling case can be obtained if the design is balanced, that is, $n_0 = n_1 = n/2$. Then if (5.37) holds and the discriminant is anti-symmetric, we have that

$$E\left[\hat{\varepsilon}_{n/2,n/2}^{r,01}\right] = E\left[\hat{\varepsilon}_{n/2,n/2}^{r,0}\right] = E\left[\hat{\varepsilon}_{n/2,n/2}^{r,1}\right], \tag{5.40}$$

regardless of the value of the prior probabilities c_0 and c_1. Hence, it suffices to study one of the population-specific expected error rates, irrespective of the prior probabilities.

5.5.3 Leave-One-Out Error

The development here follows much of that in the previous section, except when it comes to the "near-unbiasedness" theorem of leave-one-out cross-validation.

By taking expectations on both sides of (5.9), it follows that the expected leave-one-out population-specific error rates are given by

$$\begin{aligned} E[\hat{\varepsilon}^{l,0}] &= P(W^{(1)}(\mathbf{X}_1) \leq 0 \mid Y_1 = 0), \\ E[\hat{\varepsilon}^{l,1}] &= P(W^{(1)}(\mathbf{X}_1) > 0 \mid Y_1 = 1). \end{aligned} \tag{5.41}$$

The previous expressions are valid for both the mixture- and separate-sampling schemes. The expected error rates under the two sampling mechanisms are related by

$$\begin{aligned} E\left[\hat{\varepsilon}^{l,0}_{n_0,n_1}\right] &= E\left[\hat{\varepsilon}^{l,0}_{n} \mid N_0 = n_0\right], \\ E\left[\hat{\varepsilon}^{l,1}_{n_0,n_1}\right] &= E\left[\hat{\varepsilon}^{l,1}_{n} \mid N_0 = n_0\right]. \end{aligned} \tag{5.42}$$

As before, it follows that

$$\begin{aligned} E\left[\hat{\varepsilon}^{l,0}_{n}\right] &= \sum_{n_0=0}^{n} \binom{n}{n_0} c_0^{n_0} c_1^{n_1} E\left[\hat{\varepsilon}^{l,0}_{n_0,n_1}\right], \\ E\left[\hat{\varepsilon}^{l,1}_{n}\right] &= \sum_{n_0=0}^{n} \binom{n}{n_0} c_0^{n_0} c_1^{n_1} E\left[\hat{\varepsilon}^{l,1}_{n_0,n_1}\right]. \end{aligned} \tag{5.43}$$

If the prior probabilities c_0 and c_1 are known, then the estimator in (5.10) can be used. Taking expectations on both sides of that equation yields

$$E[\hat{\varepsilon}^{l,01}] = c_0 E[\hat{\varepsilon}^{l,0}] + c_1 E[\hat{\varepsilon}^{l,1}]. \tag{5.44}$$

As before, the mixture- and separate-sampling cases are related by

$$E\left[\hat{\varepsilon}^{l,01}_{n_0,n_1}\right] = E\left[\hat{\varepsilon}^{l,01}_{n} \mid N_0 = n_0\right] \tag{5.45}$$

and

$$\begin{aligned} E\left[\hat{\varepsilon}^{l,01}_{n}\right] &= \sum_{n_0=0}^{n} \binom{n}{n_0} c_0^{n_0} c_1^{n_1} E\left[\hat{\varepsilon}^{l,01}_{n_0,n_1}\right] \\ &= \sum_{n_0=0}^{n} \binom{n}{n_0} c_0^{n_0} c_1^{n_1} \left(c_0 E\left[\hat{\varepsilon}^{l,0}_{n_0,n_1}\right] + c_1 E\left[\hat{\varepsilon}^{l,1}_{n_0,n_1}\right]\right). \end{aligned} \tag{5.46}$$

In addition, it is easy to show that for the standard leave-one-out error estimator in (5.11), we have an expression similar to (5.44):

$$E\left[\hat{\varepsilon}_n^{l}\right] = c_0 E\left[\hat{\varepsilon}_n^{l,0}\right] + c_1 E\left[\hat{\varepsilon}_n^{l,1}\right]. \tag{5.47}$$

Now, in contrast to the case of resubstitution, the expected leave-one-out error rates are linked with expected true error rates, via the cross-validation "near-unbiasedness" theorem, which we now discuss.

In the mixture-sampling case, notice that

$$W^{(1)}(S_n, \mathbf{X}_1) \mid Y_1 = i \sim W(S_{n-1}, \mathbf{X}) \mid Y = i, \quad i = 0, 1, \tag{5.48}$$

where $(\mathbf{X}, Y)$ is an independent test point. From (5.23) and (5.41), it follows that

$$\begin{aligned} E\left[\hat{\varepsilon}_n^{l,0}\right] &= E\left[\varepsilon_{n-1}^{0}\right], \\ E\left[\hat{\varepsilon}_n^{l,1}\right] &= E\left[\varepsilon_{n-1}^{1}\right]. \end{aligned} \tag{5.49}$$

Concerning the estimators in (5.10) or (5.11), this implies that

$$\begin{aligned} E\left[\hat{\varepsilon}_n^{l,01}\right] = E\left[\hat{\varepsilon}_n^{l}\right] &= c_0 E\left[\hat{\varepsilon}_n^{l,0}\right] + c_1 E\left[\hat{\varepsilon}_n^{l,1}\right] \\ &= c_0 E\left[\varepsilon_{n-1}^{0}\right] + c_1 E\left[\varepsilon_{n-1}^{1}\right] \\ &= E[\varepsilon_{n-1}]. \end{aligned} \tag{5.50}$$

The previous expression is the well-known "near-unbiasedness" theorem of leave-one-out error estimation.

Now, let us consider separate sampling. In this case,

$$\begin{aligned} W^{(1)}(S_{n_0,n_1}, \mathbf{X}_1) \mid Y_1 = 0 &\sim W(S_{n_0-1,n_1}, \mathbf{X}) \mid Y = 0, \\ W^{(1)}(S_{n_0,n_1}, \mathbf{X}_1) \mid Y_1 = 1 &\sim W(S_{n_0,n_1-1}, \mathbf{X}) \mid Y = 1, \end{aligned} \tag{5.51}$$

where $(\mathbf{X}, Y)$ is an independent test point. Again from (5.23) and (5.41), it follows that

$$\begin{aligned} E\left[\hat{\varepsilon}_{n_0,n_1}^{l,0}\right] &= E\left[\varepsilon_{n_0-1,n_1}^{0}\right], \\ E\left[\hat{\varepsilon}_{n_0,n_1}^{l,1}\right] &= E\left[\varepsilon_{n_0,n_1-1}^{1}\right]. \end{aligned} \tag{5.52}$$

An interesting fact occurs at this point. If the prior probabilities are known, one can use the estimator (5.10) as a separate-sampling estimator of the overall error rate, as mentioned previously. However,

$$\begin{aligned} E\left[\hat{\varepsilon}_{n_0,n_1}^{l,01}\right] &= c_0 E\left[\hat{\varepsilon}_{n_0,n_1}^{l,0}\right] + c_1 E\left[\hat{\varepsilon}_{n_0,n_1}^{l,1}\right] \\ &= c_0 E\left[\varepsilon_{n_0-1,n_1}^{0}\right] + c_1 E\left[\varepsilon_{n_0,n_1-1}^{1}\right]. \end{aligned} \tag{5.53}$$

In general, this is all that can be said. The previous expression is not in general equal to $E[\varepsilon_{n_0-1,n_1}]$, $E[\varepsilon_{n_0,n_1-1}]$, or $E[\varepsilon_{n_0-1,n_1-1}]$. Therefore, there is no "near-unbiasedness" theorem for the overall error rate in the separate-sampling case even if the prior probabilities are known.

Let us investigate this with further assumptions. If (5.26) holds, and in addition the discriminant is anti-symmetric, then it was shown previously that $E[\varepsilon^0_{n_0,n_1}] = E[\varepsilon^1_{n_1,n_0}]$ and $E[\varepsilon^1_{n_0,n_1}] = E[\varepsilon^0_{n_1,n_0}]$. Now, if in addition the design is balanced, $n_0 = n_1 = n/2$, then it follows that

$$E\left[\hat{\varepsilon}^{l,01}_{n/2,n/2}\right] = E\left[\hat{\varepsilon}^{l,0}_{n/2,n/2}\right] = E\left[\hat{\varepsilon}^{l,1}_{n/2,n/2}\right] = E\left[\varepsilon^{0}_{n/2-1,n/2}\right] = E\left[\varepsilon^{1}_{n/2,n/2-1}\right]. \tag{5.54}$$

This is all that can be said without much stronger additional assumptions. For example, even if the additional assumption of equally likely populations is made, $c_0 = c_1 = 0.5$ — in which case, as shown previously, $E[\varepsilon_{n,n-1}] = E[\varepsilon_{n-1,n}]$ — it is not possible to get the identity $E[\hat{\varepsilon}^{l,01}_{n,n}] = E[\varepsilon_{n,n-1}] = E[\varepsilon_{n-1,n}]$. The difficulty here is that the standard leave-one-out estimator provides no way to estimate the error rate $E[\varepsilon^0_{n,n-1}] = E[\varepsilon^1_{n-1,n}]$ in a nearly unbiased manner. The problem is that the leave-one-out estimator is inherently asymmetric between the classes in the separate-sampling case. We will see in the next section how this can be fixed by using the definition of separate-sampling cross-validation given in Section 5.4.3.

5.5.4 Cross-Validation Error

In the mixture-sampling case, one can start from (5.12) and use the fact that

$$W^{(U)}(S_n, \mathbf{X}_i) \mid Y_i = j \sim W(S_{n-k}, \mathbf{X}) \mid Y = j, \quad i \in U, j = 0, 1, \tag{5.55}$$

where $(\mathbf{X}, Y)$ is an independent test point to show that $E[\hat{\varepsilon}^{\mathrm{cv}(k)}_n] = E[\varepsilon_{n-k}]$, the "near-unbiasedness" theorem of cross-validation.

Let us consider now the separate sampling version of cross-validation introduced in Section 5.4.3. Since

$$\begin{aligned} W^{(U,V)}(S_{n_0,n_1}, \mathbf{X}_i) &\sim W(S_{n_0-k_0,n_1-k_1}, \mathbf{X}) \mid Y = 0, \quad i \in U, \\ W^{(U,V)}(S_{n_0,n_1}, \mathbf{X}_i) &\sim W(S_{n_0-k_0,n_1-k_1}, \mathbf{X}) \mid Y = 1, \quad i \in V, \end{aligned} \tag{5.56}$$

from (5.14),

$$\begin{aligned} E\left[\hat{\varepsilon}^{\mathrm{cv}(k_0,k_1),0}_{n_0,n_1}\right] &= E\left[\varepsilon^0_{n_0-k_0,n_1-k_1}\right], \\ E\left[\hat{\varepsilon}^{\mathrm{cv}(k_0,k_1),1}_{n_0,n_1}\right] &= E\left[\varepsilon^1_{n_0-k_0,n_1-k_1}\right]. \end{aligned} \tag{5.57}$$

Further, if the prior probabilities are known, one can use (5.15) as an estimator of the overall error rate, in which case

$$E\left[\hat{\varepsilon}_{n_0,n_1}^{cv(k_0,k_1),01}\right] = c_0 E\left[\hat{\varepsilon}_{n_0,n_1}^{cv(k_0,k_1),0}\right] + c_1 E\left[\hat{\varepsilon}_{n_0,n_1}^{cv(k_0,k_1),1}\right]$$
$$= c_0 E\left[\varepsilon^0_{n_0-k_0,n_1-k_1}\right] + c_1 E\left[\varepsilon^1_{n_0-k_0,n_1-k_1}\right]$$
$$= E[\varepsilon_{n_0-k_0,n_1-k_1}]. \tag{5.58}$$

This provides the "near-unbiasedness" theorem for cross-validation in the separate-sampling case. For example, for the *separate-sampling leave-(1,1)-out estimator* of the overall error rate, given by the estimator in (5.15) with $k_0 = k_1 = 1$, we have $E[\hat{\varepsilon}_{n_0,n_1}^{cv(1,1),01}] = E[\varepsilon_{n_0-1,n_1-1}]$. Contrast this to the expectation of the leave-one-out estimator $\hat{\varepsilon}_{n_0,n_1}^{l,01}$ in (5.53).

5.5.5 Bootstrap Error

In the mixture-sampling case, from (5.18),

$$E\left[\hat{\varepsilon}_n^{\,\text{zero}}\right] = E\left[E\left[\hat{\varepsilon}_n^{\mathbf{C}} \mid S_n\right]\right] = E\left[\hat{\varepsilon}_n^{\mathbf{C}}\right] = \sum_{\mathbf{C}} E\left[\hat{\varepsilon}_n^{\mathbf{C}} \mid \mathbf{C}\right] P(\mathbf{C}), \tag{5.59}$$

where $P(\mathbf{C})$ is given by (5.16). Now, note that (5.17) can be rewritten as

$$\hat{\varepsilon}_n^{\mathbf{C}} = \frac{n_0(\mathbf{C})}{n(\mathbf{C})}\left(\frac{1}{n_0(\mathbf{C})}\sum_{i=1}^{n} I_{W(S_n^{\mathbf{C}},\mathbf{X}_i)\le 0} I_{Y_i=0} I_{C_i=0}\right)$$
$$+ \frac{n_1(\mathbf{C})}{n(\mathbf{C})}\left(\frac{1}{n_1(\mathbf{C})}\sum_{i=1}^{n} I_{W(S_n^{\mathbf{C}},\mathbf{X}_i)> 0} I_{Y_i=1} I_{C_i=0}\right), \tag{5.60}$$

where $n_0(\mathbf{C}) = \sum_{i=1}^n I_{C_i=0} I_{Y_i=0}$ and $n_1(\mathbf{C}) = \sum_{i=1}^n I_{C_i=0} I_{Y_i=1}$ are the number of zeros in $\mathbf{C}$ corresponding to points with class label 0 and 1, respectively.

For ease of notation in the remainder of this section, we assume that, given $N_0 = n_0$, the sample indices are sorted such that $Y_1 = \cdots = Y_{n_0} = 0$ and $Y_{n_0+1} = \cdots = Y_{n_0+n_1} = 1$, without loss of generality. Taking conditional expectations on both sides of (5.60) yields

$$E\left[\hat{\varepsilon}_n^{\mathbf{C}} \mid N_0 = n_0, \mathbf{C}\right] = \frac{n_0(\mathbf{C})}{n(\mathbf{C})} P\left(W\left(S_n^{\mathbf{C}},\mathbf{X}\right) \le 0 \mid Y = 0, N_0 = n_0, \mathbf{C}\right)$$
$$+ \frac{n_1(\mathbf{C})}{n(\mathbf{C})} P\left(W\left(S_n^{\mathbf{C}},\mathbf{X}\right) > 0 \mid Y = 1, N_0 = n_0, \mathbf{C}\right)$$
$$= \frac{n_0(\mathbf{C})}{n(\mathbf{C})} E\left[\varepsilon_{n_0,n_1}^{\mathbf{C}(1:n_0),\mathbf{C}(n_0+1:n),0} \mid \mathbf{C}\right]$$
$$+ \frac{n_1(\mathbf{C})}{n(\mathbf{C})} E\left[\varepsilon_{n_0,n_1}^{\mathbf{C}(1:n_0),\mathbf{C}(n_0+1:n),1} \mid \mathbf{C}\right], \tag{5.61}$$

where $\mathbf{C}(1:n_0)$ and $\mathbf{C}(n_0+1:n)$ are the vectors containing the first n_0 and last n_1 elements of $\mathbf{C}$, respectively, and $\varepsilon_{n_0,n_1}^{\mathbf{C}(1:n_0),\mathbf{C}(n_0+1:n),0}$ and $\varepsilon_{n_0,n_1}^{\mathbf{C}(1:n_0),\mathbf{C}(n_0+1:n),1}$ are obtained from the error rates

$$\varepsilon_{n_0,n_1}^{\mathbf{C}_0,\mathbf{C}_1,0} = P\left(W\left(S_{n_0,n_1}^{\mathbf{C}_0,\mathbf{C}_1},\mathbf{X}\right) \leq 0 \mid Y=0, S_{n_0,n_1}^{\mathbf{C}_0,\mathbf{C}_1}, \mathbf{C}_0, \mathbf{C}_1\right),$$
$$\varepsilon_{n_0,n_1}^{\mathbf{C}_0,\mathbf{C}_1,1} = P\left(W\left(S_{n_0,n_1}^{\mathbf{C}_0,\mathbf{C}_1},\mathbf{X}\right) > 0 \mid Y=1, S_{n_0,n_1}^{\mathbf{C}_0,\mathbf{C}_1}, \mathbf{C}_0, \mathbf{C}_1\right), \tag{5.62}$$

by letting $\mathbf{C}_0 = \mathbf{C}(1:n_0)$ and $\mathbf{C}_1 = \mathbf{C}(n_0+1:n)$. Note that the error rates in (5.62) are the true population-specific errors of a classifier designed on the bootstrap sample $S_{n_0,n_1}^{\mathbf{C}_0,\mathbf{C}_1}$. One important caveat here is that $\mathbf{C}(1:n_0)$ and $\mathbf{C}(n_0+1:n)$ need not sum to n_0 and n_1, respectively. All error rates are well-defined, nevertheless. From (5.59), it follows that

$$\begin{aligned}
E\left[\hat{\varepsilon}_n^{\,\mathrm{zero}}\right] &= \sum_{C} E\left[\hat{\varepsilon}_n^{\,\mathbf{C}} \mid \mathbf{C}\right] P(\mathbf{C}) \\
&= \sum_{\mathbf{C}} \left(\sum_{n_0=0}^{n} E\left[\hat{\varepsilon}_n^{\,\mathbf{C}} \mid N_0 = n_0, \mathbf{C}\right] P(N_0 = n_0) \right) P(\mathbf{C}) \\
&= \sum_{\mathbf{C}} \left[\sum_{n_0=0}^{n} \binom{n}{n_0} c_0^{n_0} c_1^{n_1} \left(\frac{n_0(\mathbf{C})}{n(\mathbf{C})} E\left[\varepsilon_{n_0,n_1}^{\mathbf{C}(1:n_0),\mathbf{C}(n_0+1:n),0} \mid \mathbf{C}\right] \right.\right. \\
&\qquad \left.\left. + \frac{n_1(\mathbf{C})}{n(\mathbf{C})} E\left[\varepsilon_{n_0,n_1}^{\mathbf{C}(1:n_0),\mathbf{C}(n_0+1:n),1} \mid \mathbf{C}\right] \right) \right] P(\mathbf{C}).
\end{aligned} \tag{5.63}$$

where

$$E\left[\varepsilon_{n_0,n_1}^{\mathbf{C}(1:n_0),\mathbf{C}(n_0+1:n),0} \mid \mathbf{C}\right] = P\left(W\left(S_{n_0,n_1}^{\mathbf{C}(1:n_0),\mathbf{C}(n_0+1:n)},\mathbf{X}\right) \leq 0 \mid Y=0, \mathbf{C}\right),$$
$$E\left[\varepsilon_{n_0,n_1}^{\mathbf{C}(1:n_0),\mathbf{C}(n_0+1:n),1} \mid \mathbf{C}\right] = P\left(W\left(S_{n_0,n_1}^{\mathbf{C}(1:n_0),\mathbf{C}(n_0+1:n)},\mathbf{X}\right) > 0 \mid Y=1, \mathbf{C}\right). \tag{5.64}$$

Notice that inside the inner summation in (5.63), the values of $n_0(\mathbf{C})$, $n_1(\mathbf{C})$, $\mathbf{C}(1:n_0)$, and $\mathbf{C}(n_0+1:n)$ change as n_0 changes. According to (5.63), to find the mixture-sampling expected zero bootstrap error rate, it suffices to have expressions for $E[\varepsilon_{n_0,n_1}^{\mathbf{C}(1:n_0),\mathbf{C}(n_0+1:n),0} \mid \mathbf{C}]$ and $E[\varepsilon_{n_0,n_1}^{\mathbf{C}(1:n_0),\mathbf{C}(n_0+1:n),1} \mid \mathbf{C}]$. Exact expressions that allow the computation of these expectations will be given in Chapters 6 and 7 in the Gaussian LDA case.

As an application, the availability of $E[\hat{\varepsilon}_n^{\,\mathrm{zero}}]$, in addition to $E[\hat{\varepsilon}_n^{\,r}]$ and $E[\varepsilon_n]$, allows one to compute the required weight for unbiased convex bootstrap error estimation. As discussed in Section 2.7, we can define a convex bootstrap estimator via

$$\hat{\varepsilon}_n^{\,\mathrm{conv}} = (1-w)\,\hat{\varepsilon}_n^{\,r} + w\,\hat{\varepsilon}_n^{\,\mathrm{zero}}. \tag{5.65}$$

Our goal here is to determine a weight $w = w^*$ to achieve an unbiased estimator. The weight is heuristically set to $w = 0.632$ in Efron (1983) without rigorous justification. The bias of the convex estimator in (5.65) is given by

$$E\left[\hat{\varepsilon}_n^{\,\text{conv}} - \varepsilon_n\right] = (1-w)E\left[\hat{\varepsilon}_n^{\,r}\right] + wE\left[\hat{\varepsilon}_n^{\,\text{zero}}\right] - E[\varepsilon_n]. \tag{5.66}$$

Setting this to zero yields the exact weight

$$w^* = \frac{E\left[\hat{\varepsilon}_n^{\,r}\right] - E[\varepsilon_n]}{E\left[\hat{\varepsilon}_n^{\,r}\right] - E\left[\hat{\varepsilon}_n^{\,\text{zero}}\right]} \tag{5.67}$$

that produces an unbiased error estimator. The value of w^* will be plotted in Chapters 6 and 7 in the Gaussian LDA case.

Finally, in the separate-sampling case, it follows from (5.20) that

$$\begin{aligned} E\left[\hat{\varepsilon}_{n_0,n_1}^{\,\text{zero},i}\right] &= E\left[E\left[\hat{\varepsilon}_{n_0,n_1}^{\,\mathbf{C}_0,\mathbf{C}_1,i} \mid S_{n_0,n_1}\right]\right] = E\left[\hat{\varepsilon}_{n_0,n_1}^{\,\mathbf{C}_0,\mathbf{C}_1,i}\right] \\ &= \sum_{\mathbf{C}_0,\mathbf{C}_1} E\left[\hat{\varepsilon}_{n_0,n_1}^{\,\mathbf{C}_0,\mathbf{C}_1,i} \mid \mathbf{C}_0,\mathbf{C}_1\right] P(\mathbf{C}_0)P(\mathbf{C}_1), \quad i = 0,1. \end{aligned} \tag{5.68}$$

Now, taking expectations on both sides of (5.19) yields

$$\begin{aligned} E\left[\hat{\varepsilon}_{n_0,n_1}^{\,\mathbf{C}_0,\mathbf{C}_1,0} \mid \mathbf{C}_0,\mathbf{C}_1\right] &= P\left(W\left(S_{n_0,n_1}^{\mathbf{C}_0,\mathbf{C}_1},\mathbf{X}\right) \le 0 \mid Y=0,\mathbf{C}_0,\mathbf{C}_1\right) \\ &= E\left[\varepsilon_{n_0,n_1}^{\mathbf{C}_0,\mathbf{C}_1,0} \mid \mathbf{C}_0,\mathbf{C}_1\right], \\ E\left[\hat{\varepsilon}_{n_0,n_1}^{\,\mathbf{C}_0,\mathbf{C}_1,1} \mid \mathbf{C}_0,\mathbf{C}_1\right] &= P\left(W\left(S_{n_0,n_1}^{\mathbf{C}_0,\mathbf{C}_1},\mathbf{X}\right) > 0 \mid Y=1,\mathbf{C}_0,\mathbf{C}_1\right) \\ &= E\left[\varepsilon_{n_0,n_1}^{\mathbf{C}_0,\mathbf{C}_1,1} \mid \mathbf{C}_0,\mathbf{C}_1\right], \end{aligned} \tag{5.69}$$

where $(\mathbf{X}, Y)$ is an independent test point and $\varepsilon_{n_0,n_1}^{\mathbf{C}_0,\mathbf{C}_1,0}$ and $\varepsilon_{n_0,n_1}^{\mathbf{C}_0,\mathbf{C}_1,1}$ are the population-specific true error rates associated with the discriminant $W(S_{n_0,n_1}^{\mathbf{C}_0,\mathbf{C}_1}, \mathbf{X})$. Combining (5.68) and (5.69) leads to the final result:

$$E\left[\hat{\varepsilon}_{n_0,n_1}^{\,\text{zero},i}\right] = \sum_{\mathbf{C}_0,\mathbf{C}_1} E\left[\varepsilon_{n_0,n_1}^{\mathbf{C}_0,\mathbf{C}_1,i} \mid \mathbf{C}_0,\mathbf{C}_1\right] P(\mathbf{C}_0)P(\mathbf{C}_1), \quad i = 0,1. \tag{5.70}$$

Hence, to find the population-specific expected zero bootstrap error rates, it suffices to have expressions for $E[\varepsilon_{n_0,n_1}^{\mathbf{C}_0,\mathbf{C}_1,i}]$, for $i = 0, 1$, and perform the summation in (5.70). Finally, if the prior probabilities are known and (5.21) is used as an estimator of the overall error rate, then

$$E\left[\hat{\varepsilon}_{n_0,n_1}^{\,\text{zero},01}\right] = c_0 E\left[\hat{\varepsilon}_{n_0,n_1}^{\,\text{zero},0}\right] + c_1 E\left[\hat{\varepsilon}_{n_0,n_1}^{\,\text{zero},1}\right]. \tag{5.71}$$

Exact expressions for computing $E[\varepsilon_{n_0,n_1}^{\mathbf{C}_0,\mathbf{C}_1,0} \mid \mathbf{C}_0, \mathbf{C}_1]$ and $E[\varepsilon_{n_0,n_1}^{\mathbf{C}_0,\mathbf{C}_1,1} \mid \mathbf{C}_0, \mathbf{C}_1]$ will be given in Chapters 6 and 7 in the Gaussian case.

5.6 HIGHER-ORDER MOMENTS OF ERROR RATES

Higher-order moments (including mixed moments) of the true and estimated error rates are needed for, among other things, computing the variance and RMS of error estimators, as seen in Chapter 2.

In this section we limit ourselves to describing the higher-order moments of the true error rate and of the resubstitution and leave-one-out estimated error rates, as well as the sampling distributions of resubstitution and leave-one-out.

5.6.1 True Error

From (5.4),

$$\begin{aligned} E[\varepsilon^2] &= E\left[(c_0\varepsilon^0 + c_1\varepsilon^1)^2\right] \\ &= c_0^2 E[(\varepsilon^0)^2] + 2c_0c_1E[\varepsilon^0\varepsilon^1] + c_1^2E[(\varepsilon^1)^2]. \end{aligned} \tag{5.72}$$

With $(\mathbf{X}^{(1)}, Y^{(1)})$ and $(\mathbf{X}^{(2)}, Y^{(2)})$ denoting independent test points, we obtain

$$\begin{aligned} E[(\varepsilon^0)^2] &= E[P(W(\mathbf{X}^{(1)}) \le 0 \mid Y^{(1)} = 0, S)\, P(W(\mathbf{X}^{(2)}) \le 0 \mid Y^{(2)} = 0, S)] \\ &= E[P(W(\mathbf{X}^{(1)}) \le 0, W(\mathbf{X}^{(2)}) \le 0 \mid Y^{(1)} = Y^{(2)} = 0, S)] \\ &= P(W(\mathbf{X}^{(1)}) \le 0, W(\mathbf{X}^{(2)}) \le 0 \mid Y^{(1)} = Y^{(2)} = 0). \end{aligned} \tag{5.73}$$

Similarly, one can show that

$$\begin{aligned} E[\varepsilon^0\varepsilon^1] &= P(W(\mathbf{X}^{(1)}) \le 0, W(\mathbf{X}^{(2)}) > 0 \mid Y^{(1)} = 0, Y^{(2)} = 1), \\ E[(\varepsilon^1)^2] &= P(W(\mathbf{X}^{(1)}) > 0, W(\mathbf{X}^{(2)}) > 0 \mid Y^{(1)} = Y^{(2)} = 1). \end{aligned} \tag{5.74}$$

The previous expressions are valid for both the mixture- and separate-sampling schemes, the difference between them being the distinct sampling distributions of the discriminants.

In general, the moment of ε of order k can be obtained from knowledge of the joint sampling distribution of $W(\mathbf{X}^{(1)}), \ldots, W(\mathbf{X}^{(k)}) \mid Y^{(1)}, \ldots, Y^{(k)}$, where $(\mathbf{X}^{(1)}, Y^{(1)}), \ldots, (\mathbf{X}^{(k)}, Y^{(k)})$ are independent test points. This result is summarized in the following theorem, the proof of which follows the pattern of the case $k = 2$ above, and is thus omitted. The result holds for both mixture and separate sampling.

Theorem 5.1 *The moment of order k of the true error rate is given by*

$$E[\varepsilon^k] = \sum_{l=0}^{k} \binom{k}{l} c_0^{k-l} c_1^l\, E[(\varepsilon^0)^{k-l}(\varepsilon^1)^l], \tag{5.75}$$

where

$$E[(\varepsilon^0)^{k-l}(\varepsilon^1)^l] = P(W(\mathbf{X}^{(1)}) \leq 0, \dots, W(\mathbf{X}^{(k-l)}) \leq 0, W(\mathbf{X}^{(k-l+1)}) > 0, \dots, \\ W(\mathbf{X}^{(k)}) > 0 \mid Y^{(1)} = \cdots = Y^{(k-l)} = 0, Y^{(k-l+1)} = \cdots = Y^{(k)} = 1), \tag{5.76}$$

for independent test points $(\mathbf{X}^{(1)}, Y^{(1)}), \dots, (\mathbf{X}^{(k)}, Y^{(k)})$.

5.6.2 Resubstitution Error

For the separate-sampling case, it follows from (5.6) that

$$\begin{aligned} E\left[\left(\hat{\varepsilon}^{r,0}_{n_0,n_1}\right)^2\right] &= \frac{1}{n_0} P(W(\mathbf{X}_1) \leq 0 \mid Y_1 = 0) \\ &\quad + \frac{n_0 - 1}{n_0} P(W(\mathbf{X}_1) \leq 0, W(\mathbf{X}_2) \leq 0 \mid Y_1 = 0, Y_2 = 0), \\ E\left[\hat{\varepsilon}^{r,0}_{n_0,n_1} \hat{\varepsilon}^{r,1}_{n_0,n_1}\right] &= P(W(\mathbf{X}_1) \leq 0, W(\mathbf{X}_2) > 0 \mid Y_1 = 0, Y_2 = 1), \\ E\left[\left(\hat{\varepsilon}^{r,1}_{n_0,n_1}\right)^2\right] &= \frac{1}{n_1} P(W(\mathbf{X}_1) > 0 \mid Y_1 = 1) \\ &\quad + \frac{n_1 - 1}{n_1} P(W(\mathbf{X}_1) > 0, W(\mathbf{X}_2) > 0 \mid Y_1 = 1, Y_2 = 1). \end{aligned} \tag{5.77}$$

If the prior probabilities c_0 and c_1 are known, then one can use (5.7) as an estimator of the overall error rate, in which case,

$$\begin{aligned} E\left[\left(\hat{\varepsilon}^{r,01}_{n_0,n_1}\right)^2\right] &= E\left[\left(c_0 \hat{\varepsilon}^{r,0}_{n_0,n_1} + c_1 \hat{\varepsilon}^{r,1}_{n_0,n_1}\right)^2\right] \\ &= c_0^2 E\left[\left(\hat{\varepsilon}^{r,0}_{n_0,n_1}\right)^2\right] + 2c_0 c_1 E\left[\hat{\varepsilon}^{r,0}_{n_0,n_1} \hat{\varepsilon}^{r,1}_{n_0,n_1}\right] + c_1^2 E\left[\left(\hat{\varepsilon}^{r,1}_{n_0,n_1}\right)^2\right]. \end{aligned} \tag{5.78}$$

In general, the moment of $\hat{\varepsilon}^{r,01}_{n_0,n_1}$ of order k, for $k \leq \min\{n_0, n_1\}$, can be obtained using the following theorem, the proof of which follows the general pattern of the case $k = 2$ above.

Theorem 5.2 *Assume that the prior probabilities are known and the resubstitution estimator of the overall error rate (5.7) is used. Its separate-sampling moment of order k, for* $k \leq \min\{n_0, n_1\}$, *is given by*

$$E\left[\left(\hat{\varepsilon}^{r,01}_{n_0,n_1}\right)^k\right] = \sum_{l=0}^{k} \binom{k}{l} c_0^{k-l} c_1^l \, E\left[\left(\hat{\varepsilon}^{r,0}_{n_0,n_1}\right)^{k-l} \left(\hat{\varepsilon}^{r,1}_{n_0,n_1}\right)^l\right], \tag{5.79}$$

with

$$E\left[\left(\hat{\varepsilon}^{r,0}_{n_0,n_1}\right)^{k-l}\left(\hat{\varepsilon}^{r,1}_{n_0,n_1}\right)^{l}\right] = \frac{1}{n_0^{k-l}}\frac{1}{n_1^{l}}\sum_{i=1}^{k-l}\sum_{j=1}^{l} d_{n_0}(k-l,i)d_{n_1}(l,j) \times$$
$$P(W(\mathbf{X}_1) \leq 0, \ldots, W(\mathbf{X}_i) \leq 0,$$
$$W(\mathbf{X}_{i+1}) > 0, \ldots, W(\mathbf{X}_{i+j}) > 0 \mid Y_1 = \cdots = Y_i = 0,$$
$$Y_{i+1} = \cdots = Y_{i+j} = 1), \tag{5.80}$$

where $d_n(k,m)$ is the number of k-digit numbers in base n (including any leading zeros) that have exactly m unique digits, for $m = 1, \ldots, k$.

It is possible, with some effort, to give a general formula to compute $d_n(k,m)$. We will instead illustrate it through examples. First, it is clear that

$$d_n(k,1) = n \quad \text{and} \quad d_n(k,k) = \frac{n!}{(n-k)!}. \tag{5.81}$$

With $k = 2$, this gives $d_n(2,1) = n$ and $d_n(2,2) = n(n-1)$, which leads to the formulas in (5.77) and (5.78). With $k = 3$, we obtain $d_n(3,1) = n$, $d_n(3,2) = 3n(n-1)$, and $d_n(3,3) = n(n-1)(n-2)$, which gives

$$E\left[\left(\hat{\varepsilon}^{r,0}_{n_0,n_1}\right)^{3}\right] = \frac{1}{n_0^2}P(W(\mathbf{X}_1) \leq 0 \mid Y_1 = 0)$$
$$+ \frac{3(n_0-1)}{n_0^2}P(W(\mathbf{X}_1) \leq 0, W(\mathbf{X}_2) \leq 0 \mid Y_1 = Y_2 = 0)$$
$$+ \frac{(n_0-1)(n_0-2)}{n_0^2}P(W(\mathbf{X}_1) \leq 0, W(\mathbf{X}_2) \leq 0,$$
$$W(\mathbf{X}_3) \leq 0 \mid Y_1 = Y_2 = Y_3 = 0). \tag{5.82}$$

With $k = 4$, one has $d_n(4,1) = n$, $d_n(4,2) = 7n(n-1)$, $d_n(4,2) = 6n(n-1)(n-2)$, and $d_n(3,3) = n(n-1)(n-2)(n-3)$, so that

$$E\left[\left(\hat{\varepsilon}^{r,0}_{n_0,n_1}\right)^{4}\right] = \frac{1}{n_0^3}P(W(\mathbf{X}_1) \leq 0 \mid Y_1 = 0)$$
$$+ \frac{7(n_0-1)}{n_0^2}P(W(\mathbf{X}_1) \leq 0, W(\mathbf{X}_2) \leq 0 \mid Y_1 = Y_2 = 0)$$
$$+ \frac{6(n_0-1)(n_0-2)}{n_0^2}P(W(\mathbf{X}_1) \leq 0, W(\mathbf{X}_2) \leq 0,$$
$$W(\mathbf{X}_3) \leq 0 \mid Y_1 = Y_2 = Y_3 = 0)$$

$$+\frac{(n_0-1)(n_0-2)(n_0-3)}{n_0^2}P(W(\mathbf{X}_1)\leq 0, W(\mathbf{X}_2)\leq 0,$$
$$W(\mathbf{X}_3)\leq 0, W(\mathbf{X}_4)\leq 0 \mid Y_1=Y_2=Y_3=Y_4=0). \tag{5.83}$$

The corresponding expressions for mixture sampling can be obtained by conditioning on $N_0 = n_0$ and using the previous results for the separate-sampling case.

The mixed moments of the true and resubstitution error rates are needed to compute the estimator deviation variance, RMS, and correlation with the true error. It can be shown that

$$\begin{aligned}
E[\varepsilon^0\hat{\varepsilon}^{r,0}] &= P(W(\mathbf{X})\leq 0, W(\mathbf{X}_1)\leq 0 \mid Y=0, Y_1=0),\\
E[\varepsilon^0\hat{\varepsilon}^{r,1}] &= P(W(\mathbf{X})\leq 0, W(\mathbf{X}_1)> 0 \mid Y=0, Y_1=1),\\
E[\varepsilon^1\hat{\varepsilon}^{r,0}] &= P(W(\mathbf{X})> 0, W(\mathbf{X}_1)\leq 0 \mid Y=1, Y_1=0),\\
E[\varepsilon^1\hat{\varepsilon}^{r,1}] &= P(W(\mathbf{X})> 0, W(\mathbf{X}_1)> 0 \mid Y=1, Y_1=1),
\end{aligned} \tag{5.84}$$

where $(\mathbf{X}, Y)$ is an independent test point. If the prior probabilities are known and the estimator in (5.7) is used, then

$$E[\varepsilon\hat{\varepsilon}^{r,01}] = c_0^2E[\varepsilon^0\hat{\varepsilon}^{r,0}] + c_0c_1E[\varepsilon^0\hat{\varepsilon}^{r,1}] + c_1c_0E[\varepsilon^1\hat{\varepsilon}^{r,0}] + c_1^2E[\varepsilon^1\hat{\varepsilon}^{r,1}]. \tag{5.85}$$

These results apply to both mixture and separate sampling.

5.6.3 Leave-One-Out Error

It is straightforward to verify that the results for leave-one-out are obtained from those for resubstitution by replacing $W(\mathbf{X}_i)$ by $W^{(i)}(\mathbf{X}_i)$ everywhere. Therefore, we have the following theorem.

Theorem 5.3 *Assume that the prior probabilities are known and the leave-one-out estimator of the overall error rate (5.10) is used. Its separate-sampling moment of order k, for $k \leq \min\{n_0, n_1\}$, is given by*

$$E\left[\left(\hat{\varepsilon}^{l,01}_{n_0,n_1}\right)^k\right] = \sum_{l=0}^{k}\binom{k}{l}c_0^{k-l}c_1^l\,E\left[\left(\hat{\varepsilon}^{l,0}_{n_0,n_1}\right)^{k-l}\left(\hat{\varepsilon}^{l,1}_{n_0,n_1}\right)^l\right], \tag{5.86}$$

with

$$E\left[\left(\hat{\varepsilon}^{l,0}_{n_0,n_1}\right)^{k-l}\left(\hat{\varepsilon}^{l,1}_{n_0,n_1}\right)^l\right] = \frac{1}{n_0^{k-l}}\frac{1}{n_1^l}\sum_{i=1}^{k-l}\sum_{j=1}^{l} d_{n_0}(k-l,i)d_{n_1}(l,j)\times$$
$$P(W^{(1)}(\mathbf{X}_1)\leq 0,\ldots,W^{(i)}(\mathbf{X}_i)\leq 0,$$
$$W^{(i+1)}(\mathbf{X}_{i+1})>0,\ldots,W^{(i+j)}(\mathbf{X}_{i+j})>0 \mid Y_1=\cdots=Y_i=0,$$
$$Y_{i+1}=\cdots=Y_{i+j}=1), \tag{5.87}$$

where $d_n(k, m)$ is the number of k-digit numbers in base n (including any leading zeros) that have exactly m unique digits, for $m = 1, \ldots, k$.

Similarly, for the mixed moments:

$$\begin{aligned}
E[\varepsilon^0 \hat{\varepsilon}^{l,0}] &= P(W(\mathbf{X}) \leq 0, W^{(1)}(\mathbf{X}_1) \leq 0 \mid Y = 0, Y_1 = 0), \\
E[\varepsilon^0 \hat{\varepsilon}^{l,1}] &= P(W(\mathbf{X}) \leq 0, W^{(1)}(\mathbf{X}_1) > 0 \mid Y = 0, Y_1 = 1), \\
E[\varepsilon^1 \hat{\varepsilon}^{l,0}] &= P(W(\mathbf{X}) > 0, W^{(1)}(\mathbf{X}_1) \leq 0 \mid Y = 1, Y_1 = 0), \\
E[\varepsilon^1 \hat{\varepsilon}^{l,1}] &= P(W(\mathbf{X}) > 0, W^{(1)}(\mathbf{X}_1) > 0 \mid Y = 1, Y_1 = 1).
\end{aligned} \tag{5.88}$$

If the prior probabilities are known and the estimator in (5.10) is used, then

$$E[\varepsilon \hat{\varepsilon}^{l,01}] = c_0^2 E[\varepsilon^0 \hat{\varepsilon}^{l,0}] + c_0 c_1 E[\varepsilon^0 \hat{\varepsilon}^{l,1}] + c_1 c_0 E[\varepsilon^1 \hat{\varepsilon}^{l,0}] + c_1^2 E[\varepsilon^1 \hat{\varepsilon}^{l,1}]. \tag{5.89}$$

These results apply to both mixture and separate sampling.

5.7 SAMPLING DISTRIBUTION OF ERROR RATES

Going beyond the moments, the full distribution of the error rates is also of interest. In this section, we provide general results on the distributions of the classical mixture-sampling resubstitution and leave-one-out estimators of the overall error rate. It may not be possible to provide similar general results on the marginal distribution of the true error rate or the joint distribution of the true and estimated error rates, which are key to describing completely the performance of an error estimator, as discussed in Chapter 2. However, results can be obtained for specific distributions. In the next two chapters this is done for the specific case of LDA discriminants over Gaussian populations. (Joint distributions were also computed in Chapter 4 for discrete classifiers using complete enumeration.)

5.7.1 Resubstitution Error

First note that

$$P\left(\hat{\varepsilon}_n^r = \frac{k}{n}\right) = \sum_{n_0=0}^{n} P\left(\hat{\varepsilon}_n^r = \frac{k}{n} \,\middle|\, N_0 = n_0\right) P(N_0 = n_0), \quad k = 0, 1, \ldots, n, \tag{5.90}$$

where

$$P\left(\hat{\varepsilon}_n^r = \frac{k}{n} \,\middle|\, N_0 = n_0\right) = \sum_{l=0}^{k} P(U = l,\ V = k - l \mid N_0 = n_0), \quad k = 0, 1, \ldots, n, \tag{5.91}$$

with U and V representing the numbers of errors committed in populations Π_0 and Π_1, respectively. Now, it is straightforward to establish that

$$\begin{aligned} P(U = l,\, V = k - l \mid N_0 = n_0) = \binom{n_0}{l}\binom{n_1}{k-l} & P(W(\mathbf{X}_1) \le 0, \dots, W(\mathbf{X}_l) \le 0, \\ & W(\mathbf{X}_{l+1}) > 0, \dots, W(\mathbf{X}_{n_0+k-l}) > 0, \\ & W(\mathbf{X}_{n_0+k-l+1}) \le 0, \dots, W(\mathbf{X}_n) \le 0 \mid \\ & Y_1 = \cdots = Y_l = 0, Y_{l+1} = \cdots = Y_{n_0+k-l} = 1, \\ & Y_{n_0+k-l+1} = \cdots = Y_n = 0), \end{aligned} \tag{5.92}$$

for $k = 0, 1, \dots, n$. Putting everything together establishes the following theorem.

Theorem 5.4 *The sampling distribution of the resubstitution estimator of the overall error rate in the mixture-sampling case in (5.8) is given by*

$$\begin{aligned} P\left(\hat{\varepsilon}_n^r = \frac{k}{n}\right) = \sum_{n_0=0}^{n} \binom{n}{n_0} c_0^{n_0} c_1^{n_1} \sum_{l=0}^{k} \binom{n_0}{l}\binom{n_1}{k-l} \\ \times P(W(\mathbf{X}_1) \le 0, \dots, W(\mathbf{X}_l) \le 0, \\ W(\mathbf{X}_{l+1}) > 0, \dots, W(\mathbf{X}_{n_0+k-l}) > 0, \\ W(\mathbf{X}_{n_0+k-l+1}) \le 0, \dots, W(\mathbf{X}_n) \le 0 \mid \\ Y_1 = \cdots = Y_l = 0, Y_{l+1} = \cdots = Y_{n_0+k-l} = 1, \\ Y_{n_0+k-l+1} = \cdots = Y_n = 0), \end{aligned} \tag{5.93}$$

for $k = 0, 1, \dots, n$.

We conclude that computation of the resubstitution error rate distribution requires the joint distribution of the entire vector $W(\mathbf{X}_1), \dots, W(\mathbf{X}_n)$. It is interesting to contrast this to the expressions for the k-th order error rate moments given previously, which require the joint distribution of a smaller k-dimensional vector of discriminants, $k \le \min\{n_0, n_1\}$.

5.7.2 Leave-One-Out Error

It is straightforward to show that the corresponding results for the leave-one-out estimator are obtained from those for resubstitution by replacing $W(\mathbf{X}_i)$ by $W^{(i)}(\mathbf{X}_i)$ everywhere. In particular, we have the following theorem.

Theorem 5.5 *The sampling distribution of the leave-one-out estimator of the overall error rate in the mixture-sampling case in (5.11) is given by*

$$
\begin{aligned}
P\left(\hat{\varepsilon}_n^{\,l} = \frac{k}{n}\right) = \sum_{n_0=0}^{n} \binom{n}{n_0} c_0^{n_0} c_1^{n_1} \sum_{l=0}^{k} \binom{n_0}{l} \binom{n_1}{k-l} \\
\times P(W^{(1)}(\mathbf{X}_1) \leq 0, \ldots, W^{(l)}(\mathbf{X}_l) \leq 0, \\
W^{(l+1)}(\mathbf{X}_{l+1}) > 0, \ldots, W^{(n_0+k-l)}(\mathbf{X}_{n_0+k-l}) > 0, \\
W^{(n_0+k-l+1)}(\mathbf{X}_{n_0+k-l+1}) \leq 0, \ldots, W^{(n)}(\mathbf{X}_n) \leq 0 \\
\mid Y_1 = \cdots = Y_l = 0, Y_{l+1} = \cdots = Y_{n_0+k-l} = 1, \\
Y_{n_0+k-l+1} = \cdots = Y_n = 0), \qquad (5.94)
\end{aligned}
$$

for $k = 0, 1, \ldots, n$.

EXERCISES

5.1 Redo Figure 5.1 using the model M2 of Exercise 1.8, for the QDA, 3NN, and linear SVM classification rules, and for $d = 6$, $\sigma = 1$, $\rho = 0.2, 0.8$, $\varepsilon_{\text{bay}} = 0.2$.

5.2 Redo Figure 5.2 using the model M2 of Exercise 1.8, for the QDA, 3NN, and linear SVM classification rules, and for $d = 6$, $\sigma = 1$, $\rho = 0.2, 0.8$, $\varepsilon_{\text{bay}} = 0.2$.

5.3 Redo Figure 5.3 with 6-fold cross-validation using the model M2 of Exercise 1.8, for the LDA, QDA, 3NN, and linear SVM classification rules, and for $d = 6$, $\sigma = 1$, $\rho = 0.2, 0.8$, $\varepsilon_{\text{bay}} = 0.2$.

5.4 Redo Figure 5.3 using the model M2 of Exercise 1.8, for the LDA, QDA, 3NN, and linear SVM classification rules, and for $d = 6$, $\sigma = 1$, $\rho = 0.2, 0.8$, $\varepsilon_{\text{bay}} = 0.2$, except now use the .632 bootstrap estimator.

5.5 Redo Figure 5.3 using the model M2 of Exercise 1.8, for the LDA, QDA, 3NN, and linear SVM classification rules, and for $d = 6$, $\sigma = 1$, $\rho = 0.2, 0.8$, $\varepsilon_{\text{bay}} = 0.2$, except now use separate-sampling cross-validation with $k_0 = k_1 = 3$.

5.6 Redo Figure 5.3 using the model M2 of Exercise 1.8, for the LDA QDA, 3NN, and linear SVM classification rules, and for $d = 6$, $\sigma = 1$, $\rho = 0.2, 0.8$, $\varepsilon_{\text{bay}} = 0.2$, except now use the separate-sampling .632 bootstrap.

5.7 As we see from (5.53), there is no "near-unbiasedness" theorem for leave-one-out with separate sampling. Using QDA, the model M2 of Exercise 1.8 with $d = 6$, $\sigma = 1$, $\rho = 0.2, 0.8$, $\varepsilon_{\text{bay}} = 0.2$, and $c_0 = 0.5$, plot $E[\hat{\varepsilon}^{\,l,01}_{n_0,n_1}]$, $E[\varepsilon_{n_0-1,n_1}]$, and $E[\varepsilon_{n_0,n_1-1}]$ for $n_0, n_1 = 5, 10, 15, \ldots, 50$ (keeping one sample size fixed and varying the other), and compare the plots.

5.8 Compute the unbiasedness weight of (5.67) using the model M2 of Exercise 1.8, for the QDA, 3NN, and linear SVM classification rules, and for $d = 6$, $\sigma = 1$, $\rho = 0.2, 0.8$, $\varepsilon_{\text{bay}} = 0.2$, and $n = 5, 10, 15, \ldots, 50$.

5.9 Prove (5.74).

5.10 Prove Theorem 5.1.

5.11 Prove Theorem 5.2.

5.12 Prove (5.84).

CHAPTER 6

GAUSSIAN DISTRIBUTION THEORY: UNIVARIATE CASE

This chapter and the next present a survey of salient results in the distributional study of error rates of Linear Discriminant Analysis (LDA) and Quadratic Discriminant Analysis (QDA) under an assumption of Gaussian populations. There is a vast body of literature on this topic; excellent surveys can be found in Anderson (1984), McLachlan (1992), Raudys and Young (2004), Wyman et al. (1990). Our purpose is not to be comprehensive, rather to present a systematic account of the classical results in the field, as well as recent developments not covered in other surveys.

In Sections 6.3 and 6.4, we concentrate on results for the separate-sampling population-specific error rates. This is the case that has been historically assumed in studies on error rate moments. Furthermore, as we showed in Chapter 5, distributional results for mixture-sampling moments can be derived from results for separate-sampling moments by conditioning on the population-specific sample sizes. In Section 6.5 we consider the sampling distribution of the classical mixture-sampling resubstitution and leave-one-out error estimators, including the joint distribution and joint density with the true error rate.

A characteristic feature of the univariate case is the availability of exact results for the heteroskedastic case (unequal population variances) — which is generally difficult to obtain in the multivariate case. All the results presented in this chapter are valid for the heteroskedastic case (the homoskedastic special cases of those results are included at some junctures for historical or didactic reasons).

Error Estimation for Pattern Recognition, First Edition. Ulisses M. Braga-Neto and Edward R. Dougherty.

6.1 HISTORICAL REMARKS

The classical approach to study the small-sample performance of sample-based statistics is the distributional approach, which was developed early in the history of modern mathematical statistics by R.A. Fisher, when he derived the sampling distribution of the correlation coefficient for Gaussian data in a series of seminal papers (Fisher, 1915, 1921, 1928). Fisher's work allowed one to express all the properties of the statistic (such as bias, variance, higher-order moments, and so on), for a population sample of finite size in terms of the parameters of the population distribution. We quote from H. Cramér's classical text, "Mathematical Methods of Statistics" (Cramér, 1946):

> It is clear that a knowledge of the *exact* form of a sampling distribution would be of a far greater value than [...] a limiting expression for large values of n. Especially when we are dealing with *small samples*, as is often the case in the applications, the asymptotic expressions are sometimes grossly inadequate, and a knowledge of the exact form of the distribution would then be highly desirable.

This approach was vigorously followed in the early literature on pattern recognition — known as "discriminant analysis" in the statistical community. LDA has a long history, having been originally based on an idea by Fisher (the "Fisher discriminant") (Fisher, 1936), developed by A. Wald (1944), and given the form known today by T.W. Anderson (1951). There is a wealth of results on the properties of the classification error of LDA. For example, the exact distribution and expectation of the true classification error were determined by S. John in the univariate case and in the multivariate case by assuming that the covariance matrix Σ is known in the formulation of the discriminant (John, 1961). The case when Σ is not known and the sample covariance matrix S is used in discrimination is very difficult, and John gave only an asymptotic approximation to the distribution of the true error (John, 1961). In publications that appeared in the same year, A. Bowker provided a statistical representation of the LDA discriminant (Bowker, 1961), whereas R. Sitgreaves gave the exact distribution of the discriminant in terms of an infinite series when the sample sizes of each class are equal (Sitgreaves, 1961). Several classical papers have studied the distribution and moments of the true classification error of LDA under a parametric Gaussian assumption, using exact and approximate methods (Anderson, 1973; Bowker and Sitgreaves, 1961; Harter, 1951; Hills, 1966; Kabe, 1963; Okamoto, 1963; Raudys, 1972; Sayre, 1980; Sitgreaves, 1951; Teichroew and Sitgreaves, 1961). G. McLachlan (1992) and T. Anderson (1984) provide extensive surveys of these methods, whereas Wyman and collaborators provide a compact survey and numerical comparison of several of the asymptotic results in the literature on the true error for Gaussian discrimination (Wyman et al., 1990). A survey of similar results published in the Russian literature has been given by S. Raudys and D.M. Young (Raudys and Young, 2004, Section 3).

Results involving estimates of the classification error are comparatively scarcer in the literature, and until recently were confined to moments and not distributions. As far as the authors know, the first journal paper to appear in the English language literature to carry out a systematic analytical study of error estimation was the paper by M. Hills

(1966) — even though P. Lachenbruch's Ph.D. thesis (Lachenbruch, 1965) predates it by 1 year, and results were published in Russian a few years earlier (Raudys and Young, 2004; Toussaint, 1974). Hills provided in Hills (1966) an exact formula for the expected resubstitution error estimate in the univariate case that involves the bivariate Gaussian cumulative distribution. M. Moran extended this result to the multivariate case when Σ is known in the formulation of the discriminant (Moran, 1975). Moran's result can also be seen as a generalization of a similar result given by John in John (1961) for the expectation of the true error. G. McLachlan provided an asymptotic expression for the expected resubstitution error in the multivariate case, for unknown covariance matrix (McLachlan, 1976), with a similar result having been provided by Raudys in Raudys (1978). In Foley (1972), D. Foley derived an asymptotic expression for the variance of the resubstitution error. Finally, Raudys applied a double-asymptotic approach where both sample size and dimensionality increase to infinity, which we call the "Raudys-Kolmogorov method," to obtain a simple, asymptotically exact expression for the expected resubstitution error (Raudys, 1978).

The distributional study of error estimation has waned considerably since the 1970s, and researchers have turned mostly to Monte-Carlo simulation studies. A hint why is provided in a 1978 paper by A. Jain and W. Waller, in which they write, "The exact expression for the average probability of error using the W statistic with the restriction that $N_1 = N_2 = N$ as given by Sitgreaves (Sitgreaves, 1961) is very complicated and difficult to evaluate even on a computer" (Jain and Waller, 1978). However, the distributional approach has been recently revived by the work of H. McFarland, D.Richards, and others (McFarland and Richards, 2001, 2002; Vu, 2011; Vu et al., 2014; Zollanvari, 2010; Zollanvari et al., 2009, 2010, 2011, 2012; Zollanvari and Dougherty, 2014). In addition, this approach has been facilitated by the availability of high-performance computing, which allows the evaluation of the complex expressions involved.

6.2 UNIVARIATE DISCRIMINANT

Throughout this chapter, the general case of LDA classification for arbitrary univariate Gaussian populations $\Pi_i \sim N(\mu_i, \sigma_i^2)$, for $i = 0, 1$, is referred to as the *univariate heteroskedastic Gaussian LDA model*, whereas the specialization of this model to the case $\sigma_0 = \sigma_1 = \sigma$ is referred to as the *univariate homoskedastic Gaussian LDA model*.

Given a sample $S = \{(X_1, Y_1), \ldots, (X_n, Y_n)\}$, the sample-based LDA discriminant in (1.48) becomes

$$W_L(S, x) = \frac{1}{\hat{\sigma}^2}(x - \bar{X})(\bar{X}_0 - \bar{X}_1), \tag{6.1}$$

where

$$\bar{X}_0 = \frac{1}{n_0} \sum_{i=1}^{n} X_i I_{Y_i=0} \quad \text{and} \quad \bar{X}_1 = \frac{1}{n_1} \sum_{i=1}^{n} X_i I_{Y_i=1} \tag{6.2}$$

are the sample means (in the case of mixture sampling, n_0 and n_1 are random variables), $\bar{X} = \frac{1}{2}(\bar{X}_0 + \bar{X}_1)$, and $\hat{\sigma}^2$ is the pooled sample variance,

$$\hat{\sigma}^2 = \frac{1}{n-2} \sum_{i=1}^{n} [(X_i - \bar{X}_0)(X_i - \bar{X}_0) I_{Y_i=0} + (X_i - \bar{X}_1)(X_i - \bar{X}_1) I_{Y_1=1}] \quad (6.3)$$

Assuming a threshold of zero, the term $\hat{\sigma}^2$ can be dropped, since it is always positive and only the sign of W_L matters in this case. The LDA discriminant can thus equivalently be written as

$$W_L(S, x) = (x - \bar{X})(\bar{X}_0 - \bar{X}_1). \quad (6.4)$$

This simplified form renders the discriminant a function only of the sample means. The LDA classification rule is

$$\Psi_L(S, x) = \begin{cases} 0, & (x - \bar{X})(\bar{X}_0 - \bar{X}_1) > 0, \\ 1, & \text{otherwise.} \end{cases} \quad (6.5)$$

The preceding formulas apply regardless of whether separate or mixture sampling is assumed. As in Chapter 5, we will write $W_L(x)$ whenever there is no danger of confusion.

The simplicity of the LDA discriminant and classification rule in the univariate case allows a wealth of exact distributional results, even in the heteroskedastic case, as we will see in the sequel.

6.3 EXPECTED ERROR RATES

In this section, we will give results for the first moment of the true error and several error estimators for the separate-sampling case. As demonstrated in Chapter 5, this is enough to obtain the corresponding expected error rates for the mixture-sampling case, by conditioning on the population-specific sample sizes.

6.3.1 True Error

For Gaussian populations of equal and unit variance, Hills provided simple formulas for the expectation of the true error (Hills, 1966). Let

$$\Phi(u) = \int_{-\infty}^{u} \frac{1}{2\pi} e^{-\frac{x^2}{2}} \, dx \quad (6.6)$$

be the cumulative distribution function of a zero-mean, unit-variance Gaussian distribution, and

$$\Phi(u, v; \rho) = \int_{-\infty}^{u} \int_{-\infty}^{v} \frac{1}{2\pi\sqrt{1-\rho^2}} \exp\left\{ -\frac{x^2 + y^2 - 2\rho xy}{2(1-\rho^2)} \right\} dx\, dy \quad (6.7)$$

be the cumulative distribution function of a zero-mean, unit-variance bivariate Gaussian distribution, with correlation ρ (c.f. Section A.11). This is a good place to remark that all expressions given in this chapter involve integration of (at times, singular) multivariate Gaussian densities over rectangular regions. Thankfully, fast and accurate methods to perform these integrations are available; see for example Genz (1992), Genz and Bretz (2002).

The following is a straightforward generalization of Hills' formulas in Hills (1966) for the univariate homoskedastic Gaussian LDA model:

$$E\left[\varepsilon^0_{n_0,n_1}\right] = \Phi(a) + \Phi(b) - 2\Phi(a,b;\rho_e), \tag{6.8}$$

where

$$a = -\frac{1}{2}\frac{\delta}{\sqrt{1+h}}, \quad b = -\frac{1}{2}\frac{\delta}{\sqrt{h}},$$
$$\rho_e = \frac{-\frac{1}{4}\left(\frac{1}{n_0}-\frac{1}{n_1}\right)}{\sqrt{h(1+h)}}, \quad h = \frac{1}{4}\left(\frac{1}{n_0}+\frac{1}{n_1}\right), \tag{6.9}$$

with $\delta = (\mu_1 - \mu_0)/\sigma$. The formula for $E[\varepsilon^1_{n_0,n_1}]$ is obtained by interchanging n_0 and n_1, and μ_0 and μ_1. We remark that a small typo in Hills (1966) was fixed in the previous formula.

Hills' formulas are a special case of a more general result given by John for the general heteroskedastic case (John, 1961). An equivalent formulation of John's expressions is given in the next theorem, where the expressions $\mathbf{Z} \leq \mathbf{0}$ and $\mathbf{Z} \geq \mathbf{0}$ mean that all components of $\mathbf{Z}$ are nonpositive and nonnegative, respectively.

Theorem 6.1 *(Zollanvari et al., 2012) Under the univariate heteroskedastic Gaussian LDA model,*

$$E\left[\varepsilon^0_{n_0,n_1}\right] = P(\mathbf{Z} \leq \mathbf{0}) + P(\mathbf{Z} \geq \mathbf{0}), \tag{6.10}$$

where $\mathbf{Z}$ *is a Gaussian bivariate vector with mean and covariance matrix given by*

$$\mu_{\mathbf{Z}} = \begin{bmatrix} \frac{\mu_0-\mu_1}{2} \\ \mu_1 - \mu_0 \end{bmatrix}, \quad \Sigma_{\mathbf{Z}} = \begin{pmatrix} \left(1+\frac{1}{4n_0}\right)\sigma_0^2 + \frac{\sigma_1^2}{4n_1} & \frac{\sigma_0^2}{2n_0} - \frac{\sigma_1^2}{2n_1} \\ \cdot & \frac{\sigma_0^2}{n_0} + \frac{\sigma_1^2}{n_1} \end{pmatrix}. \tag{6.11}$$

The result for $E[\varepsilon^1_{n_0,n_1}]$ *can be obtained by exchanging all indices 0 and 1.*

Proof: It follows from the expression for W_L in (6.4) that

$$\begin{aligned} E\left[\varepsilon^0_{n_0,n_1}\right] &= P(W_L(X) \le 0 \mid Y = 0) \\ &= P(X - \bar{X} < 0, \bar{X}_0 - \bar{X}_1 > 0 \mid Y = 0) \\ &+ P(X - \bar{X} \ge 0, \bar{X}_0 - \bar{X}_1 < 0 \mid Y = 0). \end{aligned} \tag{6.12}$$

For simplicity of notation, let the sample indices be sorted such that $Y_1 = \cdots = Y_{n_0} = 0$ and $Y_{n_0+1} = \cdots = Y_{n_0+n_1} = 1$. Expanding $\bar{X}$, $\bar{X}_1$, and $\bar{X}_2$ results in

$$P(W_L(X) \le 0 \mid Y = 0) = P(\mathbf{Z}^{\mathrm{I}} < 0) + P(\mathbf{Z}^{\mathrm{I}} \ge 0), \tag{6.13}$$

where $\mathbf{Z}^{\mathrm{I}} = A\mathbf{U}$, in which $\mathbf{U} = [X, X_1, \ldots, X_{n_0}, X_{n_0+1}, \ldots, X_{n_0+n_1}]^T$ and

$$A = \begin{pmatrix} 1 & -\frac{1}{2n_0} & -\frac{1}{2n_0} & \cdots & -\frac{1}{2n_0} & -\frac{1}{2n_1} & \cdots & -\frac{1}{2n_1} \\ 0 & -\frac{1}{n_0} & -\frac{1}{n_0} & \cdots & -\frac{1}{n_0} & \frac{1}{n_1} & \cdots & \frac{1}{n_1} \end{pmatrix}. \tag{6.14}$$

Therefore, $\mathbf{Z}^{\mathrm{I}}$ is a Gaussian random vector with mean $A\boldsymbol{\mu}_{\mathbf{U}}$ and covariance matrix $A\Sigma_{\mathbf{U}}A^T$, where

$$\boldsymbol{\mu}_{\mathbf{U}} = \left[\mu_0 \mathbf{1}_{n_0+1}, \mu_1 \mathbf{1}_{n_1}\right]^T, \tag{6.15}$$

$$\Sigma_{\mathbf{U}} = \operatorname{diag}\left(\sigma_0^2 \mathbf{1}_{n_0+1}, \sigma_1^2 \mathbf{1}_{n_1}\right), \tag{6.16}$$

which leads to the expression stated in the theorem. ■

By using the properties of the bivariate Gaussian CDF in Section A.11, it is easy to show that, when $\sigma_0 = \sigma_1 = \sigma$, the expression for $E[\varepsilon^0_{n_0,n_1}]$ given in the previous theorem coincides with Hills' formulas (6.8) and (6.9).

Asymptotic behavior is readily derived from the previous finite-sample formulas. Using Theorem 6.1, it is clear that

$$\lim_{n_0,n_1\to\infty} E[\varepsilon_{n_0,n_1}] = c_0\Phi\left(-\frac{1}{2}\frac{|\mu_1 - \mu_0|}{\sigma_0}\right) + c_1\Phi\left(-\frac{1}{2}\frac{|\mu_1 - \mu_0|}{\sigma_1}\right). \tag{6.17}$$

In the case $\sigma_0 = \sigma_1 = \sigma$, this further reduces to

$$\lim_{n_0,n_1\to\infty} E[\varepsilon_{n_0,n_1}] = \Phi\left(-\frac{1}{2}\frac{|\mu_1 - \mu_0|}{\sigma}\right) = \varepsilon_{\mathrm{bay}}. \tag{6.18}$$

This shows that LDA is a consistent classification rule in this case.

6.3.2 Resubstitution Error

Under the univariate homoskedastic Gaussian LDA model, with the additional constraint of unit variance, and assuming without loss of generality that $\delta = \mu_1 - \mu_0 > 0$, Hills showed that $E[\hat{\varepsilon}^{r,0}_{n_0,n_1}]$ can be expressed as (Hills, 1966)

$$E\left[\hat{\varepsilon}^{r,0}_{n_0,n_1}\right] = \Phi(c) + \Phi(b) - 2\Phi(c,b;\rho_r), \tag{6.19}$$

where

$$c = -\frac{1}{2}\frac{\delta}{\sqrt{1+h-\frac{1}{n_0}}}, \qquad b = -\frac{1}{2}\frac{\delta}{\sqrt{h}},$$

$$\rho_r = \sqrt{\frac{h}{1+h-\frac{1}{n_0}}}, \qquad h = \frac{1}{4}\left(\frac{1}{n_0}+\frac{1}{n_1}\right). \tag{6.20}$$

The corresponding result for $E[\hat{\varepsilon}^{r,1}_{n_0,n_1}]$ is obtained by interchanging n_0 and n_1, and μ_0 and μ_1. We remark that a small typo in the formulas of Hills (1966) has been fixed above.

Hills' result can be readily extended to the case of populations of arbitrary unknown variance.

Theorem 6.2 *(Zollanvari et al., 2012) Under the univariate heteroskedastic Gaussian LDA model,*

$$E\left[\hat{\varepsilon}^{r,0}_{n_0,n_1}\right] = P(\mathbf{Z} \leq \mathbf{0}) + P(\mathbf{Z} \geq \mathbf{0}), \tag{6.21}$$

where $\mathbf{Z}$ *is a Gaussian bivariate vector with mean and covariance matrix given by*

$$\mu_{\mathbf{Z}} = \begin{bmatrix} \frac{\mu_0-\mu_1}{2} \\ \mu_1-\mu_0 \end{bmatrix}, \quad \Sigma_{\mathbf{Z}} = \begin{pmatrix} \left(1-\frac{3}{4n_0}\right)\sigma_0^2 + \frac{\sigma_1^2}{4n_1} & -\frac{\sigma_0^2}{2n_0} - \frac{\sigma_1^2}{2n_1} \\ \cdot & \frac{\sigma_0^2}{n_0}+\frac{\sigma_1^2}{n_1} \end{pmatrix}. \tag{6.22}$$

The result for $E[\hat{\varepsilon}^{r,1}_{n_0,n_1}]$ *can be obtained by exchanging all indices 0 and 1.*

Proof: The proof is based on using (6.4) and following the same approach as in the proof of Theorem 6.1. ■

By using the properties of the bivariate Gaussian CDF in Section A.11, it is easy to show that, when $\sigma_0 = \sigma_1 = 1$, the expression for $E[\hat{\varepsilon}^{r,0}_{n_0,n_1}]$ given in the previous theorem coincides with Hills' formulas (6.19) and (6.20).

Let us illustrate the usefulness of the previous results by determining the bias of the separate-sampling resubstitution estimator of the overall error rate (5.7), assuming that it can be used with knowledge of c_0 and c_1. First note that the LDA discriminant (6.4) satisfies the anti-symmetry property defined in (5.3). Furthermore, it is easy to establish that the invariance properties (5.26) and (5.37) are satisfied for the LDA discriminant under homoskedastic Gaussian populations. With further assumptions of equal, unit population variances and balanced design $n_0 = n_1 = n/2$, a compact expression for the bias can be obtained by using Hills' expressions for $E[\varepsilon^0_{n_0,n_1}]$ and $E[\hat{\varepsilon}^{r,0}_{n_0,n_1}]$:

$$\text{Bias}\left(\hat{\varepsilon}^{r,01}_{n/2,n/2}\right) = \Phi(c) - \Phi(a) - 2\left[\Phi(c, b; \rho_r) - \Phi(a)\Phi(b)\right], \tag{6.23}$$

where

$$a = -\frac{1}{2}\frac{\delta}{\sqrt{1+\frac{1}{n}}}, \quad b = -\frac{1}{2}\frac{\delta}{\sqrt{\frac{1}{n}}}, \quad c = -\frac{1}{2}\frac{\delta}{\sqrt{1-\frac{1}{n}}}, \quad \rho_r = \sqrt{\frac{1}{n-1}}. \tag{6.24}$$

One can observe that as $n \to \infty$, we have $b \to \infty$ and $c \to a$, so that $\text{Bias}(\hat{\varepsilon}^{r,01}_{n_0,n_1}) \to 0$, for any value of $\delta > 0$, establishing that resubstitution is asymptotically unbiased in this case. Larger sample sizes are needed to get small bias for smaller values of δ (i.e., overlapping populations) than for larger values of δ (i.e., well-separated populations).

6.3.3 Leave-One-Out Error

It follows from (5.52) that the first moment of the leave-one-out error estimator can be obtained by using Theorem 6.1, while replacing n_0 by $n_0 - 1$ and n_1 by $n_1 - 1$ at the appropriate places.

6.3.4 Bootstrap Error

In this section, we build on the results derived in Chapter 5 regarding the bootstrapped error rate. As defined in Section 5.4.4 for the separate-sampling case, let $S^{\mathbf{C}_0,\mathbf{C}_1}_{n_0,n_1}$ be a bootstrap sample corresponding to multinomial vectors $\mathbf{C}_0 = (C_{01}, \dots, C_{0n_0})$ and $\mathbf{C}_1 = (C_{11}, \dots, C_{1n_1})$. We assume a more general case where $\mathbf{C}_0$ and $\mathbf{C}_1$ do not necessarily add up to n_0 and n_1 separately, but together add up to the total sample size n. This corresponds to the case where $\mathbf{C}_0$ and $\mathbf{C}_1$ are partitions of size n_0 and n_1 from a multinomial vector $\mathbf{C}$ of size n. This allows one to apply the results in this section to the computation of the expectations $E[\varepsilon^{\mathbf{C}(1:n_0),\mathbf{C}(n_0+1:n),0}_{n_0,n_1} \mid \mathbf{C}]$ and $E[\varepsilon^{\mathbf{C}(1:n_0),\mathbf{C}(n_0+1:n),1}_{n_0,n_1} \mid \mathbf{C}]$, needed in (5.63) to obtain $E[\hat{\varepsilon}^{\text{zero}}_n]$, in the mixture-sampling univariate case.

To facilitate notation, let the sample indices be sorted such that $Y_1 = \cdots = Y_{n_0} = 0$ and $Y_{n_0+1} = \cdots = Y_{n_0+n_1} = 1$, without loss of generality. The univariate LDA discriminant (6.4) applied to $S_{n_0,n_1}^{\mathbf{C}_0,\mathbf{C}_1}$ can be written as

$$W_L\left(S_{n_0,n_1}^{\mathbf{C}_0,\mathbf{C}_1}, x\right) = (x - \bar{X}^{\mathbf{C}_0,\mathbf{C}_1})\left(\bar{X}_0^{\mathbf{C}_0} - \bar{X}_1^{\mathbf{C}_1}\right), \tag{6.25}$$

where

$$\bar{X}_0^{\mathbf{C}_0} = \frac{\sum_{i=1}^{n_0} C_{0i}X_i}{\sum_{i=1}^{n_0} C_{0i}} \quad \text{and} \quad \bar{X}_1^{\mathbf{C}_1} = \frac{\sum_{i=1}^{n_1} C_{1i}X_{n_0+i}}{\sum_{i=1}^{n_1} C_{1i}} \tag{6.26}$$

are *bootstrap sample means*, and $\bar{X}^{\mathbf{C}_0,\mathbf{C}_1} = \frac{1}{2}(\bar{X}_0^{\mathbf{C}_0} + \bar{X}_1^{\mathbf{C}_1})$.

The next theorem extends Theorem 6.1 for the univariate expected true error rate to the case of the bootstrapped LDA discriminant defined by (6.25) and (6.26).

Theorem 6.3 *(Vu et al., 2014) Under the univariate heteroskedastic Gaussian LDA model,*

$$E\left[\varepsilon_{n_0,n_1}^{\mathbf{C}_0,\mathbf{C}_1,0} \mid \mathbf{C}_0, \mathbf{C}_1\right] = P(\mathbf{Z} \leq \mathbf{0}) + P(\mathbf{Z} \geq \mathbf{0}), \tag{6.27}$$

where $\mathbf{Z}$ *is a Gaussian bivariate vector with mean and covariance matrix given by*

$$\mu_{\mathbf{Z}} = \begin{bmatrix} \frac{\mu_0-\mu_1}{2} \\ \mu_1 - \mu_0 \end{bmatrix}, \quad \Sigma_{\mathbf{Z}} = \begin{pmatrix} \left(1 + \frac{s_0}{4}\right)\sigma_0^2 + \frac{s_1\sigma_1^2}{4} & \frac{s_0\sigma_0^2}{2} - \frac{s_1\sigma_1^2}{2} \\ \cdot & s_0\sigma_0^2 + s_1\sigma_1^2 \end{pmatrix}, \tag{6.28}$$

with

$$s_0 = \frac{\sum_{i=1}^{n_0} C_{0i}^2}{\left(\sum_{i=1}^{n_0} C_{0i}\right)^2} \quad \text{and} \quad s_1 = \frac{\sum_{i=1}^{n_1} C_{1i}^2}{\left(\sum_{i=1}^{n_1} C_{1i}\right)^2}. \tag{6.29}$$

The corresponding result for $E[\varepsilon_{n_0,n_1}^{\mathbf{C}_0,\mathbf{C}_1,1} \mid \mathbf{C}_0, \mathbf{C}_1]$ *is obtained by interchanging all indices 0 and 1.*

Proof: For ease of notation, we omit the conditioning on $\mathbf{C}_0, \mathbf{C}_1$ below. We write

$$\begin{aligned} E\left[\varepsilon_{n_0,n_1}^{\mathbf{C}_0,\mathbf{C}_1,0}\right] &= P\left(W_L\left(S_{n_0,n_1}^{\mathbf{C}_0,\mathbf{C}_1}, X\right) \leq 0 \mid Y = 0\right) \\ &= P\left(\bar{X}_1^{\mathbf{C}_1} > \bar{X}_0^{\mathbf{C}_0}, X > \bar{X}^{\mathbf{C}_0,\mathbf{C}_1} \middle| Y = 0\right) \\ &\quad + P\left(\bar{X}_1^{\mathbf{C}_1} \leq \bar{X}_0^{\mathbf{C}_0}, X \leq \bar{X}^{\mathbf{C}_0,\mathbf{C}_1} \middle| Y = 0\right) \\ &= P(UV > 0 \mid Y = 0), \end{aligned} \tag{6.30}$$

where (X, Y) is an independent test point, $U = \bar{X}_1^{\mathbf{C}_1} - \bar{X}_0^{\mathbf{C}_0}$, and $V = X - \bar{X}^{\mathbf{C}_0,\mathbf{C}_1}$. From (6.26), it is clear that $\bar{X}_0^{\mathbf{C}_0}$ and $\bar{X}_1^{\mathbf{C}_1}$ are independent Gaussian random variables, such that $\bar{X}_i^{\mathbf{C}_i} \sim N(\mu_i, s_i\sigma_i^2)$, for $i = 0, 1$, where s_0 and s_1 are defined in (6.29). It follows that U and V are jointly Gaussian random variables, with the following parameters:

$$
\begin{aligned}
E[U \mid Y = 0] &= \mu_1 - \mu_0, \quad \mathrm{Var}(U \mid Y = 0) = s_0\sigma_0^2 + s_1\sigma_1^2, \\
E[V \mid Y = 0] &= \frac{\mu_0 - \mu_1}{2}, \quad \mathrm{Var}(V \mid Y = 0) = \left(1 + \frac{s_0}{4}\right)\sigma_0^2 + \frac{s_1}{4}\sigma_1^2, \\
\mathrm{Cov}(U, V \mid Y = 0) &= \frac{s_0\sigma_0^2 - s_1\sigma_1^2}{2}.
\end{aligned} \tag{6.31}
$$

The result then follows after some algebraic manipulation. By symmetry, to obtain $E[\varepsilon_{n_0,n_1}^{\mathbf{C}_0,\mathbf{C}_1,1} \mid \mathbf{C}_0, \mathbf{C}_1]$ one needs only to interchange all indices 0 and 1. ■

It is easy to check that the result in Theorem 6.3 reduces to the one in Theorem 6.1 when $\mathbf{C}_0 = \mathbf{1}_{n_0}$ and $\mathbf{C}_1 = \mathbf{1}_{n_1}$, that is, the bootstrap sample equals the original sample.

As an illustration of the results in this chapter so far, let us compute the weight w^* for mixture-sampling unbiased convex bootstrap error estimation, given by (5.67). Given a univariate Gaussian model, one can compute $E[\varepsilon_n]$ by using (5.25) and Theorem 6.1; similarly, one can obtain $E[\hat{\varepsilon}_n^r]$ by using (5.32), (5.36), and Theorem 6.2; finally, one can find $E[\hat{\varepsilon}_n^{\mathrm{zero}}]$ by using (5.63) and Theorem 6.3.

An important fact that can be gleaned from the previous expressions is that, in the homoskedastic, balanced-sampling, univariate Gaussian case, all the expected error rates become functions only of n and δ, and therefore so does the weight w^*. Since the Bayes classification error in this case is $\varepsilon_{\mathrm{bay}} = \Phi(-\delta/2)$, where $\delta > 0$, there is a one-to-one correspondence between Bayes error and the Mahalanobis distance. Therefore, in the homoskedastic, balanced-sampling case, the weight w^* is a function only of the Bayes error $\varepsilon_{\mathrm{bay}}$ and the sample size n, being insensitive to the prior probabilities c_0 and c_1.

Figure 6.1 displays the exact value of w^* thus computed, in the homoskedastic, balanced-sampling, univariate Gaussian case for several sample sizes and Bayes errors. One can see in Figure 6.1(a) that w^* varies wildly and can be very far from the heuristic 0.632 weight; however, as the sample size increases, w^* appears to settle around an asymptotic fixed value. This asymptotic value is approximately 0.675, which is slightly larger than 0.632. In addition, Figure 6.1(b) allows one to see that convergence to the asymptotic value is faster for smaller Bayes errors. These facts help explain the good performance of the original convex .632 bootstrap error estimator with moderate sample sizes and small Bayes errors.

6.4 HIGHER-ORDER MOMENTS OF ERROR RATES

6.4.1 True Error

The following result gives an exact expression for the second-order moments of the true error. This result, together with (5.72), allows for the computation of $E[\varepsilon_{n_0,n_1}^2]$ and, therefore, of $\mathrm{Var}(\varepsilon_{n_0,n_1})$.

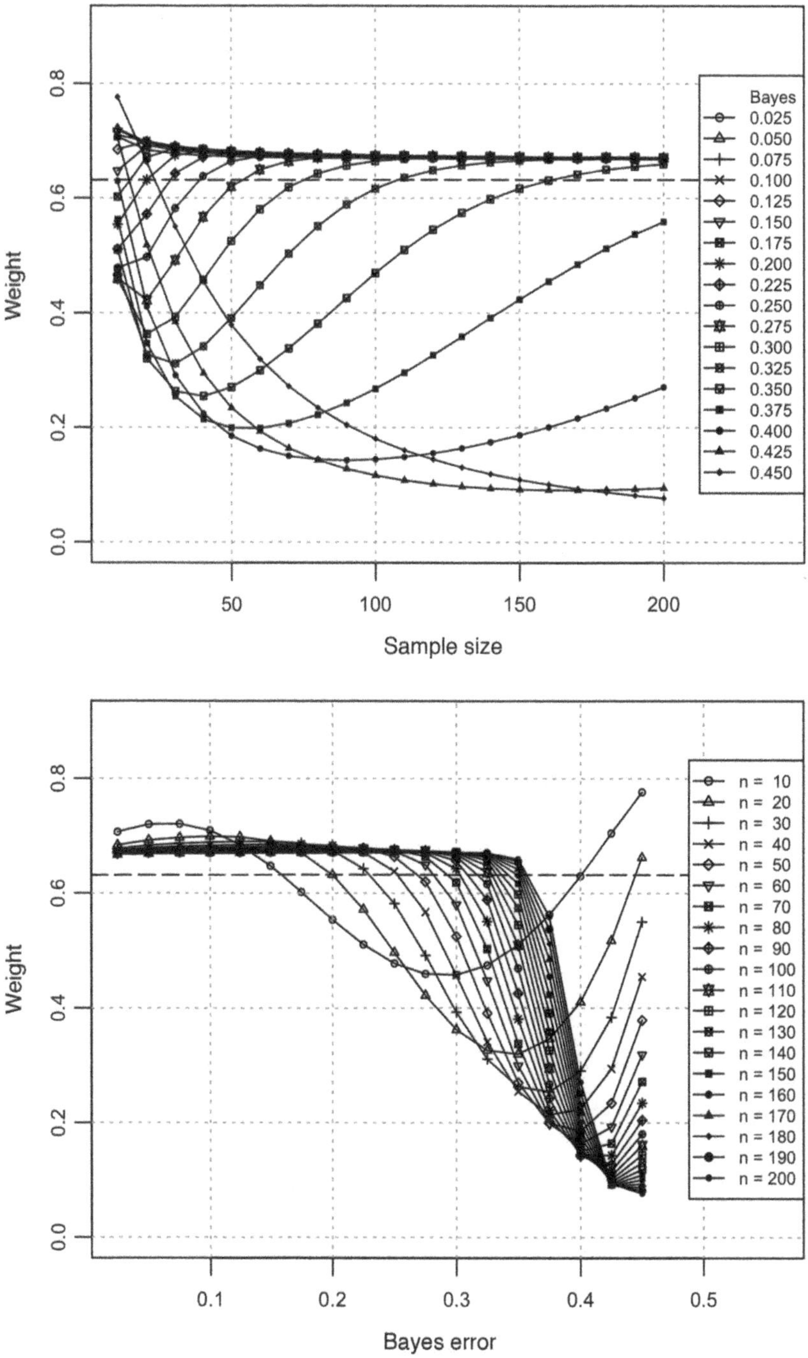

FIGURE 6.1 Required weight w^* for unbiased convex bootstrap estimation plotted against sample size and Bayes error in the homoskedastic, balanced-sampling, univariate Gaussian case.

Theorem 6.4 *(Zollanvari et al., 2012) Under the univariate heteroskedastic Gaussian LDA model,*

$$
\begin{aligned}
E\left[\left(\varepsilon^0_{n_0,n_1}\right)^2\right] &= P(\mathbf{Z}^I \le \mathbf{0}) + P(\mathbf{Z}^I \ge \mathbf{0}), \\
E\left[\varepsilon^0_{n_0,n_1}\varepsilon^1_{n_0,n_1}\right] &= P(\mathbf{Z}^{II} \le \mathbf{0}) + P(\mathbf{Z}^{II} \ge \mathbf{0}), \\
E\left[\left(\varepsilon^1_{n_0,n_1}\right)^2\right] &= P(\mathbf{Z}^{III} \le \mathbf{0}) + P(\mathbf{Z}^{III} \ge \mathbf{0}),
\end{aligned} \tag{6.32}
$$

where $\mathbf{Z}^j$, *for* $j = \mathrm{I}, \mathrm{II}, \mathrm{III}$, *are 3-variate Gaussian random vectors with means and covariance matrices given by*

$$
\boldsymbol{\mu}_{\mathbf{Z}^{\mathrm{I}}} = \begin{bmatrix} \frac{\mu_0-\mu_1}{2} \\ \mu_1 - \mu_0 \\ \frac{\mu_0-\mu_1}{2} \end{bmatrix}, \tag{6.33}
$$

$$
\Sigma_{\mathbf{Z}^{\mathrm{I}}} = \begin{pmatrix} \left(1+\frac{1}{4n_0}\right)\sigma_0^2 + \frac{\sigma_1^2}{4n_1} & \frac{\sigma_0^2}{2n_0} - \frac{\sigma_1^2}{2n_1} & \frac{\sigma_0^2}{4n_0} + \frac{\sigma_1^2}{4n_1} \\ \cdot & \frac{\sigma_0^2}{n_0} + \frac{\sigma_1^2}{n_1} & \frac{\sigma_0^2}{2n_0} - \frac{\sigma_1^2}{2n_1} \\ \cdot & \cdot & \left(1+\frac{1}{4n_0}\right)\sigma_0^2 + \frac{\sigma_1^2}{4n_1} \end{pmatrix}, \tag{6.34}
$$

$$
\Sigma_{\mathbf{Z}^{\mathrm{II}}} = \begin{pmatrix} \left(1+\frac{1}{4n_0}\right)\sigma_0^2 + \frac{\sigma_1^2}{4n_1} & \frac{\sigma_0^2}{2n_0} - \frac{\sigma_1^2}{2n_1} & -\frac{\sigma_0^2}{4n_0} - \frac{\sigma_1^2}{4n_1} \\ \cdot & \frac{\sigma_0^2}{n_0} + \frac{\sigma_1^2}{n_1} & -\frac{\sigma_0^2}{2n_0} + \frac{\sigma_1^2}{2n_1} \\ \cdot & \cdot & \frac{\sigma_0^2}{4n_0} + \left(1+\frac{1}{4n_1}\right)\sigma_1^2 \end{pmatrix}, \tag{6.35}
$$

with $\boldsymbol{\mu}_{\mathbf{Z}^I} = \boldsymbol{\mu}_{\mathbf{Z}^{II}} = \boldsymbol{\mu}_{\mathbf{Z}^{III}}$, *and* $\Sigma_{\mathbf{Z}^{III}}$ *obtained from* $\Sigma_{\mathbf{Z}^I}$ *by exchanging all indices 0 and 1.*

Proof: We prove the result for $E[(\varepsilon^0_{n_0,n_1})^2]$, the other cases being similar. It follows from the expression for W_L in (6.4) and (5.73) that

$$
\begin{aligned}
E\left[\left(\varepsilon^0_{n_0,n_1}\right)^2\right] &= P\left(W_L(X^{(1)}) \le 0, W_L(X^{(2)}) \le 0 \mid Y^{(1)} = Y^{(2)} = 0\right) \\
&= P\left(X^{(1)} - \bar{X} \ge 0, \bar{X}_0 - \bar{X}_1 < 0, X^{(2)} - \bar{X} \ge 0, \bar{X}_0 - \bar{X}_1 < 0 \right. \\
&\qquad\qquad \left. \mid Y^{(1)} = Y^{(2)} = 0\right) \\
&+ P\left(X^{(1)} - \bar{X} < 0, \bar{X}_0 - \bar{X}_1 > 0, X^{(2)} - \bar{X} < 0, \bar{X}_0 - \bar{X}_1 \ge 0 \right. \\
&\qquad\qquad \left. \mid Y^{(1)} = Y^{(2)} = 0\right)
\end{aligned}
$$

$$
\begin{aligned}
&+ P\left(X^{(1)} - \bar{X} \geq 0, \bar{X}_0 - \bar{X}_1 < 0, X^{(2)} - \bar{X} < 0, \bar{X}_0 - \bar{X}_1 \geq 0 \right. \\
&\qquad\qquad\qquad\qquad\qquad\qquad \left. \mid Y^{(1)} = Y^{(2)} = 0\right) \\
&+ P\left(X^{(1)} - \bar{X} < 0, \bar{X}_0 - \bar{X}_1 \geq 0, X^{(2)} - \bar{X} \geq 0, \bar{X}_0 - \bar{X}_1 < 0 \right. \\
&\qquad\qquad\qquad\qquad\qquad\qquad \left. \mid Y^{(1)} = Y^{(2)} = 0\right) \\
&= P\left(X^{(1)} - \bar{X} \geq 0, \bar{X}_0 - \bar{X}_1 < 0, X^{(2)} - \bar{X} \geq 0 \right. \\
&\qquad\qquad\qquad\qquad\qquad\qquad \left. \mid Y^{(1)} = Y^{(2)} = 0\right) \\
&+ P\left(X^{(1)} - \bar{X} < 0, \bar{X}_0 - \bar{X}_1 > 0, X^{(2)} - \bar{X} < 0 \right. \\
&\qquad\qquad\qquad\qquad\qquad\qquad \left. \mid Y^{(1)} = Y^{(2)} = 0\right),
\end{aligned}
\tag{6.36}
$$

where $(X^{(1)}, Y^{(1)})$ and $(X^{(2)}, Y^{(2)})$ denote independent test points. For simplicity of notation, let the sample indices be sorted such that $Y_1 = \cdots = Y_{n_0} = 0$ and $Y_{n_0+1} = \cdots = Y_{n_0+n_1} = 1$. Expanding $\bar{X}_0$, $\bar{X}_1$, and $\bar{X}$ results in

$$
\begin{aligned}
&P\left(W_L(X^{(1)}) \leq 0, W_L(S, X^{(2)}) \leq 0 \mid Y^{(1)} = Y^{(2)} = 0\right) \\
&\quad = P(\mathbf{Z}^{\mathrm{I}} < \mathbf{0}) + P(\mathbf{Z}^{\mathrm{I}} \geq \mathbf{0}),
\end{aligned}
\tag{6.37}
$$

where $\mathbf{Z}^{\mathrm{I}} = A\mathbf{U}$ in which $\mathbf{U} = [X, X^{(2)}, X_1, \ldots, X_{n_0}, X_{n_0+1}, \ldots, X_{n_0+n_1}]^T$ and

$$
A = \begin{pmatrix}
1 & 0 & -\frac{1}{2n_0} & -\frac{1}{2n_0} & \cdots & -\frac{1}{2n_0} & -\frac{1}{2n_1} & \cdots & -\frac{1}{2n_1} \\
0 & 0 & -\frac{1}{n_0} & -\frac{1}{n_0} & \cdots & -\frac{1}{n_0} & \frac{1}{n_1} & \cdots & \frac{1}{n_1} \\
0 & 1 & -\frac{1}{2n_0} & -\frac{1}{2n_0} & \cdots & -\frac{1}{2n_0} & -\frac{1}{2n_1} & \cdots & -\frac{1}{2n_1}
\end{pmatrix}.
\tag{6.38}
$$

Therefore, $\mathbf{Z}^{\mathrm{I}}$ is a Gaussian random vector with mean $A\boldsymbol{\mu}_{\mathbf{U}}$ and covariance matrix $A\Sigma_{\mathbf{U}}A^T$, where $\boldsymbol{\mu}_{\mathbf{U}} = [\mu_0 \mathbf{1}_{n_0+2}, \mu_1 \mathbf{1}_{n_1}]^T$ and $\Sigma_{\mathbf{U}} = \mathrm{diag}(\sigma_0^2 \mathbf{1}_{n_0+2}, \sigma_1^2 \mathbf{1}_{n_1})$, which leads to the expression stated in the theorem. ■

6.4.2 Resubstitution Error

The second-order moments of the resubstitution error estimator, needed to compute the estimator variance, can be found by using the next theorem.

Theorem 6.5 *(Zollanvari et al., 2012) Under the univariate heteroskedastic Gaussian LDA model,*

$$
\begin{aligned}
E\left[\left(\hat{\epsilon}^{r,0}_{n_0,n_1}\right)^2\right] &= \frac{1}{n_0}\left[P(\mathbf{Z}^{I} < \mathbf{0}) + P(\mathbf{Z}^{I} \geq \mathbf{0})\right] \\
&\quad + \frac{n_0 - 1}{n_0}\left[P(\mathbf{Z}^{III} < \mathbf{0}) + P(\mathbf{Z}^{III} \geq \mathbf{0})\right],
\end{aligned}
$$

$$
\begin{aligned}
E\left[\left(\hat{\varepsilon}^{r,1}_{n_0,n_1}\right)^2\right] &= \frac{1}{n_1}\left[P(\mathbf{Z}^{II} < \mathbf{0}) + P(\mathbf{Z}^{II} \geq \mathbf{0})\right] \\
&\quad + \frac{n_1 - 1}{n_1}\left[P(\mathbf{Z}^{IV} < \mathbf{0}) + P(\mathbf{Z}^{IV} \geq \mathbf{0})\right], \\
E\left[\hat{\varepsilon}^{r,0}_{n_0,n_1}\hat{\varepsilon}^{r,1}_{n_0,n_1}\right] &= P(\mathbf{Z}^{V} < \mathbf{0}) + P(\mathbf{Z}^{V} \geq \mathbf{0}),
\end{aligned}
\tag{6.39}
$$

where $\mathbf{Z}^I$ is distributed as $\mathbf{Z}$ in Theorem 6.2, $\mathbf{Z}^{II}$ is distributed as $\mathbf{Z}$ in Theorem 6.2 by exchanging all indices 0 and 1, and $\mathbf{Z}^j$ for $j = III, IV, V$ are 3-variate Gaussian random vectors with means and covariance matrices given by

$$
\boldsymbol{\mu}_{\mathbf{Z}^{III}} = \begin{bmatrix} \frac{\mu_0-\mu_1}{2} \\ \mu_1 - \mu_0 \\ \frac{\mu_0-\mu_1}{2} \end{bmatrix},
\tag{6.40}
$$

$$
\Sigma_{\mathbf{Z}^{III}} = \begin{pmatrix} \left(1 - \frac{3}{4n_0}\right)\sigma_0^2 + \frac{\sigma_1^2}{4n_1} & -\frac{\sigma_0^2}{2n_0} - \frac{\sigma_1^2}{2n_1} & -\frac{3\sigma_0^2}{4n_0} + \frac{\sigma_1^2}{4n_1} \\ \cdot & \frac{\sigma_0^2}{n_0} + \frac{\sigma_1^2}{n_1} & -\frac{\sigma_0^2}{2n_0} - \frac{\sigma_1^2}{2n_1} \\ \cdot & \cdot & \left(1 - \frac{3}{4n_0}\right)\sigma_0^2 + \frac{\sigma_1^2}{4n_1} \end{pmatrix},
\tag{6.41}
$$

$$
\Sigma_{\mathbf{Z}^{V}} = \begin{pmatrix} \left(1 - \frac{3}{4n_0}\right)\sigma_0^2 + \frac{\sigma_1^2}{4n_1} & -\frac{\sigma_0^2}{2n_0} - \frac{\sigma_1^2}{2n_1} & \frac{\sigma_0^2}{4n_0} + \frac{\sigma_1^2}{4n_1} \\ \cdot & \frac{\sigma_0^2}{n_0} + \frac{\sigma_1^2}{n_1} & -\frac{\sigma_0^2}{2n_0} - \frac{\sigma_1^2}{2n_1} \\ \cdot & \cdot & \frac{\sigma_0^2}{4n_0} + \left(1 - \frac{3}{4n_1}\right)\sigma_1^2 \end{pmatrix},
\tag{6.42}
$$

with $\boldsymbol{\mu}_{\mathbf{Z}^{III}} = \boldsymbol{\mu}_{\mathbf{Z}^{IV}} = \boldsymbol{\mu}_{\mathbf{Z}^{V}}$, and $\Sigma_{\mathbf{Z}^{IV}}$ obtained from $\Sigma_{\mathbf{Z}^{III}}$ by exchanging all indices 0 and 1.

Proof: We prove the result for $E[(\hat{\varepsilon}^{r,0}_{n_0,n_1})^2]$, the other cases being similar. By (5.77) and the result in Theorem 6.2, it suffices to show that

$$
P(W_L(X_1) \leq 0, W_L(X_2) \leq 0 \mid Y_1 = Y_2 = 0) = P(\mathbf{Z}^{\mathrm{III}} < \mathbf{0}) + P(\mathbf{Z}^{\mathrm{III}} \geq \mathbf{0}).
\tag{6.43}
$$

From the expression for W_L in (6.4),

$$
\begin{aligned}
&P(W_L(X_1) \leq 0, W_L(X_2) \leq 0 \mid Y_1 = Y_2 = 0) \\
&\quad = P(X_1 - \bar{X} \geq 0, \bar{X}_0 - \bar{X}_1 < 0, X_2 - \bar{X} \geq 0, \bar{X}_0 - \bar{X}_1 < 0 \mid Y_1 = Y_2 = 0) \\
&\qquad + P(X_1 - \bar{X} < 0, \bar{X}_0 - \bar{X}_1 > 0, X_2 - \bar{X} < 0, \bar{X}_0 - \bar{X}_1 \geq 0 \mid Y_1 = Y_2 = 0)
\end{aligned}
$$

$$\begin{aligned}
&+P(X_1-\bar{X}\geq 0,\bar{X}_0-\bar{X}_1<0,X_2-\bar{X}<0,\bar{X}_0-\bar{X}_1\geq 0\mid Y_1=Y_2=0)\\
&+P(X_1-\bar{X}<0,\bar{X}_0-\bar{X}_1\geq 0,X_2-\bar{X}\geq 0,\bar{X}_0-\bar{X}_1<0\mid Y_1=Y_2=0)\\
=&\ P(X_1-\bar{X}\geq 0,\bar{X}_0-\bar{X}_1<0,X_2-\bar{X}\geq 0\mid Y_1=Y_2=0)\\
&+P(X_1-\bar{X}<0,\bar{X}_0-\bar{X}_1>0,X_2-\bar{X}<0\mid Y_1=Y_2=0).
\end{aligned} \tag{6.44}$$

Expanding $\bar{X}_0$, $\bar{X}_1$, and $\bar{X}$ yields the desired result. ■

The mixed moments of the resubstitution error estimator with the true error, which are needed for computation of the RMS, are given by the following theorem.

Theorem 6.6 *Under the univariate heteroskedastic Gaussian LDA model,*

$$\begin{aligned}
E\left[\varepsilon^0_{n_0,n_1}\,\hat{\varepsilon}^{r,0}_{n_0,n_1}\right]&=P(\mathbf{Z}^I<\mathbf{0})+P(\mathbf{Z}^I\geq\mathbf{0}),\\
E\left[\varepsilon^0_{n_0,n_1}\,\hat{\varepsilon}^{r,1}_{n_0,n_1}\right]&=P(\mathbf{Z}^{II}<\mathbf{0})+P(\mathbf{Z}^{II}\geq\mathbf{0}),\\
E\left[\varepsilon^1_{n_0,n_1}\,\hat{\varepsilon}^{r,0}_{n_0,n_1}\right]&=P(\mathbf{Z}^{III}<\mathbf{0})+P(\mathbf{Z}^{III}\geq\mathbf{0}),\\
E\left[\varepsilon^1_{n_0,n_1}\,\hat{\varepsilon}^{r,1}_{n_0,n_1}\right]&=P(\mathbf{Z}^{IV}<\mathbf{0})+P(\mathbf{Z}^{IV}\geq\mathbf{0}),
\end{aligned} \tag{6.45}$$

where $\mathbf{Z}^j$, for $j=I,\ldots,IV$, are 3-variate Gaussian random vectors with means and covariances given by

$$\boldsymbol{\mu}_{\mathbf{Z}^I}=\begin{bmatrix}\frac{\mu_0-\mu_1}{2}\\ \mu_1-\mu_0\\ \frac{\mu_0-\mu_1}{2}\end{bmatrix},\quad \boldsymbol{\mu}_{\mathbf{Z}^{III}}=\begin{bmatrix}\frac{\mu_1-\mu_0}{2}\\ \mu_0-\mu_1\\ \frac{\mu_1-\mu_0}{2}\end{bmatrix}, \tag{6.46}$$

$$\Sigma_{\mathbf{Z}^I}=\begin{pmatrix}\left(1+\frac{1}{4n_0}\right)\sigma_0^2+\frac{\sigma_1^2}{4n_1} & \frac{\sigma_0^2}{2n_0}-\frac{\sigma_1^2}{2n_1} & -\frac{\sigma_0^2}{4n_0}+\frac{\sigma_1^2}{4n_1}\\ \cdot & \frac{\sigma_0^2}{n_0}+\frac{\sigma_1^2}{n_1} & -\frac{\sigma_0^2}{2n_0}-\frac{\sigma_1^2}{2n_1}\\ \cdot & \cdot & \left(1-\frac{3}{4n_0}\right)\sigma_0^2+\frac{\sigma_1^2}{4n_1}\end{pmatrix}, \tag{6.47}$$

$$\Sigma_{\mathbf{Z}^{II}}=\begin{pmatrix}\left(1+\frac{1}{4n_0}\right)\sigma_0^2+\frac{\sigma_1^2}{4n_1} & \frac{\sigma_0^2}{2n_0}-\frac{\sigma_1^2}{2n_1} & -\frac{\sigma_0^2}{4n_0}+\frac{\sigma_1^2}{4n_1}\\ \cdot & \frac{\sigma_0^2}{n_0}+\frac{\sigma_1^2}{n_1} & -\frac{\sigma_0^2}{2n_0}-\frac{\sigma_1^2}{2n_1}\\ \cdot & \cdot & \frac{\sigma_0^2}{4n_0}+\left(1-\frac{3}{4n_1}\right)\sigma_1^2\end{pmatrix}, \tag{6.48}$$

with $\boldsymbol{\mu}_{\mathbf{Z}^I}=\boldsymbol{\mu}_{\mathbf{Z}^{II}}$, $\boldsymbol{\mu}_{\mathbf{Z}^{III}}=\boldsymbol{\mu}_{\mathbf{Z}^{IV}}$, and $\Sigma_{\mathbf{Z}^{III}}$ and $\Sigma_{\mathbf{Z}^{IV}}$ obtained from $\Sigma_{\mathbf{Z}^{II}}$ and $\Sigma_{\mathbf{Z}^I}$, respectively, by exchanging all indices 0 and 1.

Proof: We prove the result for $E[\varepsilon^{0}_{n_0,n_1}\hat{\varepsilon}^{r,0}_{n_0,n_1}]$, the other cases being similar. By (5.84), it suffices to show that

$$P(W_L(X) \le 0, W_L(X_1) \le 0 \mid Y = 0, Y_1 = 0) = P(\mathbf{Z}^{\mathrm{I}} < \mathbf{0}) + P(\mathbf{Z}^{\mathrm{I}} \ge \mathbf{0}). \quad (6.49)$$

From the expression for W_L in (6.4),

$$\begin{aligned}
&P(W_L(X) \le 0, W_L(X_1) \le 0 \mid Y = 0, Y_1 = 0)\\
&\quad = P(X - \bar{X} \ge 0, \bar{X}_0 - \bar{X}_1 < 0, X_1 - \bar{X} \ge 0, \bar{X}_0 - \bar{X}_1 < 0 \mid Y = 0, Y_1 = 0)\\
&\qquad + P(X - \bar{X} < 0, \bar{X}_0 - \bar{X}_1 > 0, X_1 - \bar{X} < 0, \bar{X}_0 - \bar{X}_1 \ge 0 \mid Y = 0, Y_1 = 0)\\
&\qquad + P(X - \bar{X} \ge 0, \bar{X}_0 - \bar{X}_1 < 0, X_1 - \bar{X} < 0, \bar{X}_0 - \bar{X}_1 \ge 0 \mid Y = 0, Y_1 = 0)\\
&\qquad + P(X - \bar{X} < 0, \bar{X}_0 - \bar{X}_1 \ge 0, X_1 - \bar{X} \ge 0, \bar{X}_0 - \bar{X}_1 < 0 \mid Y = 0, Y_1 = 0)\\
&\quad = P(X - \bar{X} \ge 0, \bar{X}_0 - \bar{X}_1 < 0, X_1 - \bar{X} \ge 0 \mid Y = 0, Y_1 = 0)\\
&\qquad + P(X - \bar{X} < 0, \bar{X}_0 - \bar{X}_1 > 0, X_1 - \bar{X} < 0 \mid Y = 0, Y_1 = 0). \qquad (6.50)
\end{aligned}$$

Expanding $\bar{X}_0$, $\bar{X}_1$, and $\bar{X}$ results in the Gaussian vector $\mathbf{Z}^I$ with the mean and covariance matrix stated in the theorem. ■

6.4.3 Leave-One-Out Error

The second moment of the leave-one-out error estimator, needed to compute its variance, can be obtained by using the following theorem.

Theorem 6.7 *(Zollanvari et al., 2012) Under the univariate heteroskedastic Gaussian LDA model,*

$$\begin{aligned}
E\left[\left(\hat{\varepsilon}^{l,0}_{n_0,n_1}\right)^2\right] &= \frac{1}{n_0}\left[P(\mathbf{Z}^I < \mathbf{0}) + P(\mathbf{Z}^I \ge \mathbf{0})\right]\\
&\quad + \frac{n_0 - 1}{n_0}\left[P(\mathbf{Z}_0^{III} < \mathbf{0}) + P(\mathbf{Z}_0^{III} \ge \mathbf{0}) + P(\mathbf{Z}_1^{III} < \mathbf{0}) + P(\mathbf{Z}_1^{III} \ge \mathbf{0})\right],\\
E\left[\left(\hat{\varepsilon}^{l,1}_{n_0,n_1}\right)^2\right] &= \frac{1}{n_1}\left[P(\mathbf{Z}^{II} < \mathbf{0}) + P(\mathbf{Z}^{II} \ge \mathbf{0})\right]\\
&\quad + \frac{n_1 - 1}{n_1}\left[P(\mathbf{Z}_0^{IV} < \mathbf{0}) + P(\mathbf{Z}_0^{IV} \ge \mathbf{0}) + P(\mathbf{Z}_1^{IV} < \mathbf{0}) + P(\mathbf{Z}_1^{IV} \ge \mathbf{0})\right],\\
E\left[\hat{\varepsilon}^{l,0}_{n_0,n_1}\hat{\varepsilon}^{l,1}_{n_0,n_1}\right] &= P(\mathbf{Z}_0^{V} < \mathbf{0}) + P(\mathbf{Z}_0^{V} \ge \mathbf{0}) + P(\mathbf{Z}_1^{V} < \mathbf{0}) + P(\mathbf{Z}_1^{V} \ge \mathbf{0}), \qquad (6.51)
\end{aligned}$$

where $\mathbf{Z}^I$ *is distributed as* $\mathbf{Z}$ *in Theorem 6.1 with* n_0 *and* n_1 *replaced by* $n_0 - 1$ *and* $n_1 - 1$*, respectively,* $\mathbf{Z}^{II}$ *is distributed as* $\mathbf{Z}$ *in Theorem 6.1 by exchanging all* *indices 0 and 1 and replacing* n_0 *and* n_1 *by* $n_0 - 1$ *and* $n_1 - 1$*, respectively, while* $\mathbf{Z}_i^j$*, for* $i = 0, 1$*,* $j = III, IV, V$*, are 4-variate Gaussian random vectors with means and covariance matrices given by*

$$\boldsymbol{\mu}_{\mathbf{Z}_0^{III}} = \begin{bmatrix} \frac{\mu_0-\mu_1}{2} \\ \mu_1 - \mu_0 \\ \frac{\mu_0-\mu_1}{2} \\ \mu_1 - \mu_0 \end{bmatrix}, \quad \boldsymbol{\mu}_{\mathbf{Z}_1^{III}} = \begin{bmatrix} \frac{\mu_0-\mu_1}{2} \\ \mu_1 - \mu_0 \\ \frac{\mu_1-\mu_0}{2} \\ \mu_0 - \mu_1 \end{bmatrix}, \tag{6.52}$$

$$\Sigma_{\mathbf{Z}_0^{III}} = \begin{pmatrix} \left(1+\frac{1}{4(n_0-1)}\right)\sigma_0^2 + \frac{\sigma_1^2}{4n_1} & \frac{\sigma_0^2}{2(n_0-1)} - \frac{\sigma_1^2}{2n_1} & \frac{(-3n_0+2)\sigma_0^2}{4(n_0-1)^2} + \frac{\sigma_1^2}{4n_1} & -\frac{n_0\sigma_0^2}{2(n_0-1)^2} - \frac{\sigma_1^2}{2n_1} \\ \cdot & \frac{\sigma_0^2}{n_0-1} + \frac{\sigma_1^2}{n_1} & -\frac{n_0\sigma_0^2}{2(n_0-1)^2} - \frac{\sigma_1^2}{2n_1} & \frac{(n_0-2)\sigma_0^2}{(n_0-1)^2} + \frac{\sigma_1^2}{n_1} \\ \cdot & \cdot & \left(1+\frac{1}{4(n_0-1)}\right)\sigma_0^2 + \frac{\sigma_1^2}{4n_1} & \frac{\sigma_0^2}{2(n_0-1)} - \frac{\sigma_1^2}{2n_1} \\ \cdot & \cdot & \cdot & \frac{\sigma_0^2}{n_0-1} + \frac{\sigma_1^2}{n_1} \end{pmatrix}, \tag{6.53}$$

$$\Sigma_{\mathbf{Z}_1^{III}} = \begin{pmatrix} \left(1+\frac{1}{4(n_0-1)}\right)\sigma_0^2 + \frac{\sigma_1^2}{4n_1} & \frac{\sigma_0^2}{2(n_0-1)} - \frac{\sigma_1^2}{2n_1} & -\frac{(-3n_0+2)\sigma_0^2}{4(n_0-1)^2} - \frac{\sigma_1^2}{4n_1} & \frac{n_0\sigma_0^2}{2(n_0-1)^2} + \frac{\sigma_1^2}{2n_1} \\ \cdot & \frac{\sigma_0^2}{n_0-1} + \frac{\sigma_1^2}{n_1} & \frac{n_0\sigma_0^2}{2(n_0-1)^2} + \frac{\sigma_1^2}{2n_1} & -\frac{(n_0-2)\sigma_0^2}{(n_0-1)^2} - \frac{\sigma_1^2}{n_1} \\ \cdot & \cdot & \left(1+\frac{1}{4(n_0-1)}\right)\sigma_0^2 + \frac{\sigma_1^2}{4n_1} & \frac{\sigma_0^2}{2(n_0-1)} - \frac{\sigma_1^2}{2n_1} \\ \cdot & \cdot & \cdot & \frac{\sigma_0^2}{n_0-1} + \frac{\sigma_1^2}{n_1} \end{pmatrix}, \tag{6.54}$$

$$\Sigma_{\mathbf{Z}_0^{V}} = \begin{pmatrix} \left(1+\frac{1}{4(n_0-1)}\right)\sigma_0^2 + \frac{\sigma_1^2}{4n_1} & \frac{\sigma_0^2}{2(n_0-1)} - \frac{\sigma_1^2}{2n_1} & \frac{\sigma_0^2}{4n_0} + \frac{\sigma_1^2}{4n_1} & -\frac{\sigma_0^2}{2n_0} - \frac{\sigma_1^2}{2n_1} \\ \cdot & \frac{\sigma_0^2}{n_0-1} + \frac{\sigma_1^2}{n_1} & -\frac{\sigma_0^2}{2n_0} - \frac{\sigma_1^2}{2n_1} & \frac{\sigma_0^2}{n_0} + \frac{\sigma_1^2}{n_1} \\ \cdot & \cdot & \frac{\sigma_0^2}{4n_0} + \left(1+\frac{1}{4(n_1-1)}\right)\sigma_1^2 & -\frac{\sigma_0^2}{2n_0} + \frac{\sigma_1^2}{2(n_1-1)} \\ \cdot & \cdot & \cdot & \frac{\sigma_0^2}{n_0} + \frac{\sigma_1^2}{n_1-1} \end{pmatrix}, \tag{6.55}$$

$$
\Sigma_{\mathbf{Z}_1^V} =
\begin{pmatrix}
\left(1+\frac{1}{4(n_0-1)}\right)\sigma_0^2+\frac{\sigma_1^2}{4n_1} & \frac{\sigma_0^2}{2(n_0-1)}-\frac{\sigma_1^2}{2n_1} & -\frac{\sigma_0^2}{4n_0}-\frac{\sigma_1^2}{4n_1} & \frac{\sigma_0^2}{2n_0}+\frac{\sigma_1^2}{2n_1} \\
\cdot & \frac{\sigma_0^2}{n_0-1}+\frac{\sigma_1^2}{n_1} & \frac{\sigma_0^2}{2n_0}+\frac{\sigma_1^2}{2n_1} & -\frac{\sigma_0^2}{n_0}-\frac{\sigma_1^2}{n_1} \\
\cdot & \cdot & \frac{\sigma_0^2}{4n_0}+\left(1+\frac{1}{4(n_1-1)}\right)\sigma_1^2 & -\frac{\sigma_0^2}{2n_0}+\frac{\sigma_1^2}{2(n_1-1)} \\
\cdot & \cdot & \cdot & \frac{\sigma_0^2}{n_0}+\frac{\sigma_1^2}{n_1-1}
\end{pmatrix},
\tag{6.56}
$$

with $\boldsymbol{\mu}_{\mathbf{Z}_0^{III}} = \boldsymbol{\mu}_{\mathbf{Z}_0^{IV}} = \boldsymbol{\mu}_{\mathbf{Z}_0^{V}}$, $\boldsymbol{\mu}_{\mathbf{Z}_1^{III}} = \boldsymbol{\mu}_{\mathbf{Z}_1^{IV}} = \boldsymbol{\mu}_{\mathbf{Z}_1^{V}}$, *and* $\Sigma_{\mathbf{Z}_0^{IV}}$ *and* $\Sigma_{\mathbf{Z}_1^{IV}}$ *obtained from* $\Sigma_{\mathbf{Z}_0^{III}}$ *and* $\Sigma_{\mathbf{Z}_1^{III}}$, *respectively, by exchanging all indices 0 and 1.*

Proof: We prove the result for $E[(\hat{\varepsilon}^{l,0}_{n_0,n_1})^2]$, the other cases being similar. First note that $E[(\hat{\varepsilon}^{l,0}_{n_0,n_1})^2]$ can be written as $E[(\hat{\varepsilon}^{r,0}_{n_0,n_1})^2]$ in (5.77), by replacing $W_L(X_i)$ by $W_L^{(i)}(X_i)$, for $i = 1, 2$. Given the fact that $P(W_L^{(1)}(X_1) \le 0 \mid Y_1 = 0)$ can be written by replacing $\mathbf{Z}$, n_0, and n_1 by $\mathbf{Z}^{\mathrm{I}}$, $n_0 - 1$ and $n_1 - 1$, respectively, in the result of Theorem 6.1, it suffices to show that

$$
\begin{aligned}
&P\left(W_L^{(1)}(X_1) \le 0, W_L^{(2)}(X_2) \le 0 \mid Y_1 = Y_2 = 0\right) \\
&\quad = P\left(\mathbf{Z}_0^{\mathrm{III}} < \mathbf{0}\right) + P\left(\mathbf{Z}_0^{\mathrm{III}} \ge \mathbf{0}\right) + P\left(\mathbf{Z}_1^{\mathrm{III}} < \mathbf{0}\right) + P\left(\mathbf{Z}_1^{\mathrm{III}} \ge \mathbf{0}\right).
\end{aligned}
\tag{6.57}
$$

From the expression for W_L in (6.4),

$$
\begin{aligned}
&P\left(W_L^{(1)}(X_1) \le 0, W_L^{(2)}(X_2) \le 0 \mid Y_1 = Y_2 = 0\right) \\
&\quad = P\Big(X_1 - \bar{X}^{(1)} \ge 0, \bar{X}_0^{(1)} - \bar{X}_1 < 0, X_2 - \bar{X}^{(2)} \ge 0, \bar{X}_0^{(2)} - \bar{X}_1 < 0 \\
&\qquad\qquad \mid Y_1 = Y_2 = 0\Big) \\
&\quad + P\Big(X_1 - \bar{X}^{(1)} < 0, \bar{X}_0^{(1)} - \bar{X}_1 \ge 0, X_2 - \bar{X}^{(2)} < 0, \bar{X}_0^{(2)} - \bar{X}_1 \ge 0 \\
&\qquad\qquad \mid Y_1 = Y_2 = 0\Big) \\
&\quad + P\Big(X_1 - \bar{X}^{(1)} \ge 0, \bar{X}_0^{(1)} - \bar{X}_1 < 0, X_2 - \bar{X}^{(2)} < 0, \bar{X}_0^{(2)} - \bar{X}_1 \ge 0 \\
&\qquad\qquad \mid Y_1 = Y_2 = 0\Big) \\
&\quad + P\Big(X_1 - \bar{X}^{(1)} < 0, \bar{X}_0^{(1)} - \bar{X}_1 \ge 0, X_2 - \bar{X}^{(2)} \ge 0, \bar{X}_0^{(2)} - \bar{X}_1 < 0 \\
&\qquad\qquad \mid Y_1 = Y_2 = 0\Big).
\end{aligned}
\tag{6.58}
$$

Expanding $\bar{X}_0^{(j)}$, $\bar{X}_1$, and $\bar{X}^{(j)}$ results in the Gaussian vectors $\mathbf{Z}_0^{III}$ and $\mathbf{Z}_1^{III}$ with the means and covariance matrices stated in the theorem. ■

The mixed moments of the leave-one-out error estimator with the true error, which are needed for computation of the RMS, are given by the following theorem.

Theorem 6.8 *Under the univariate heteroskedastic Gaussian LDA model,*

$$
\begin{aligned}
E\left[\varepsilon^{0}_{n_0,n_1}\hat{\varepsilon}^{l,0}_{n_0,n_1}\right] &= P\left(\mathbf{Z}_0^{I} < \mathbf{0}\right) + P\left(\mathbf{Z}_0^{I} \geq \mathbf{0}\right) + P\left(\mathbf{Z}_1^{I} < \mathbf{0}\right) + P\left(\mathbf{Z}_1^{I} \geq \mathbf{0}\right),\\
E\left[\varepsilon^{0}_{n_0,n_1}\hat{\varepsilon}^{l,1}_{n_0,n_1}\right] &= P\left(\mathbf{Z}_0^{II} < \mathbf{0}\right) + P\left(\mathbf{Z}_0^{II} \geq \mathbf{0}\right) + P\left(\mathbf{Z}_1^{II} < \mathbf{0}\right) + P\left(\mathbf{Z}_1^{II} \geq \mathbf{0}\right),\\
E\left[\varepsilon^{1}_{n_0,n_1}\hat{\varepsilon}^{l,0}_{n_0,n_1}\right] &= P\left(\mathbf{Z}_0^{III} < \mathbf{0}\right) + P\left(\mathbf{Z}_0^{III} \geq \mathbf{0}\right) + P\left(\mathbf{Z}_1^{III} < \mathbf{0}\right) + P\left(\mathbf{Z}_1^{III} \geq \mathbf{0}\right),\\
E\left[\varepsilon^{1}_{n_0,n_1}\hat{\varepsilon}^{l,1}_{n_0,n_1}\right] &= P\left(\mathbf{Z}_0^{IV} < \mathbf{0}\right) + P\left(\mathbf{Z}_0^{IV} \geq \mathbf{0}\right) + P\left(\mathbf{Z}_1^{IV} < \mathbf{0}\right) + P\left(\mathbf{Z}_1^{IV} \geq \mathbf{0}\right),
\end{aligned}
\tag{6.59}
$$

where $\mathbf{Z}_i^j$*, for* $i = 0, 1$ *and* $j = I, \ldots, IV$*, are 4-variate Gaussian random vectors with means and covariance matrices given by*

$$
\boldsymbol{\mu}_{\mathbf{Z}_0^I} = \begin{bmatrix} \frac{\mu_0-\mu_1}{2} \\ \mu_1 - \mu_0 \\ \frac{\mu_0-\mu_1}{2} \\ \mu_1 - \mu_0 \end{bmatrix}, \quad \boldsymbol{\mu}_{\mathbf{Z}_1^I} = \begin{bmatrix} \frac{\mu_0-\mu_1}{2} \\ \mu_1 - \mu_0 \\ \frac{\mu_1-\mu_0}{2} \\ \mu_0 - \mu_1 \end{bmatrix}, \tag{6.60}
$$

$$
\boldsymbol{\mu}_{\mathbf{Z}_0^{III}} = \begin{bmatrix} \frac{\mu_1-\mu_0}{2} \\ \mu_0 - \mu_1 \\ \frac{\mu_0-\mu_1}{2} \\ \mu_1 - \mu_0 \end{bmatrix}, \quad \boldsymbol{\mu}_{\mathbf{Z}_1^{III}} = \begin{bmatrix} \frac{\mu_1-\mu_0}{2} \\ \mu_0 - \mu_1 \\ \frac{\mu_1-\mu_0}{2} \\ \mu_0 - \mu_1 \end{bmatrix}, \tag{6.61}
$$

$$
\Sigma_{\mathbf{Z}_0^I} =
\begin{pmatrix}
\left(1+\frac{1}{4n_0}\right)\sigma_0^2 + \frac{\sigma_1^2}{4n_1} & \frac{\sigma_0^2}{2n_0} - \frac{\sigma_1^2}{2n_1} & -\frac{\sigma_0^2}{4n_0} + \frac{\sigma_1^2}{4n_1} & \frac{\sigma_0^2}{2n_0} - \frac{\sigma_1^2}{2n_1} \\
\cdot & \frac{\sigma_0^2}{n_0} + \frac{\sigma_1^2}{n_1} & -\frac{\sigma_0^2}{2n_0} - \frac{\sigma_1^2}{2n_1} & \frac{\sigma_0^2}{n_0} + \frac{\sigma_1^2}{n_1} \\
\cdot & \cdot & \left(1+\frac{1}{4(n_0-1)}\right)\sigma_0^2 + \frac{\sigma_1^2}{4n_1} & \frac{\sigma_0^2}{2(n_0-1)} - \frac{\sigma_1^2}{2n_1} \\
\cdot & \cdot & \cdot & \frac{\sigma_0^2}{(n_0-1)} + \frac{\sigma_1^2}{n_1}
\end{pmatrix},
\tag{6.62}
$$

$$\Sigma_{\mathbf{Z}_1^I} =$$
$$\begin{pmatrix} \left(1+\frac{1}{4n_0}\right)\sigma_0^2+\frac{\sigma_1^2}{4n_1} & \frac{\sigma_0^2}{2n_0}-\frac{\sigma_1^2}{2n_1} & \frac{\sigma_0^2}{4n_0}-\frac{\sigma_1^2}{4n_1} & -\frac{\sigma_0^2}{2n_0}+\frac{\sigma_1^2}{2n_1} \\ \cdot & \frac{\sigma_0^2}{n_0}+\frac{\sigma_1^2}{n_1} & \frac{\sigma_0^2}{2n_0}+\frac{\sigma_1^2}{2n_1} & -\frac{\sigma_0^2}{n_0}-\frac{\sigma_1^2}{n_1} \\ \cdot & \cdot & \left(1+\frac{1}{4(n_0-1)}\right)\sigma_0^2+\frac{\sigma_1^2}{4n_1} & \frac{\sigma_0^2}{2(n_0-1)}-\frac{\sigma_1^2}{2n_1} \\ \cdot & \cdot & \cdot & \frac{\sigma_0^2}{(n_0-1)}+\frac{\sigma_1^2}{n_1} \end{pmatrix}, \tag{6.63}$$

$$\Sigma_{\mathbf{Z}_0^{II}} =$$
$$\begin{pmatrix} \left(1+\frac{1}{4n_0}\right)\sigma_0^2+\frac{\sigma_1^2}{4n_1} & \frac{\sigma_0^2}{2n_0}-\frac{\sigma_1^2}{2n_1} & -\frac{\sigma_0^2}{4n_0}+\frac{\sigma_1^2}{4n_0 1} & \frac{\sigma_0^2}{2n_0}-\frac{\sigma_1^2}{2n_1} \\ \cdot & \frac{\sigma_0^2}{n_0}+\frac{\sigma_1^2}{n_1} & -\frac{\sigma_0^2}{2n_0}-\frac{\sigma_1^2}{2n_1} & \frac{\sigma_0^2}{n_0}+\frac{\sigma_1^2}{n_1} \\ \cdot & \cdot & \frac{\sigma_0^2}{4n_0}+\left(1+\frac{1}{4(n_1-1)}\right)\sigma_1^2 & -\frac{\sigma_0^2}{2n_0}+\frac{\sigma_1^2}{2(n_1-1)} \\ \cdot & \cdot & \cdot & \frac{\sigma_0^2}{n_0}+\frac{\sigma_1^2}{n_1-1} \end{pmatrix}, \tag{6.64}$$

$$\Sigma_{\mathbf{Z}_1^{II}} =$$
$$\begin{pmatrix} \left(1+\frac{1}{4n_0}\right)\sigma_0^2+\frac{\sigma_1^2}{4n_1} & \frac{\sigma_0^2}{2n_0}-\frac{\sigma_1^2}{2n_1} & \frac{\sigma_0^2}{4n_0}-\frac{\sigma_1^2}{4n_1} & -\frac{\sigma_0^2}{2n_0}+\frac{\sigma_1^2}{2n_1} \\ \cdot & \frac{\sigma_0^2}{n_0}+\frac{\sigma_1^2}{n_1} & \frac{\sigma_0^2}{2n_0}+\frac{\sigma_1^2}{2n_1} & -\frac{\sigma_0^2}{n_0}-\frac{\sigma_1^2}{n_1} \\ \cdot & \cdot & \frac{\sigma_0^2}{4n_0}+\left(1+\frac{1}{4(n_1-1)}\right)\sigma_1^2 & -\frac{\sigma_0^2}{2n_0}+\frac{\sigma_1^2}{2(n_1-1)} \\ \cdot & \cdot & \cdot & \frac{\sigma_0^2}{n_0}+\frac{\sigma_1^2}{n_1-1} \end{pmatrix}, \tag{6.65}$$

with $\boldsymbol{\mu}_{\mathbf{Z}_i^I} = \boldsymbol{\mu}_{\mathbf{Z}_i^{II}}$, $\boldsymbol{\mu}_{\mathbf{Z}_i^{III}} = \boldsymbol{\mu}_{\mathbf{Z}_i^{IV}}$, *and* $\Sigma_{\mathbf{Z}_i^{III}}$ *and* $\Sigma_{\mathbf{Z}_i^{IV}}$ *obtained from* $\Sigma_{\mathbf{Z}_{1-i}^{II}}$ *and* $\Sigma_{\mathbf{Z}_{1-i}^{I}}$, *respectively, by exchanging all indices 0 and 1, for* $i = 0, 1$.

Proof: We prove the result for $E[\varepsilon^0_{n_0,n_1}\hat{\varepsilon}^{l,0}_{n_0,n_1}]$, the other cases being similar. By (5.88), it suffices to show that

$$\begin{aligned} &P\left(W_L(X) \le 0, W_L^{(1)}(X_1) \le 0 \mid Y = 0, Y_1 = 0\right) \\ &\quad = P\left(\mathbf{Z}_0^I < \mathbf{0}\right) + P\left(\mathbf{Z}_0^I \ge \mathbf{0}\right) + P\left(\mathbf{Z}_1^I < \mathbf{0}\right) + P\left(\mathbf{Z}_1^I \ge \mathbf{0}\right). \end{aligned} \tag{6.66}$$

From the expression for W_L in (6.4),

$$\begin{aligned}
&P\left(W_L(X) \le 0, W_L^{(1)}(X_1) \le 0 \mid Y = 0, Y_1 = 0\right) \\
&= P\left(X - \bar{X} \ge 0, \bar{X}_0 - \bar{X}_1 < 0, X_1 - \bar{X}^{(1)} \ge 0, \bar{X}_0^{(1)} - \bar{X}_1 < 0 \mid Y = 0, Y_1 = 0\right) \\
&+ P\left(X - \bar{X} < 0, \bar{X}_0 - \bar{X}_1 \ge 0, X_1 - \bar{X}^{(1)} < 0, \bar{X}_0^{(1)} - \bar{X}_1 \ge 0 \mid Y = 0, Y_1 = 0\right) \\
&+ P\left(X - \bar{X} \ge 0, \bar{X}_0 - \bar{X}_1 < 0, X_1 - \bar{X}^{(1)} < 0, \bar{X}_0^{(1)} - \bar{X}_1 \ge 0 \mid Y = 0, Y_1 = 0\right) \\
&+ P\left(X - \bar{X} < 0, \bar{X}_0 - \bar{X}_1 \ge 0, X_1 - \bar{X}^{(1)} \ge 0, \bar{X}_0^{(1)} - \bar{X}_1 < 0 \mid Y = 0, Y_1 = 0\right).
\end{aligned} \tag{6.67}$$

Expanding $\bar{X}_0, \bar{X}_0^{(j)}, \bar{X}_1, \bar{X}$, and $\bar{X}^{(j)}$ results in the Gaussian vectors $\mathbf{Z}_0^I$ and $\mathbf{Z}_1^I$ with the means and covariance matrices stated in the theorem. ■

6.4.4 Numerical Example

The expressions provided by the preceding several theorems allow one to compute the RMS of the resubstitution and leave-one-out error estimators. This is illustrated in Figure 6.2, which displays the RMS of the separate-sampling resubstitution and leave-one-out estimators of the overall error rate $\hat{\varepsilon}^{r,01}_{n_0,n_1}$ and $\hat{\varepsilon}^{l,01}_{n_0,n_1}$, respectively, by assuming that the prior probabilities $c_0 = c_1 = 0.5$ are known. The RMS in each case is plotted against the Bayes error for different sample sizes and balanced design, $n_0 = n_1 = n$, under equally likely univariate Gaussian distributions with means $\mu_1 = -\mu_0 = 1$ and $\sigma_0 = \sigma_1 = 1$. We see that the RMS is an increasing function

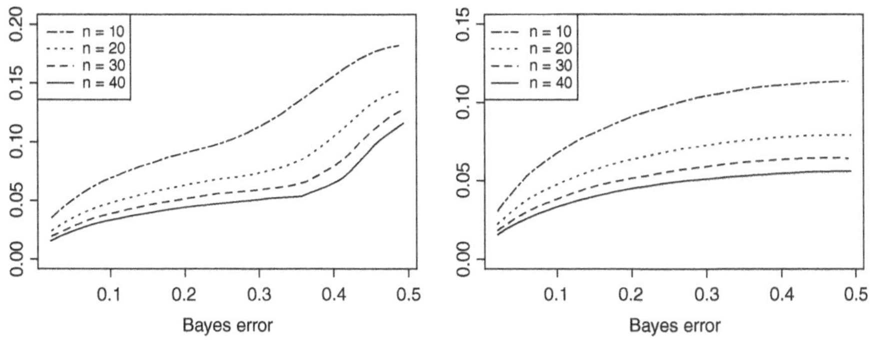

FIGURE 6.2 RMS of separate-sampling error estimators of the overall error rate versus Bayes error in a univariate homoskedastic Gaussian LDA model. Left panel: leave-one-out. Right panel: resubstitution.

of the Bayes error $\varepsilon_{\rm bay}$. Implicit in Figure 6.2 is the determination of a required sample size to achieve a desired RMS. For example, it allows one to obtain the bounds $\text{RMS}[\hat{\varepsilon}^{l,01}_{n_0,n_1}] \leq 0.145$ and $\text{RMS}[\hat{\varepsilon}^{r,01}_{n_0,n_1}] \leq 0.080$ for $n = 20$, and $\text{RMS}[\hat{\varepsilon}^{l,01}_{n_0,n_1}] \leq 0.127$ and $\text{RMS}[\hat{\varepsilon}^{r,01}_{n_0,n_1}] \leq 0.065$ for $n = 30$.

6.5 SAMPLING DISTRIBUTIONS OF ERROR RATES

In this section, we present exact expressions for the marginal sampling distributions of the classical mixture-sampling resubstitution and leave-one-out estimators using the results in Zollanvari et al. (2009). We also discuss the joint sampling distribution and density of each of these error estimators and the true error in the univariate Gaussian case, and some of its applications, using results published in Zollanvari et al. (2010). The lemmas and theorems appearing in the next two subsections are based on the results of Zollanvari et al. (2009) and Section 5.7.

6.5.1 Marginal Distribution of Resubstitution Error

We are interested in computing the sampling distribution of the mixture-sampling resubstitution estimator

$$P\left(\hat{\varepsilon}^r_n = \frac{k}{n}\right), \quad k = 0, 1, \ldots, n. \tag{6.68}$$

We begin with the following result, which gives the probability of perfect separation between the classes in the training data.

Lemma 6.1 *Under the univariate heteroskedastic Gaussian LDA model,*

$$\begin{aligned} P\left(\hat{\varepsilon}^r_n = 0\right) = & \sum_{n_0=0}^{n} \binom{n}{n_0} c_0^{n_0} c_1^{n_1} P(W_L(X_1) \geq 0, \ldots, W_L(X_{n_0}) \geq 0, \\ & W_L(X_{n_0+1}) < 0, \ldots, W_L(X_n) < 0 \\ & \quad \mid Y_1 = \cdots = Y_{n_0} = 0, Y_{n_0+1} = \cdots = Y_{n_0+n_1} = 1) \\ = & \sum_{n_0=0}^{n} \binom{n}{n_0} c_0^{n_0} c_1^{n_1} [P(\mathbf{Z} \geq \mathbf{0}) + P(\mathbf{Z} < \mathbf{0})], \end{aligned} \tag{6.69}$$

where $\mathbf{Z}$ *is a Gaussian random vector of size* $n + 1$*, with mean* $\boldsymbol{\mu}_{\mathbf{Z}}$ *given by*

$$\boldsymbol{\mu}_{\mathbf{Z}} = (\mu_0 - \mu_1)\,\mathbf{1}_{n+1} \tag{6.70}$$

and covariance matrix $\Sigma_{\mathbf{Z}}$ defined by

$$(\Sigma_{\mathbf{Z}})_{ij} = \begin{cases} (4n_0 - 3)\frac{\sigma_0^2}{n_0} + \frac{\sigma_1^2}{n_1}, & i,j = 1, \ldots, n_0, i = j, \\ -3\frac{\sigma_0^2}{n_0} + \frac{\sigma_1^2}{n_1}, & i,j = 1, \ldots, n_0, i \neq j, \\ \frac{\sigma_0^2}{n_0} + (4n_1 - 3)\frac{\sigma_1^2}{n_1}, & i,j = n_0 + 1, \ldots, n, i = j, \\ \frac{\sigma_0^2}{n_0} - 3\frac{\sigma_1^2}{n_1}, & i,j = n_0 + 1, \ldots, n, i \neq j, \\ \frac{\sigma_0^2}{n_0} + \frac{\sigma_1^2}{n_1}, & \text{otherwise.} \end{cases} \tag{6.71}$$

Proof: For conciseness, we omit conditioning on $Y_1 = \cdots = Y_{n_0} = 0, Y_{n_0+1} = \cdots = Y_{n_0+n_1} = 1$ in the equations below. It follows from (6.4) that

$$\begin{aligned} & P(W_L(X_1) \geq 0, \ldots, W_L(X_{n_0}) \geq 0, W_L(X_{n_0+1}) < 0, \ldots, W_L(X_n) < 0) \\ & \quad = P(W_L(X_1) \geq 0, \ldots, W_L(X_{n_0}) \geq 0, \\ & \qquad\quad W_L(X_{n_0+1}) < 0, \ldots, W_L(X_n) < 0, \bar{X}_0 - \bar{X}_1 \geq 0) \\ & \qquad + P(W_L(X_1) \geq 0, \ldots, W_L(X_{n_0}) \geq 0, \\ & \qquad\quad W_L(X_{n_0+1}) < 0, \ldots, W_L(X_n) < 0, \bar{X}_0 - \bar{X}_1 < 0) \\ & \quad = P(X_1 - \bar{X} \geq 0, \ldots, X_{n_0} - \bar{X} \geq 0, \\ & \qquad\quad \bar{X} - X_{n_0+1} \geq 0, \ldots, \bar{X} - X_n \geq 0, \bar{X}_0 - \bar{X}_1 \geq 0) \\ & \qquad + P(X_1 - \bar{X} < 0, \ldots, X_{n_0} - \bar{X} < 0, \\ & \qquad\quad \bar{X} - X_{n_0+1} < 0, \ldots, \bar{X} - X_n < 0, \bar{X}_0 - \bar{X}_1 < 0) \\ & \quad = P(\mathbf{Z} \geq \mathbf{0}) + P(\mathbf{Z} < \mathbf{0}), \end{aligned} \tag{6.72}$$

where

$$\mathbf{Z} = [2(X_1 - \bar{X}), \ldots, 2(X_{n_0} - \bar{X}), 2(\bar{X} - X_{n_0+1}), \ldots, 2(\bar{X} - X_n), \bar{X}_0 - \bar{X}_1]^T. \tag{6.73}$$

Vector $\mathbf{Z}$ is a linear combination of the vector $\mathbf{X} = [X_1, \ldots, X_n]^T$ of observations, namely, $\mathbf{Z} = A\mathbf{X}$, where matrix A is given by

$$A_{(n+1)\times n} = \begin{pmatrix} A^1 \\ A^2 \\ A^3 \end{pmatrix}, \tag{6.74}$$

where

$$A^1_{n_0\times n} = \begin{pmatrix} \left(2-\frac{1}{n_0}\right) & -\frac{1}{n_0} & \cdots & -\frac{1}{n_0} & -\frac{1}{n_1} & \cdots & -\frac{1}{n_1} \\ -\frac{1}{n_0} & \left(2-\frac{1}{n_0}\right) & \cdots & -\frac{1}{n_0} & -\frac{1}{n_1} & \cdots & -\frac{1}{n_1} \\ \vdots & \vdots & \ddots & \vdots & & \vdots & \\ -\frac{1}{n_0} & -\frac{1}{n_0} & \cdots & \left(2-\frac{1}{n_0}\right) & -\frac{1}{n_1} & \cdots & -\frac{1}{n_1} \end{pmatrix}, \qquad (6.75)$$

$$A^2_{n_1\times n} = \begin{pmatrix} \frac{1}{n_0} & \cdots & \frac{1}{n_0} & \left(\frac{1}{n_1}-2\right) & \frac{1}{n_1} & \cdots & \frac{1}{n_1} \\ \frac{1}{n_0} & \cdots & \frac{1}{n_0} & \frac{1}{n_1} & \left(\frac{1}{n_1}-2\right) & \cdots & \frac{1}{n_1} \\ & \vdots & & \vdots & \vdots & \ddots & \\ \frac{1}{n_0} & \cdots & \frac{1}{n_0} & \frac{1}{n_1} & \frac{1}{n_1} & \cdots & \left(\frac{1}{n_1}-2\right) \end{pmatrix}, \qquad (6.76)$$

$$A^3_{1\times n} = \left(\frac{1}{n_0} \ \cdots \ \frac{1}{n_0} \ -\frac{1}{n_1} \ \cdots \ -\frac{1}{n_1}\right). \qquad (6.77)$$

Therefore, $\mathbf{Z}$ is a Gaussian random vector, with mean $\boldsymbol{\mu}_\mathbf{Z} = A\boldsymbol{\mu}_\mathbf{X}$ and covariance $\Sigma_\mathbf{Z} = A\Sigma_\mathbf{X}A^T$, where $\boldsymbol{\mu}_\mathbf{X} = [\mu_0\mathbf{1}_{n_0}, \mu_1\mathbf{1}_{n_1}]^T$ and $\Sigma_\mathbf{X} = \text{diag}(\sigma_0^2\mathbf{1}_{n_0}, \sigma_1^2\mathbf{1}_{n_1})$, which results in (6.70) and (6.71). The result now follows from Theorem 5.4. ■

Matrix $\Sigma_\mathbf{Z}$ in the previous Lemma is of rank $n-1$, that is, not full rank, so that the Gaussian density of $\mathbf{Z}$ is singular. As mentioned previously, there are accurate algorithms that can perform the required integrations (Genz, 1992; Genz and Bretz, 2002).

Using Lemma 6.1 one can prove the following result.

Theorem 6.9 *Under the univariate heteroskedastic Gaussian LDA model,*

$$P\left(\hat{\varepsilon}_n^r = \frac{k}{n}\right) = \sum_{n_0=0}^{n} \binom{n}{n_0} c_0^{n_0} c_1^{n_1} \sum_{l=0}^{k} \binom{n_0}{l}\binom{n_1}{k-l} \left[P(E_{l,k-l}\mathbf{Z} \geq \mathbf{0}) + P(E_{l,k-l}\mathbf{Z} < \mathbf{0})\right], \qquad (6.78)$$

for $k = 0, 1, \ldots, n$, *where* $\mathbf{Z}$ *is defined in Lemma 6.1, and* $E_{l,k-l}$ *is a diagonal matrix of size* $n+1$, *with diagonal elements* $(-\mathbf{1}_l, \mathbf{1}_{n_0-l}, -\mathbf{1}_{k-l}, \mathbf{1}_{n_1-(k-l-1)})$.

Proof: According to Theorem 5.4, it suffices to show that

$$\begin{aligned}
&P(W_L(X_1) \le 0, \dots, W_L(X_l) \le 0,\\
&\quad W_L(X_{l+1}) > 0, \dots, W_L(X_{n_0}) > 0, W_L(X_{n_0+1}) > 0, \dots,\\
&\quad W_L(X_{n_0+k-l}) > 0, W_L(X_{n_0+k-l+1}) \le 0, \dots, W_L(X_n) \le 0\\
&\qquad \mid Y_1 = \cdots = Y_l = 0, Y_{l+1} = \cdots = Y_{n_0+k-l} = 1,\\
&\qquad Y_{n_0+k-l+1} = \cdots = Y_n = 0),\\
&= P(E_{l,k-l}\mathbf{Z} \ge \mathbf{0}) + P(\mathbf{Z}_{l,k-l} < \mathbf{0}).
\end{aligned} \tag{6.79}$$

But this follows from Lemma 6.1 and the fact that, to obtain the probability of an error happening to point X_i, one need only multiply the corresponding component $\mathbf{Z}_i$ of the vector $\mathbf{Z}$ by -1. ∎

6.5.2 Marginal Distribution of Leave-One-Out Error

A similar approach can be taken to obtain the exact sampling distribution of the mixture-sampling leave-one-out error estimator

$$P\left(\hat{\varepsilon}_n^l = \frac{k}{n}\right), \quad k = 0, 1, \dots, n, \tag{6.80}$$

although computational complexity in this case is much higher. The following result is the counterpart of Lemma 6.1.

Lemma 6.2 *Under the univariate heteroskedastic Gaussian LDA model,*

$$\begin{aligned}
P\left(\hat{\varepsilon}_n^l = 0\right) &= \sum_{n_0=0}^{n} \binom{n}{n_0} c_0^{n_0} c_1^{n_1} P(W_L^{(1)}(X_1) \ge 0, \dots, W_L^{(n_0)}(X_{n_0}) \ge 0,\\
&\qquad W_L^{(n_0+1)}(X_{n_0+1}) < 0, \dots, W_L^{(n)}(X_n) < 0\\
&\qquad \mid Y_1 = \cdots = Y_{n_0} = 0, Y_{n_0+1} = \cdots = Y_{n_0+n_1} = 1)\\
&= \sum_{n_0=0}^{n} \binom{n}{n_0} c_0^{n_0} c_1^{n_1} \sum_{r=0}^{n_0} \sum_{s=0}^{n_1} \binom{n_0}{r} \binom{n_1}{s} P(E_{r,s}\mathbf{Z} \ge \mathbf{0}),
\end{aligned} \tag{6.81}$$

where $E_{r,s}$is a diagonal matrix of size $2n$, with diagonal elements

$$(-\mathbf{1}_r, \mathbf{1}_{n_0-r}, -\mathbf{1}_r, \mathbf{1}_{n_0-r}, -\mathbf{1}_s, \mathbf{1}_{n_1-s}, -\mathbf{1}_s, \mathbf{1}_{n_1-s})$$

and $\mathbf{Z}$ is a Gaussian vector with mean

$$\boldsymbol{\mu}_{\mathbf{Z}} = \begin{bmatrix} \frac{n_0-1}{n_0}(\mu_0 - \mu_1)\mathbf{1}_{2n_0} \\ \frac{n_1-1}{n_1}(\mu_0 - \mu_1)\mathbf{1}_{2n_1} \end{bmatrix} \tag{6.82}$$

and covariance matrix

$$\Sigma_{\mathbf{Z}} = \left[\begin{array}{c|c} C^1 & C^2 \\ \hline C^{2^T} & C^3 \end{array}\right], \tag{6.83}$$

where

$$(C^1)_{ij} = \begin{cases} \left(4 - \frac{7}{n_0} + \frac{3}{n_0^2}\right)\sigma_0^2 + \frac{(n_0-1)^2\sigma_1^2}{n_0^2 n_1}, & i,j = 1,\dots,n_0, i = j, \\ (\frac{-3}{n_0} + \frac{2}{n_0^2})\sigma_0^2 + \frac{(n_0-1)^2\sigma_1^2}{n_0^2 n_1}, & i,j = 1,\dots,n_0, i \neq j, \\ \frac{(n_0-1)\sigma_0^2}{n_0^2} + \frac{(n_0-1)^2\sigma_1^2}{n_1 n_0^2}, & i,j = n_0+1,\dots,2n_0, i = j, \\ \frac{(n_0-2)\sigma_0^2}{n_0^2} + \frac{(n_0-1)^2\sigma_1^2}{n_1 n_0^2}, & i,j = n_0+1,\dots,2n_0, i \neq j, \\ \frac{-(n_0-1)\sigma_0^2}{n_0^2} + \frac{(n_0-1)^2\sigma_1^2}{n_1 n_0^2}, & \begin{cases} i = n_0+1,\dots,2n_0, j = i - n_0, \\ j = n_0+1,\dots,2n_0, i = j - n_0, \end{cases} \\ \frac{\sigma_0^2}{n_0} + \frac{(n_0-1)^2\sigma_1^2}{n_1 n_0^2}, & \text{otherwise}, \end{cases} \tag{6.84}$$

and

$$C^2 = \left[\frac{(n_1-1)(n_0-1)\sigma_0^2}{n_0^2 n_1} + \frac{(n_1-1)(n_0-1)\sigma_1^2}{n_1^2 n_0}\right]\mathbf{1}_{2n_0\times 2n_1}, \tag{6.85}$$

with C^3 being determined by interchanging the roles of σ_0^2 and σ_1^2 and of n_0 and n_1 in the expression for C^1.

Proof: For conciseness, we omit conditioning on $Y_1 = \cdots = Y_{n_0} = 0, Y_{n_0+1} = \cdots = Y_{n_0+n_1} = 1$ in the equations below. The univariate discriminant where the i-th sample point is left out is given by

$$W_L^{(i)}(x) = (x - \bar{X}^{(i)})(\bar{X}_0^{(i)} - \bar{X}_1^{(i)}), \tag{6.86}$$

where $\bar{X}_0^{(i)}$ and $\bar{X}_1^{(i)}$ are the sample means, when the i-th sample point is left out and $\bar{X}^{(i)}$ is their average. We have

$$P\Big(W_L^{(1)}(X_1) \geq 0, \dots, W_L^{(n_0)}(X_{n_0}) \geq 0,$$
$$W_L^{(n_0+1)}(X_{n_0+1}) < 0, \dots, W_L^{(n)}(X_n) < 0\Big)$$

$$
\begin{aligned}
&= P\Big(X_1 - \bar{X}^{(1)} \ge 0, \bar{X}_0^{(1)} - \bar{X}_1^{(1)} \ge 0, X_2 - \bar{X}^{(2)} \ge 0,\\
&\quad \bar{X}_0^{(2)} - \bar{X}_1^{(2)} \ge 0, \ldots, X_{n_0} - \bar{X}^{(n_0)} \ge 0, \bar{X}_0^{(n_0)} - \bar{X}_1^{(n_0)} \ge 0,\\
&\quad \bar{X}^{(n_0+1)} - X_{n_0+1} \ge 0, \bar{X}_0^{(n_0+1)} - \bar{X}_1^{(n_0+1)} \ge 0, \ldots,\\
&\quad \bar{X}^{(n)} - X_n \ge 0, \bar{X}_0^{(n)} - \bar{X}_1^{(n)} \ge 0\Big)\\
&+ P\Big(X_1 - \bar{X}^{(1)} < 0, \bar{X}_0^{(1)} - \bar{X}_1^{(1)} < 0, X_2 - \bar{X}^{(2)} \ge 0,\\
&\quad \bar{X}_0^{(2)} - \bar{X}_1^{(2)} \ge 0, \ldots, X_{n_0} - \bar{X}^{(n_0)} \ge 0, \bar{X}_0^{(n_0)} - \bar{X}_1^{(n_0)} \ge 0,\\
&\quad \bar{X}^{(n_0+1)} - X_{n_0+1} \ge 0, \bar{X}_0^{(n_0+1)} - \bar{X}_1^{(n_0+1)} \ge 0, \ldots,\\
&\quad \bar{X}^{(n)} - X_n \ge 0, \bar{X}_0^{(n)} - \bar{X}_1^{(n)} \ge 0\Big)\\
&\quad \vdots\\
&+ P\Big(X_1 - \bar{X}^{(1)} < 0, \bar{X}_0^{(1)} - \bar{X}_1^{(1)} < 0, X_2 - \bar{X}^{(2)} < 0,\\
&\quad \bar{X}_0^{(2)} - \bar{X}_1^{(2)} < 0, \ldots, X_{n_0} - \bar{X}^{(n_0)} < 0, \bar{X}_0^{(n_0)} - \bar{X}_1^{(n_0)} < 0,\\
&\quad \bar{X}^{(n_0+1)} - X_{n_0+1} < 0, \bar{X}_0^{(n_0+1)} - \bar{X}_1^{(n_0+1)} < 0, \ldots,\\
&\quad \bar{X}^{(n)} - X_n < 0, \bar{X}_0^{(n)} - \bar{X}_1^{(n)} < 0\Big),
\end{aligned}
\tag{6.87}
$$

where the total number of joint probabilities that should be computed is 2^n. Simplification by grouping repeated probabilities results in

$$
P\left(\hat{\varepsilon}^{l}_{n_0,n_1} = 0\right) = \sum_{r=0}^{n_0} \sum_{s=0}^{n_1} \binom{n_0}{r} \binom{n_1}{s} P(\mathbf{Z}_{r,s} \ge \mathbf{0}), \tag{6.88}
$$

with $\mathbf{Z}_{r,s} = E_{r,s}\mathbf{Z}$, where $E_{r,s}$ is given in the statement of the theorem and $\mathbf{Z}$ is a linear combination of the vector $\mathbf{X} = [X_1, \ldots, X_n]^T$ of observations, namely, $\mathbf{Z} = A\mathbf{X}$, for the matrix A given by

$$
A_{2n\times n} = \begin{pmatrix} A^1 \\ A^2 \\ A^3 \\ A^4 \end{pmatrix}, \tag{6.89}
$$

where

$$A^1_{n_0 \times n} =$$

$$\begin{pmatrix} 2\left(1-\frac{1}{n_0}\right) & -\frac{1}{n_0} & \dots & -\frac{1}{n_0} & -\left(\frac{1}{n_1}-\frac{1}{n_0 n_1}\right) & \dots & -\left(\frac{1}{n_1}-\frac{1}{n_0 n_1}\right) \\ -\frac{1}{n_0} & 2\left(1-\frac{1}{n_0}\right) & \dots & -\frac{1}{n_0} & -\left(\frac{1}{n_1}-\frac{1}{n_0 n_1}\right) & \dots & -\left(\frac{1}{n_1}-\frac{1}{n_0 n_1}\right) \\ \vdots & \vdots & \ddots & \vdots & & \vdots & \\ -\frac{1}{n_0} & -\frac{1}{n_0} & \dots & 2\left(1-\frac{1}{n_0}\right) & -\left(\frac{1}{n_1}-\frac{1}{n_0 n_1}\right) & \dots & -\left(\frac{1}{n_1}-\frac{1}{n_0 n_1}\right) \end{pmatrix}, \tag{6.90}$$

$$A^2_{n_0 \times n} = \begin{pmatrix} 0 & \frac{1}{n_0} & \dots & \frac{1}{n_0} & -\left(\frac{1}{n_1}-\frac{1}{n_0 n_1}\right) & \dots & -\left(\frac{1}{n_1}-\frac{1}{n_0 n_1}\right) \\ \frac{1}{n_0} & 0 & \dots & \frac{1}{n_0} & -\left(\frac{1}{n_1}-\frac{1}{n_0 n_1}\right) & \dots & -\left(\frac{1}{n_1}-\frac{1}{n_0 n_1}\right) \\ & \vdots & \ddots & \vdots & & \vdots & \\ \frac{1}{n_0} & \dots & \frac{1}{n_0} & 0 & -\left(\frac{1}{n_1}-\frac{1}{n_0 n_1}\right) & \dots & -\left(\frac{1}{n_1}-\frac{1}{n_0 n_1}\right) \end{pmatrix}, \tag{6.91}$$

$$A^3_{n_1 \times n} =$$

$$\begin{pmatrix} \left(\frac{1}{n_0}-\frac{1}{n_0 n_1}\right) & \dots & \left(\frac{1}{n_0}-\frac{1}{n_0 n_1}\right) & -2\left(1-\frac{1}{n_1}\right) & \frac{1}{n_1} & \dots & \frac{1}{n_1} \\ \left(\frac{1}{n_0}-\frac{1}{n_0 n_1}\right) & \dots & \left(\frac{1}{n_0}-\frac{1}{n_0 n_1}\right) & \frac{1}{n_1} & -2\left(1-\frac{1}{n_1}\right) & \dots & \frac{1}{n_1} \\ & \vdots & & \vdots & \vdots & \ddots & \\ \left(\frac{1}{n_0}-\frac{1}{n_0 n_1}\right) & \dots & \left(\frac{1}{n_0}-\frac{1}{n_0 n_1}\right) & \frac{1}{n_1} & \dots & \frac{1}{n_1} & -2\left(1-\frac{1}{n_1}\right) \end{pmatrix}, \tag{6.92}$$

$$A^4_{n_1 \times n} = \begin{pmatrix} \left(\frac{1}{n_0}-\frac{1}{n_0 n_1}\right) & \dots & \left(\frac{1}{n_0}-\frac{1}{n_0 n_1}\right) & 0 & -\frac{1}{n_1} & -\frac{1}{n_1} & \dots & -\frac{1}{n_1} \\ \left(\frac{1}{n_0}-\frac{1}{n_0 n_1}\right) & \dots & \left(\frac{1}{n_0}-\frac{1}{n_0 n_1}\right) & -\frac{1}{n_1} & 0 & -\frac{1}{n_1} & \dots & -\frac{1}{n_1} \\ & \vdots & & \vdots & \vdots & \ddots & & \\ \left(\frac{1}{n_0}-\frac{1}{n_0 n_1}\right) & \dots & \left(\frac{1}{n_0}-\frac{1}{n_0 n_1}\right) & -\frac{1}{n_1} & -\frac{1}{n_1} & \dots & -\frac{1}{n_1} & 0 \end{pmatrix}, \tag{6.93}$$

and the result holds. ■

The covariance matrix of $E_{r,s}\mathbf{Z}$ in the previous lemma is of rank $n-1$, that is, not full rank, so that the corresponding Gaussian density is singular. However, as before, there are accurate algorithms that can perform the required integrations (Genz, 1992; Genz and Bretz, 2002).

Using Lemma 6.2 one can prove the following result.

Theorem 6.10 *Under the univariate heteroskedastic Gaussian LDA model,*

$$P\left(\hat{\varepsilon}_n^l = \frac{k}{n}\right) = \sum_{n_0=0}^{n} \binom{n}{n_0} c_0^{n_0} c_1^{n_1} \sum_{l=0}^{k} \binom{n_0}{l} \binom{n_1}{k-l} \sum_{p=0}^{l} \sum_{q=0}^{k-l} \binom{l}{p} \binom{k-l}{q} \sum_{r=0}^{n_0-l} \sum_{s=0}^{n_1-(k-l)} \binom{n_0-l}{r} \binom{n_1-(k-l)}{s} P(E_{r,s,p,q,k,l}\mathbf{Z} \geq \mathbf{0}), \tag{6.94}$$

for $k = 0, 1, \ldots, n$, where $\mathbf{Z}$ is defined in Lemma 6.2 and

$$E_{r,s,p,q,k,l} = E_{r,s,k,l} E_{p,q,k,l}, \tag{6.95}$$

such that $E_{r,s,k,l}$ and $E_{p,q,k,l}$ are diagonal matrices of size $2n$ with diagonal elements

$$(-\mathbf{1}_p, \mathbf{1}_{n_0}, -\mathbf{1}_{l-p}, \mathbf{1}_{n_0-l}, -\mathbf{1}_q, \mathbf{1}_{n_1}, -\mathbf{1}_{k-l-q}, -\mathbf{1}_{n_1-k+l})$$

and

$$(-\mathbf{1}_l, \mathbf{1}_r, -\mathbf{1}_{n_0-r}, \mathbf{1}_r, -\mathbf{1}_{n_0-l-r}, -\mathbf{1}_{k-l}, \mathbf{1}_s, -\mathbf{1}_{n_1-s}, \mathbf{1}_s, -\mathbf{1}_{n_1-k+l-s}),$$

respectively.

Proof: According to Theorem 5.5, it suffices to show that

$$\begin{aligned}
&P\Big(W_L^{(1)}(X_1) \leq 0, \ldots, W_L^{(l)}(X_l) \leq 0, W_L^{(l+1)}(X_{l+1}) > 0, \ldots, \\
&\quad W_L^{(n_0)}(X_{n_0}) > 0, W_L^{(n_0+1)}(X_{n_0+1}) > 0, \ldots, \\
&\quad W_L^{(n_0+k-l)}(X_{n_0+k-l}) > 0, W_L^{(n_0+k-l+1)}(X_{n_0+k-l+1}) \leq 0, \ldots, \\
&\quad W_L^{(n)}(X_n) \leq 0 \mid Y_1 = \cdots = Y_l = 0, Y_{l+1} = \cdots = Y_{n_0+k-l} = 1, \\
&\qquad Y_{n_0+k-l+1} = \cdots = Y_n = 0\Big), \\
&= \sum_{p=0}^{l} \sum_{q=0}^{k-l} \binom{l}{p} \binom{k-l}{q} \sum_{r=0}^{n_0-l} \sum_{s=0}^{n_1-(k-l)} \binom{n_0-l}{r} \binom{n_1-(k-l)}{s} \\
&\qquad \times P(E_{r,s,p,q,k,l}\mathbf{Z} \geq \mathbf{0}).
\end{aligned} \tag{6.96}$$

But this follows from Lemma 6.2 and the fact that, to obtain the probability of an error happening to point X_i, one only needs to multiply the appropriate components of the vectors $\mathbf{Z}_{r,s}$ in (6.88) by -1. ■

6.5.3 Joint Distribution of Estimated and True Errors

The joint sampling distribution between the true and estimated error rates completely characterizes error estimation behavior. In other words, this distribution contains all the information regarding performance of the error estimator (i.e., how well it "tracks" the true error). Bias, deviation variance, RMS, deviation distributions, and more, can be computed with knowledge of the joint sampling distribution between the true and estimated error rates.

More specifically, for a general error estimator $\hat{\varepsilon}_n$ one is interested in exact expressions for the joint probability $P(\hat{\varepsilon}_n = \frac{k}{n}, \varepsilon_n < z)$ and the joint probability density $p(\hat{\varepsilon}_n = \frac{k}{n}, \varepsilon_n = z)$, for $k = 0, 1, \dots, n$ and $0 \le z \le 1$. Even though we are using the terminology "density," note that $p(\hat{\varepsilon}_n = \frac{k}{n}, \varepsilon_n = z)$ is in fact a combination of a density in ε_n and a probability mass function in $\hat{\varepsilon}_n$.

Once again, conditioning on the sample size N_0 proves useful. Clearly,

$$P\left(\hat{\varepsilon}_n = \frac{k}{n}, \varepsilon_n < z\right) = \sum_{n_0=0}^{n} \binom{n}{n_0} c_0^{n_0} c_1^{n_1} P\left(\hat{\varepsilon}_n = \frac{k}{n}, \varepsilon_n < z \,\middle|\, N_0 = n_0\right) \tag{6.97}$$

and

$$p\left(\hat{\varepsilon}_n = \frac{k}{n}, \varepsilon_n = z\right) = \sum_{n_0=0}^{n} \binom{n}{n_0} c_0^{n_0} c_1^{n_1} p\left(\hat{\varepsilon}_n = \frac{k}{n}, \varepsilon_n = z \,\middle|\, N_0 = n_0\right), \tag{6.98}$$

for $k = 0, 1, \dots, n$ and $0 \le z \le 1$.

Results are given in Zollanvari et al. (2010) that allow the exact calculation of the conditional distributions on the right-hand side of the previous equations for the univariate heteroskedastic Gaussian sampling model, when $\hat{\varepsilon}_n$ is either the classical resubstitution or leave-one-out error estimators. The expressions are very long and therefore are not reproduced here.

An application of these results is the computation of confidence interval bounds on the true classification error given the observed value of the error estimator,

$$P\left(\varepsilon_n < z \,\middle|\, \hat{\varepsilon}_n = \frac{k}{n}\right) = P\left(\hat{\varepsilon}_n = \frac{k}{n}, \varepsilon_n < z\right) \Big/ P\left(\hat{\varepsilon}_n = \frac{k}{n}\right), \tag{6.99}$$

for $k = 0, 1, \dots, n$, provided that the denominator $P\left(\hat{\varepsilon}_n = \frac{k}{n}\right)$ is nonzero (this probability is determined by Theorem 6.9 for resubstitution and Theorem 6.10 for

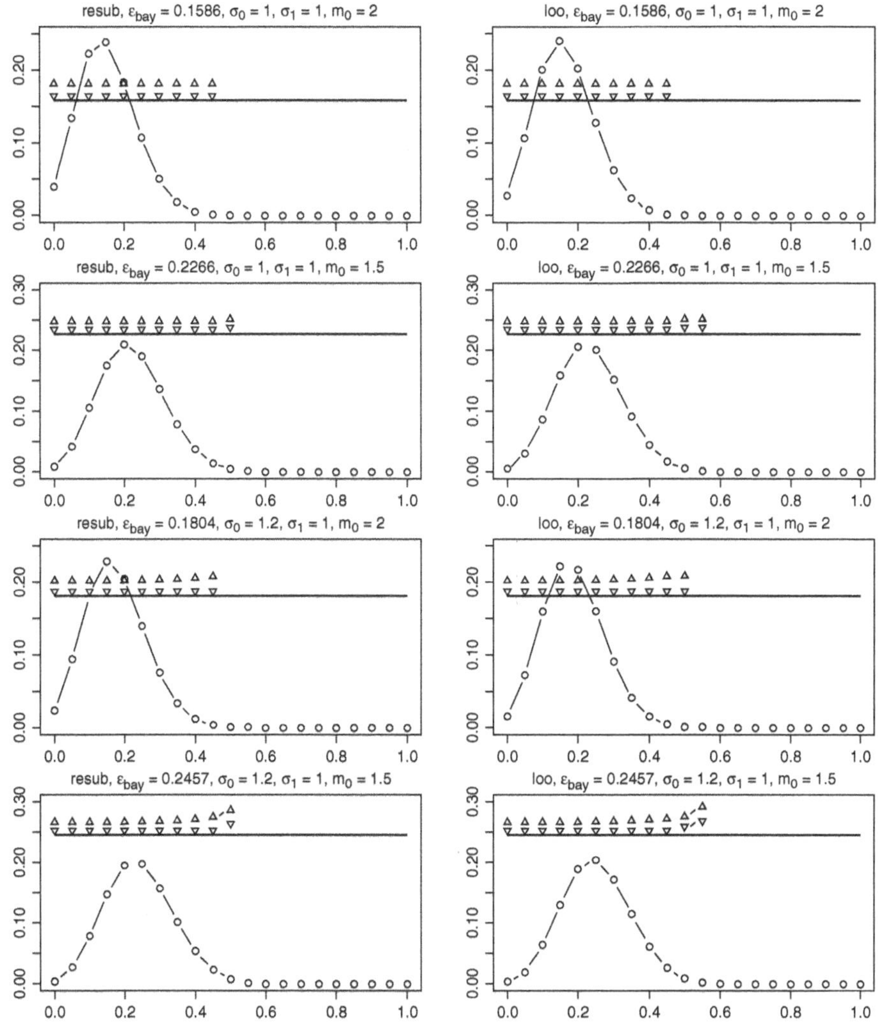

FIGURE 6.3 The 95% confidence interval upper bounds and regression of true error given the resubstitution and leave-one-out error estimates in the univariate case. In all cases, $n = 20$ and $m_1 = 0$. The horizontal solid line displays the Bayes error. The marginal probability mass function for the error estimators in each case is also plotted for reference. Legend key: 95% confidence interval upper bound (Δ), regression (∇), probability mass function ($\circ$).

leave-one-out), and then finding an exact $100(1-\alpha)\%$ upper bound on the true error given the error estimate, namely, finding z_α such that

$$P\left(\varepsilon_n < z_\alpha \middle| \hat{\varepsilon}_n = \frac{k}{n}\right) = 1 - \alpha. \tag{6.100}$$

The value z_α can be found by means of a simple one-dimensional search.

In addition, great insight can be obtained by finding the expected classification error conditioned on the observed value of the error estimator, which contains "local" information on the accuracy of the classification rule, as opposed to the "global" information contained in the unconditional expected classification error. Using the conditional distribution in (6.99), we have the regression

$$E\left(\varepsilon_n \middle| \hat{\varepsilon}_n = \frac{k}{n}\right) = \int_0^1 \left(1 - P\left(\varepsilon_n < z \middle| \hat{\varepsilon}_n = \frac{k}{n}\right)\right) dz, \tag{6.101}$$

by using the fact that $E[X] = \int P(X > z)\, dz$ for any nonnegative random variable X (c.f. Section A.7).

Figure 6.3 illustrates the exact 95% confidence interval upper bound and regression in the univariate case, using the exact expressions for the joint distribution between true and estimated errors obtained in Zollanvari et al. (2010). In the figure, m_0 and m_1 are the means and σ_0 and σ_1 are the standard deviations of the populations. The total number of sample points is kept to 20 to facilitate computation. In all examples, in order to avoid numerical instability issues, the bounds and regression are calculated for only those values of the error estimate that have a significant probability mass, for example, $P(\hat{\varepsilon}_n = \frac{k}{n}) > \tau$, where τ is a small positive number. Note that the aforementioned probability mass function is displayed in the plots to show the concentration of mass of the error estimates.

EXERCISES

6.1 Consider that every point in a sample S of size n is doubled up, producing an augmented sample S' of size $2n$. Use Theorem 6.1 to analyze the behavior of the univariate heteroskedastic Gaussian LDA model under this scheme. In particular, how does the expected error rate for this augmented sampling mechanism compares to that for a true random sample of size $2n$? Does the answer depend on the ratio between n_0 and n_1 or on the distributional parameters?

6.2 Show that the result in Theorem 6.1 reduces to Hill's formulas (6.8) and (6.9) when $\sigma_0 = \sigma_1 = \sigma$.

6.3 Let $\mu_0 = 0, \sigma_0 = 1, \sigma_1 = 2$, and μ_1 be chosen so that $\varepsilon_{\text{bay}} = 0.1, 0.2, 0.3$ in the univariate heteroskedastic Gaussian sampling model. Compute $E[\varepsilon^0_{n_0,n_1}]$ via Theorem 6.1 and compare this exact value to that obtained by Monte-Carlo sampling for $n_0, n_1 = 10, 20, \ldots, 50$ using $1000, 2000, \ldots, 10000$ Monte-Carlo sample points.

6.4 Fill in the details of the proof of Theorem 6.2.

6.5 Show that the result in Theorem 6.2 reduces to Hill's formulas (6.19) and (6.20) when $\sigma_0 = \sigma_1 = 1$.

6.6 Extend the resubstitution estimation bias formulas in (6.23) and (6.24) to the general case $n_0 \neq n_1$. Establish the asymptotic unbiasedness of resubstitution as both sample sizes $n_0, n_1 \to \infty$.

6.7 Repeat Exercise 6.3 except this time for Theorem 6.2 and $E[\hat{\varepsilon}^{r,0}_{n_0,n_1}]$.

6.8 Repeat Exercise 6.3 except this time for Theorem 6.4 and $E[(\varepsilon^{0}_{n_0,n_1})^2]$.

6.9 Repeat Exercise 6.3 except this time for Theorem 6.5 and $E[(\hat{\varepsilon}^{r,0}_{n_0,n_1})^2]$.

6.10 Repeat Exercise 6.3 except this time for Theorem 6.6 and $E[\varepsilon^{0}_{n_0,n_1}\hat{\varepsilon}^{r,0}_{n_0,n_1}]$.

6.11 Repeat Exercise 6.3 except this time for Theorem 6.7 and $E[(\hat{\varepsilon}^{l,0}_{n_0,n_1})^2]$.

6.12 Repeat Exercise 6.3 except this time for Theorem 6.8 and $E[\varepsilon^{0}_{n_0,n_1}\hat{\varepsilon}^{l,0}_{n_0,n_1}]$.

6.13 Prove Theorem 6.4 for the mixed moment $E[\varepsilon^{0}_{n_0,n_1}\varepsilon^{1}_{n_0,n_1}]$.

6.14 Prove Theorem 6.5 for the mixed moment $E[\hat{\varepsilon}^{r,0}_{n_0,n_1}\hat{\varepsilon}^{r,1}_{n_0,n_1}]$.

6.15 Prove Theorem 6.6 for the mixed moment $E[\varepsilon^{0}_{n_0,n_1}\hat{\varepsilon}^{r,1}_{n_0,n_1}]$.

6.16 Prove Theorem 6.7 for the mixed moment $E[\hat{\varepsilon}^{l,0}_{n_0,n_1}\hat{\varepsilon}^{l,1}_{n_0,n_1}]$.

6.17 Prove Theorem 6.8 for the mixed moment $E[\varepsilon^{0}_{n_0,n_1}\hat{\varepsilon}^{l,1}_{n_0,n_1}]$.

6.18 Using Theorem 6.9, plot the distribution $P\left(\hat{\varepsilon}^{r}_{n} = \frac{k}{n}\right)$, for $k = 0, 1, \ldots, n$, assuming the model of Exercise 6.3 and $n = 31$.

6.19 Using Theorem 6.10, plot the distribution $P(\hat{\varepsilon}^{l}_{n} = \frac{k}{n})$, for $k = 0, 1, \ldots, n$, assuming the model of Exercise 6.3 and $n = 31$.

CHAPTER 7

GAUSSIAN DISTRIBUTION THEORY: MULTIVARIATE CASE

This chapter surveys results in the distributional theory of error rates of the Linear (LDA) and Quadratic Discriminant Analysis (QDA) classification rules under multivariate Gaussian populations $\Pi_0 \sim N_d(\boldsymbol{\mu}_0, \Sigma_0)$ and $\Pi_1 \sim N_d(\boldsymbol{\mu}_1, \Sigma_1)$, with the aim of obtaining exact or approximate (accurate) expressions for the moments of true and estimated error rates, as well as the sampling distributions of error estimators.

In the univariate case, addressed in the previous chapter, direct methods lead to exact expressions for the distribution of the discriminant, the error rate moments, and their distribution, even in the heteroskedastic case. By contrast, the analysis of the multivariate case is more difficult. There are basically two approaches to the study of the multivariate case: exact small-sample (i.e., finite-sample) methods based on statistical representations and approximate large-sample (i.e., asymptotic) methods. The former do not necessarily produce exact results because approximation methods may still be required to simulate some of the random variables involved. In this chapter, we focus on these two approaches, with an emphasis on separate-sampling error rates. As was shown in Chapter 5, these can be used to derive mixture-sampling error rates by conditioning on the population sample sizes

7.1 MULTIVARIATE DISCRIMINANTS

Given a sample $S = \{(\mathbf{X}_1, Y_1), \ldots, (\mathbf{X}_n, Y_n)\}$, the sample-based LDA multivariate discriminant was introduced in Section 1.4:

$$W_L(S, \mathbf{x}) = \left(\mathbf{x} - \bar{\mathbf{X}}\right)^T \hat{\Sigma}^{-1} \left(\bar{\mathbf{X}}_0 - \bar{\mathbf{X}}_1\right), \tag{7.1}$$

Error Estimation for Pattern Recognition, First Edition. Ulisses M. Braga-Neto and Edward R. Dougherty.

where

$$\bar{\mathbf{X}}_0 = \frac{1}{n_0}\sum_{i=1}^{n}\mathbf{X}_i I_{Y_i=0} \quad \text{and} \quad \bar{\mathbf{X}}_1 = \frac{1}{n_1}\sum_{i=1}^{n}\mathbf{X}_i I_{Y_i=1} \tag{7.2}$$

are the population-specific sample means (in the case of mixture sampling, n_0 and n_1 are random variables), $\bar{\mathbf{X}} = \frac{1}{2}(\bar{\mathbf{X}}_0 + \bar{\mathbf{X}}_1)$, and

$$\hat{\Sigma} = \frac{1}{n-2}\sum_{i=1}^{n}\left[(\mathbf{X}_i - \bar{\mathbf{X}}_0)(\mathbf{X}_i - \bar{\mathbf{X}}_0)^T I_{Y_i=0} + (\mathbf{X}_i - \bar{\mathbf{X}}_1)(\mathbf{X}_i - \bar{\mathbf{X}}_1)^T I_{Y_1=1}\right] \tag{7.3}$$

is the pooled sample covariance matrix. As mentioned in Section 1.4, the previous discriminant is also known as Anderson's discriminant. If Σ is assumed to be known, it can be used in placed of $\hat{\Sigma}$ to obtain John's discriminant:

$$W_{\bar{L}}(S,\mathbf{x}) = (\mathbf{x} - \bar{\mathbf{X}})^T\Sigma^{-1}(\bar{\mathbf{X}}_0 - \bar{\mathbf{X}}_1). \tag{7.4}$$

In the general heteroskedastic Gaussian model, $\Sigma_0 \neq \Sigma_1$, the sample-based QDA discriminant introduced in Section 1.4 is

$$\begin{aligned} W_Q(S,\mathbf{x}) = &-\frac{1}{2}(\mathbf{x} - \bar{\mathbf{X}}_0)^T\hat{\Sigma}_0^{-1}(\mathbf{x} - \bar{\mathbf{X}}_0) \\ &+\frac{1}{2}(\mathbf{x} - \bar{\mathbf{X}}_1)^T\hat{\Sigma}_1^{-1}(\mathbf{x} - \bar{\mathbf{X}}_1) + \ln\frac{|\hat{\Sigma}_1|}{|\hat{\Sigma}_0|}, \end{aligned} \tag{7.5}$$

where

$$\hat{\Sigma}_0 = \frac{1}{n_0 - 1}\sum_{i=1}^{n}(\mathbf{X}_i - \bar{\mathbf{X}}_0)(\mathbf{X}_i - \bar{\mathbf{X}}_0)^T I_{Y_i=0}, \tag{7.6}$$

$$\hat{\Sigma}_1 = \frac{1}{n_1 - 1}\sum_{i=1}^{n}(\mathbf{X}_i - \bar{\mathbf{X}}_1)(\mathbf{X}_i - \bar{\mathbf{X}}_1)^T I_{Y_i=1}, \tag{7.7}$$

are the sample covariance matrices for each population.

The preceding formulas apply regardless of whether separate or mixture sampling is assumed. As in Chapter 5, we will write $W(\mathbf{x})$, where W is one of the discriminants above, whenever there is no danger of confusion.

7.2 SMALL-SAMPLE METHODS

In this section, we examine small-sample methods based on statistical representations for the discriminants defined in the previous section in the case of Gaussian populations and separate sampling. Most results are for the homoskedastic case, with the notable exception of McFarland and Richard's representation of the QDA discriminant in the heteroskedastic case.

It was seen in Chapter 5 that computation of error rate moments and distributions requires the evaluation of orthant probabilities involving the discriminant W. In the case of expected error rates, these probabilities are univariate, whereas in the case of higher-order moments and sampling distributions, multivariate probabilities are required. For example, it was seen in Chapter 5 that $E[\hat{\varepsilon}^{r,0}_{n_0,n_1}] = P(W(S_{n_0,n_1}, \mathbf{X}_1) \leq 0 \mid Y_1 = 0)$, c.f. (5.30), whereas the computation of $E[(\hat{\varepsilon}^{r,0}_{n_0,n_1})^2]$ requires probabilities of the kind $P(W(S_{n_0,n_1}, \mathbf{X}_1) \leq 0, W(S_{n_0,n_1}, \mathbf{X}_2) \leq 0 \mid Y_1 = 0, Y_2 = 0)$, c.f. (5.77). Statistical representations for the discriminant $W(S_{n_0,n_1}, \mathbf{X})$ in terms of known random variables can be useful to compute such orthant probabilities. More details on computational issues will be given in Section 7.2.2. For now, we concentrate in the next section on multivariate statistical representations of the random variable $W(S_{n_0,n_1}, \mathbf{X})$.

7.2.1 Statistical Representations

A *statistical representation* expresses a random variable in terms of other random variables, in such a way that this alternative expression has the same distribution as the original random variable. Typically, this alternative expression will involve familiar random variables that are easier to study and simulate than the original random variable.

In this section, statistical representations will be given for $W(S_{n_0,n_1}, \mathbf{X}) \mid Y = 0$ and $W(S_{n_0,n_1}, \mathbf{X}_1) \mid Y_1 = 0$ in the case of Anderson's and John's LDA discriminants, as well as the general QDA discriminant, under multivariate Gaussian populations and separate sampling. These representations are useful for computing error rates specific to population Π_0, such as $E[\varepsilon^0]$ and $E[\hat{\varepsilon}^{r,0}]$, respectively. To obtain error rates specific to population Π_1, such as $E[\varepsilon^1]$ and $E[\hat{\varepsilon}^{r,1}]$, one needs representations for $W(S_{n_0,n_1}, \mathbf{X}) \mid Y = 1$ and $W(S_{n_0,n_1}, \mathbf{X}_1) \mid Y_1 = 1$, respectively. These can be obtained easily from the previous representations if the discriminant is anti-symmetric, which is the case of all discriminants considered here. To see this, suppose that a statistical representation for $W(S_{n_0,n_1}, \mathbf{X}) \mid Y = 0$ is given (the argument is the same if $\mathbf{X}$ and Y are replaced by $\mathbf{X}_1$ and Y_1, respectively), where S_{n_0,n_1} is a sample from $F_{\mathbf{X}Y}$. Flipping all class labels in such a representation, including all distributional parameters and sample sizes (this may require some care), produces a representation of $W(\overline{S_{n_0,n_1}}, \mathbf{X}) \mid Y = 1$, where $\overline{S_{n_0,n_1}}$ is a sample of sizes n_1 and n_0 from $F_{\mathbf{X}\overline{Y}}$ (see Section 5.2 for all the relevant definitions). Anti-symmetry of W, c.f. (5.3), then implies that multiplication of this representation by -1 produces a representation for $W(S_{n_0,n_1}, \mathbf{X}) \mid Y = 1$.

The distribution of the discriminant is exchangeable, in the sense defined in (5.26) and (5.37), if it is invariant to a flip of all class labels of only the distributional parameters, without a flip of the sample sizes. To formalize this, consider that a statistical representation of $W(S_{n_0,n_1}, \mathbf{X}) \mid Y = 0$ is given — the argument is similar for a representation of $W(S_{n_0,n_1}, \mathbf{X}_1) \mid Y_1 = 0$ — where S_{n_0,n_1} is a sample from $F_{\mathbf{X}Y}$. If one flips the class labels of only the distribution parameters in this representation, without flipping sample sizes, then one obtains a representation of $W(\overline{S_{n_1,n_0}}, \mathbf{X}) \mid Y = 1$, where $\overline{S_{n_1,n_0}}$ is a sample of sizes n_0 and n_1 from $F_{\mathbf{X}\overline{Y}}$. The exchangeability property requires that $W(S_{n_0,n_1}, \mathbf{X}) \mid Y = 0 \sim W(\overline{S_{n_1,n_0}}, \mathbf{X}) \mid Y = 1)$, that is, the distribution of the discriminant is invariant to this operation. It will be shown, as a corollary of the results in this section, that these exchangeability properties, and thus all of their useful consequences, hold for both Anderson's and John's LDA discriminants under the homoskedastic Gaussian model.

7.2.1.1 *Anderson's LDA Discriminant on a Test Point* The distribution of Anderson's LDA discriminant on a test point is known to be very complicated (Jain and Waller, 1978; McLachlan, 1992). In Bowker (1961), Bowker provided a statistical representation of $W_L(S_{n_0,n_1}, \mathbf{X}) \mid Y = 0$ as a function of the coefficients of two independent Wishart matrices, one of which is noncentral. A clear and elegant proof of this result can be found in that reference.

Theorem 7.1 *(Bowker, 1961). Provided that $n > d$, Anderson's LDA discriminant applied to a test point can be represented in the separate-sampling homoskedastic Gaussian case as*

$$W_L(S_{n_0,n_1}, \mathbf{X}) \mid Y = 0 \sim a_1 \frac{Z_{12}Y_{22} - Y_{12}\sqrt{Z_{11}Z_{22} - Z_{12}^2}}{Y_{11}Y_{22} - Y_{12}^2} + a_2 \frac{Z_{11}Y_{22}}{Y_{11}Y_{22} - Y_{12}^2}, \tag{7.8}$$

where

$$a_1 = (n_0 + n_1 - 2)\sqrt{\frac{n_0 + n_1 + 1}{n_0 n_1}}, \quad a_2 = (n_0 + n_1 - 2)\frac{n_0 - n_1}{2 n_0 n_1}, \tag{7.9}$$

and

$$\begin{bmatrix} Y_{11} & Y_{12} \\ Y_{12} & Y_{22} \end{bmatrix}, \begin{bmatrix} Z_{11} & Z_{12} \\ Z_{12} & Z_{22} \end{bmatrix}, \tag{7.10}$$

are two independent random matrices; the first one has a central Wishart(I_2) distribution with $n_0 + n_1 - d$ degrees of freedom, while the second one has a noncentral Wishart(I_2, Λ) distribution with d degrees of freedom and noncentrality matrix

$$\Lambda = \begin{bmatrix} 1 & \sqrt{\frac{n_1}{n_0(n_0+n_1+1)}} \\ \sqrt{\frac{n_1}{n_0(n_0+n_1+1)}} & \frac{n_1}{n_0(n_0+n_1+1)} \end{bmatrix} \frac{n_0 n_1}{n_0 + n_1} \delta^2, \tag{7.11}$$

where δ is the Mahalanobis distance between the populations:

$$\delta = \sqrt{(\boldsymbol{\mu}_1 - \boldsymbol{\mu}_0)^T \Sigma^{-1} (\boldsymbol{\mu}_1 - \boldsymbol{\mu}_0)}. \tag{7.12}$$

This result shows that the distribution of $W_L(S_{n_0,n_1}, \mathbf{X}) \mid Y = 0$ is a function only of the Mahalanobis distance between populations δ and the sample sizes n_0 and n_1. This has two important consequences.

The first is that flipping all class labels to obtain a representation for $W_L(\overline{S_{n_0,n_1}}, \mathbf{X}) \mid Y = 1$ requires only switching the sample sizes, since the only distributional parameter δ is invariant to the switch. As described in the beginning of this section, anti-symmetry of W_L implies that a representation for $W_L(S_{n_0,n_1}, \mathbf{X}) \mid Y = 1$ can be obtained by multiplying this by -1. Therefore, we have

$$W_L(S_{n_0,n_1}, \mathbf{X}) \mid Y = 1 \sim \bar{a}_1 \frac{Z_{12}Y_{22} - Y_{12}\sqrt{Z_{11}Z_{22} - Z_{12}^2}}{Y_{11}Y_{22} - Y_{12}^2} + a_2 \frac{Z_{11}Y_{22}}{Y_{11}Y_{22} - Y_{12}^2}, \tag{7.13}$$

where $\bar{a}_1 = -a_1$, a_2 remains unchanged, and all other variables are as in Theorem 7.1 except that n_0 and n_1 are exchanged in (7.11). Both representations are only a function of δ, n_0 and n_1, and hence so are the error rates $E[\varepsilon^0_{n_0,n_1}]$, $E[\varepsilon^1_{n_0,n_1}]$, and $E[\varepsilon_{n_0,n_1}]$.

The second consequence is that the exchangeability property (5.26) is satisfied, since the only distributional parameter that appears in the representation is δ, which is invariant to class label flipping. As discussed in Section 5.5.1, this implies that $E[\varepsilon^0_{n_0,n_1}] = E[\varepsilon^1_{n_1,n_0}]$ and $E[\varepsilon^1_{n_0,n_1}] = E[\varepsilon^0_{n_1,n_0}]$ for Anderson's LDA classification rule. If, in addition, $c_0 = c_1 = 0.5$, then $E[\varepsilon_{n_0,n_1}] = E[\varepsilon_{n_1,n_0}]$. Finally, for a balanced design $n_0 = n_1 = n/2$ — in which case $a_2 = 0$ and all expressions become greatly simplified — we have that $E[\varepsilon_{n/2,n/2}] = E[\varepsilon^0_{n/2,n/2}] = E[\varepsilon^1_{n/2,n/2}]$, regardless of the values of the prior probabilities c_0 and c_1.

For the efficient computation of the distribution in Theorem 7.1, an equivalent expression can be derived. It can be shown that $\sqrt{n_0 + n_1 - d}\,(Y_{11}/Y_{22})$ is distributed as Student's t random variable $t_{n_0+n_1-d}$ with $n_0 + n_1 - d$ degrees of freedom, while $Y_{11} - Y_{12}^2/Y_{22}$ is distributed independently as a (central) chi-square random variable $\chi^2_{n_0+n_1-d-1}$ with $n_0 + n_1 - d - 1$ degrees of freedom (Bowker, 1961), so that dividing (7.8) through by Y_{22} leads to

$$W_L(S_{n_0,n_1}, \mathbf{X}) | Y = 0 \sim a_1 \frac{Z_{12} - \dfrac{t_{n_0+n_1-d}}{\sqrt{n_0 + n_1 - d}}\sqrt{Z_{11}Z_{22} - Z_{12}^2}}{\chi^2_{n_0+n_1-d-1}} + a_2 \frac{Z_{11}}{\chi^2_{n_0+n_1-d-1}}. \tag{7.14}$$

In addition, it can be shown that Z_{11}, Z_{12}, and Z_{22} can be represented in terms of independent Gaussian vectors $\mathbf{U} = (U_1, \dots, U_d)$ and $\mathbf{V} = (V_1, \dots, V_d)$ as

$$Z_{11} \sim \mathbf{U}^T\mathbf{U}, \quad Z_{12} \sim \mathbf{U}^T\mathbf{V}, \quad Z_{22} \sim \mathbf{V}^T\mathbf{V}, \tag{7.15}$$

where $\Sigma_{\mathbf{U}} = \Sigma_{\mathbf{V}} = I_d$, $E[U_i] = E[V_i] = 0$, for $i = 2, \dots, d$, and

$$E[U_1] = \sqrt{\frac{n_0 n_1}{n_0 + n_1}}\ \delta, \tag{7.16}$$

$$E[V_1] = \frac{n_1}{\sqrt{(n_0 + n_1)(n_0 + n_1 + 1)}}\ \delta. \tag{7.17}$$

If $Y = 1$, the only change is to replace V_1 by $\bar{V}_1$, which is distributed as V_1 except that

$$E[\bar{V}_1] = -\frac{n_0}{\sqrt{(n_0 + n_1)(n_0 + n_1 + 1)}}\ \delta. \tag{7.18}$$

7.2.1.2 John's LDA Discriminant on a Test Point The case where the common covariance matrix Σ is known is much simpler than the previous case. In John (1961), John gave an expression for $P(W_{\bar{L}}(S_{n_0,n_1}, \mathbf{X}) \leq 0 \mid Y = 0)$ under separate sampling that suggests that $W(S_{n_0,n_1}, \mathbf{X}) \mid Y = 0$ could be represented as a difference of two noncentral chi-square random variables. A brief derivation of John's expression was provided by Moran (1975). However, neither author provided an explicit statistical representation of $W_{\bar{L}}(S_{n_0,n_1}, \mathbf{X}) \mid Y = 0$. This is provided below, along with a detailed proof.

Theorem 7.2 *John's LDA discriminant applied to a test point can be represented in the separate-sampling homoskedastic Gaussian case as*

$$W_{\bar{L}}(S_{n_0,n_1}, \mathbf{X}) \mid Y = 0 \ \sim\ a_1\, \chi_d^2(\lambda_1) - a_2\, \chi_d^2(\lambda_2), \tag{7.19}$$

for positive constants

$$a_1 = \frac{1}{4 n_0 n_1}\left(n_0 - n_1 + \sqrt{(n_0 + n_1)(n_0 + n_1 + 4 n_0 n_1)}\right),$$

$$a_2 = \frac{1}{4 n_0 n_1}\left(n_1 - n_0 + \sqrt{(n_0 + n_1)(n_0 + n_1 + 4 n_0 n_1)}\right), \tag{7.20}$$

where $\chi_d^2(\lambda_1)$ and $\chi_d^2(\lambda_2)$ are independent noncentral chi-square random variables with d degrees of freedom and noncentrality parameters

$$\lambda_1 = \frac{1}{8a_1} \frac{\left(\sqrt{n_0+n_1} + \sqrt{n_0+n_1+4n_0n_1}\right)^2}{\sqrt{(n_0+n_1)(n_0+n_1+4n_0n_1)}} \delta^2,$$

$$\lambda_2 = \frac{1}{8a_2} \frac{\left(\sqrt{n_0+n_1} - \sqrt{n_0+n_1+4n_0n_1}\right)^2}{\sqrt{(n_0+n_1)(n_0+n_1+4n_0n_1)}} \delta^2. \quad (7.21)$$

Proof: We take advantage of the fact that John's LDA discriminant is invariant under application of a nonsingular affine transformation $X' = A\mathbf{X} + b$ to all sample vectors. That is, if $S' = \{(A\mathbf{X}_1 + b, Y_1), \ldots, (A\mathbf{X}_n + b, Y_n)\}$, then $W_{\bar{L}}(S', A\mathbf{X} + b)) = W_{\bar{L}}(S, \mathbf{X})$, as can be easily seen from (7.4). We consider such a transformation,

$$\mathbf{X}' = U\Sigma^{-\frac{1}{2}}\left(\mathbf{X} - \frac{1}{2}\left(\boldsymbol{\mu}_0 + \boldsymbol{\mu}_1\right)\right), \quad (7.22)$$

where U is an orthogonal matrix whose first row is is given by

$$\frac{\Sigma^{-\frac{1}{2}}(\boldsymbol{\mu}_0 - \boldsymbol{\mu}_1)}{||\Sigma^{-\frac{1}{2}}(\boldsymbol{\mu}_0 - \boldsymbol{\mu}_1)||} = \frac{\Sigma^{-\frac{1}{2}}(\boldsymbol{\mu}_0 - \boldsymbol{\mu}_1)}{\delta}. \quad (7.23)$$

Then the transformed Gaussian populations are given by

$$\Pi_0' \sim N(\nu_0, I_d) \quad \text{and} \quad \Pi_1' \sim N(\nu_1, I_d), \quad (7.24)$$

where

$$\nu_0 = \left(\frac{\delta}{2}, 0, \ldots, 0\right) \quad \text{and} \quad \nu_1 = \left(-\frac{\delta}{2}, 0, \ldots, 0\right). \quad (7.25)$$

Due to the invariance of John's LDA discriminant, one can proceed with these transformed populations without loss of generality (and thus we drop the prime from the superscript).

Define the random vectors

$$\mathbf{U} = \sqrt{\frac{4n_0n_1}{n_0+n_1+4n_0n_1}}\left(\mathbf{X} - \bar{\mathbf{X}}\right), \quad \mathbf{V} = \sqrt{\frac{n_0n_1}{n_0+n_1}}\left(\bar{\mathbf{X}}_0 - \bar{\mathbf{X}}_1\right). \quad (7.26)$$

It is clear that $W_{\bar{L}}(S_{n_0,n_1}, \mathbf{X}) \,|\, Y = 0$ is proportional to the product $\mathbf{U}^T\mathbf{V}$. The task is therefore to find the distribution of this product.

Clearly, **U** and **V** are Gaussian random vectors such that

$$\mathbf{U} \sim N_d\left(\sqrt{\frac{4n_0n_1}{n_0+n_1+4n_0n_1}}\,\nu_0,\, I_d\right), \quad \mathbf{V} \sim N_d\left(\sqrt{\frac{n_0n_1}{n_0+n_1}}(\nu_0-\nu_1),\, I_d\right). \tag{7.27}$$

Let $\mathbf{Z} = (\mathbf{U}, \mathbf{V})$. This is a Gaussian random vector, with covariance matrix

$$\Sigma_{\mathbf{Z}} = \begin{bmatrix} \Sigma_{\mathbf{U}} & \Sigma_{\mathbf{UV}} \\ \Sigma_{\mathbf{UV}}^T & \Sigma_{\mathbf{V}} \end{bmatrix} = \begin{bmatrix} I_d & \rho I_d \\ \rho I_d & I_d \end{bmatrix}, \tag{7.28}$$

since $\Sigma_{\mathbf{U}} = \Sigma_{\mathbf{V}} = I_d$, and

$$\begin{aligned}
\Sigma_{\mathbf{UV}} &= E[\mathbf{U}\mathbf{V}^T] - E[\mathbf{U}]E[\mathbf{V}]^T \\
&= \sqrt{\frac{4n_0n_1}{n_0+n_1+4n_0n_1}}\sqrt{\frac{n_0n_1}{n_0+n_1}}\,\{E[\mathbf{X}]E[\bar{\mathbf{X}}_0-\bar{\mathbf{X}}_1]^T \\
&\quad - E\left[\bar{\mathbf{X}}(\bar{\mathbf{X}}_0-\bar{\mathbf{X}}_1)^T\right] - E[\mathbf{X}-\bar{\mathbf{X}}]E[\bar{\mathbf{X}}_0-\bar{\mathbf{X}}_1]^T\} \\
&= \sqrt{\frac{4n_0n_1}{n_0+n_1+4n_0n_1}}\sqrt{\frac{n_0n_1}{n_0+n_1}}\,\frac{1}{2}(\Sigma_{\bar{\mathbf{X}}_1}-\Sigma_{\bar{\mathbf{X}}_0}) \\
&= \sqrt{\frac{4n_0n_1}{n_0+n_1+4n_0n_1}}\sqrt{\frac{n_0n_1}{n_0+n_1}}\,\frac{1}{2}\left(\frac{1}{n_1}-\frac{1}{n_0}\right)I_d = \rho I_d,
\end{aligned} \tag{7.29}$$

where

$$\rho \overset{\Delta}{=} \frac{n_0-n_1}{\sqrt{(n_0+n_1)(n_0+n_1+4n_0n_1)}}. \tag{7.30}$$

Now,

$$\mathbf{U}^T\mathbf{V} = \frac{1}{4}(\mathbf{U}+\mathbf{V})^T(\mathbf{U}+\mathbf{V}) - \frac{1}{4}(\mathbf{U}-\mathbf{V})^T(\mathbf{U}-\mathbf{V}). \tag{7.31}$$

It follows from the properties of Gaussian random vectors given in Section A.11 that $\mathbf{U}+\mathbf{V}$ and $\mathbf{U}-\mathbf{V}$ are Gaussian, with covariance matrices

$$\begin{aligned}
\Sigma_{\mathbf{U+V}} &= \begin{bmatrix} I_d \; I_d \end{bmatrix}\Sigma_{\mathbf{Z}}\begin{bmatrix} I_d \\ I_d \end{bmatrix} = 2(1+\rho)I_d, \\
\Sigma_{\mathbf{U-V}} &= \begin{bmatrix} I_d \; -I_d \end{bmatrix}\Sigma_{\mathbf{Z}}\begin{bmatrix} I_d \\ -I_d \end{bmatrix} = 2(1-\rho)I_d.
\end{aligned} \tag{7.32}$$

Furthermore, $\mathbf{U}+\mathbf{V}$ and $\mathbf{U}-\mathbf{V}$ are independently distributed, since

$$\Sigma_{\mathbf{U}+\mathbf{V},\mathbf{U}-\mathbf{V}} = \left[I_d \; I_d\right]\Sigma_{\mathbf{Z}}\begin{bmatrix} I_d \\ -I_d \end{bmatrix} = 0. \tag{7.33}$$

It follows that

$$R_1 = \frac{1}{2(1+\rho)}(\mathbf{U}+\mathbf{V})^T(\mathbf{U}+\mathbf{V}) \tag{7.34}$$

and

$$R_2 = \frac{1}{2(1-\rho)}(\mathbf{U}-\mathbf{V})^T(\mathbf{U}-\mathbf{V}) \tag{7.35}$$

are independent noncentral chi-square random variables, with noncentrality parameters

$$\begin{aligned}\lambda_1 &= \frac{1}{2(1+\rho)}E[\mathbf{U}+\mathbf{V}]^T E[\mathbf{U}+\mathbf{V}] \\ &= \frac{1}{2(1+\rho)}\left(E[\mathbf{U}]^T E[\mathbf{U}] + 2E[\mathbf{U}]^T E[\mathbf{V}] + E[\mathbf{V}]^T E[\mathbf{V}]\right) \\ &= \frac{n_0 n_1}{2(1+\rho)}\left(\frac{1}{\sqrt{n_0+n_1+4n_0n_1}} + \frac{1}{\sqrt{n_0+n_1}}\right)^2 \delta^2 \end{aligned} \tag{7.36}$$

and

$$\begin{aligned}\lambda_2 &= \frac{1}{2(1-\rho)}E[\mathbf{U}-\mathbf{V}]^T E[\mathbf{U}-\mathbf{V}] \\ &= \frac{1}{2(1-\rho)}\left(E[\mathbf{U}]^T E[\mathbf{U}] - 2E[\mathbf{U}]^T E[\mathbf{V}] + E[\mathbf{V}]^T E[\mathbf{V}]\right) \\ &= \frac{n_0 n_1}{2(1-\rho)}\left(\frac{1}{\sqrt{n_0+n_1+4n_0n_1}} - \frac{1}{\sqrt{n_0+n_1}}\right)^2 \delta^2, \end{aligned} \tag{7.37}$$

respectively. Finally, using the definitions of $\mathbf{U}$, $\mathbf{V}$, ρ, and (7.31) allows us to write

$$\begin{aligned} W_{\bar{L}}(S_{n_0,n_1},\mathbf{X}) \mid Y = 0 &= \frac{\sqrt{(n_0+n_1)(n_0+n_1+4n_0n_1)}}{2n_0n_1}\,\mathbf{U}^T\mathbf{V} \\ &\sim \frac{\sqrt{(n_0+n_1)(n_0+n_1+4n_0n_1)}}{2n_0n_1}\left(\frac{1+\rho}{2}R_1 - \frac{1-\rho}{2}R_2\right) \\ &= a_1R_1 - a_2R_2, \end{aligned} \tag{7.38}$$

where a_1 and a_2 are as in (7.20). The noncentrality parameters can be rewritten as in (7.21) by eliminating ρ and introducing a_1 and a_2. ■

As expected, applying the representation of $W_{\bar{L}}(S_{n_0,n_1}, \mathbf{X}) \mid Y = 0$ in the previous theorem to the computation of $E[\varepsilon^0_{n_0,n_1}]$ leads to the same expression given by John and Moran.

As in the case of Anderson's LDA discriminant, the previous representation is only a function of the Mahalanobis distance δ between populations and the sample sizes n_0 and n_1. This implies that, as before, flipping all class labels to obtain a representation for $W_{\bar{L}}(\overline{S_{n_0,n_1}}, \mathbf{X}) \mid Y = 1$ requires only switching the sample sizes. Multiplying this by -1 then produces a representation for $W_{\bar{L}}(S_{n_0,n_1}, \mathbf{X}) \mid Y = 1$, by anti-symmetry of $W_{\bar{L}}$, as described in the beginning of this section. Therefore, we have

$$W_{\bar{L}}(S_{n_0,n_1}, \mathbf{X}) \mid Y = 1 \sim \bar{a}_1\, \chi^2_d(\bar{\lambda}_1) - \bar{a}_2\, \chi^2_d(\bar{\lambda}_2), \tag{7.39}$$

where $\bar{a}_1$ and $\bar{a}_2$ are obtained from a_1 and a_2 by exchanging n_0 and n_1 and multiplying by -1, while $\bar{\lambda}_1$ and $\bar{\lambda}_2$ are obtained from λ_1 and λ_2 by exchanging n_0 and n_1. Both representations are only a function of δ, n_0, and n_1, and hence so are the error rates $E[\varepsilon^0_{n_0,n_1}]$, $E[\varepsilon^1_{n_0,n_1}]$, and $E[\varepsilon_{n_0,n_1}]$.

The exchangeability property (5.26) is satisfied, since, as was the case with Anderson's LDA discriminant, the only distributional parameter that appears in the representation is δ, which is invariant to class label flipping. Hence, $E[\varepsilon^0_{n_0,n_1}] = E[\varepsilon^1_{n_1,n_0}]$ and $E[\varepsilon^1_{n_0,n_1}] = E[\varepsilon^0_{n_1,n_0}]$ for John's LDA classification rule. If in addition $c = 0.5$, then $E[\varepsilon_{n_0,n_1}] = E[\varepsilon_{n_1,n_0}]$. Finally, for a balanced design $n_0 = n_1 = n/2$, all expressions become greatly simplified. In particular,

$$a_1 = a_2 = -\bar{a}_1 = -\bar{a}_2 = \frac{\sqrt{n+1}}{n}, \tag{7.40}$$

and $\bar{\lambda}_1 = \lambda_1$, $\bar{\lambda}_2 = \lambda_2$, so that $E[\varepsilon_{n/2,n/2}] = E[\varepsilon^0_{n/2,n/2}] = E[\varepsilon^1_{n/2,n/2}]$, regardless of the values of the prior probabilities c_0 and c_1.

7.2.1.3 John's LDA Discriminant on a Training Point

Similarly, an expression was given for $P(W_{\bar{L}}(S_{n_0,n_1}, \mathbf{X}_1) \le 0 \mid Y_1 = 0)$ under the separate-sampling case by Moran (1975), but no explicit representation was provided for $W_{\bar{L}}(S_{n_0,n_1}, \mathbf{X}_1) \mid Y_1 = 0$. Such a representation is provided below.

Theorem 7.3 (Zollanvari et al., 2009) *John's LDA discriminant applied to a training point can be represented in the separate-sampling homoskedastic Gaussian case as*

$$W_{\bar{L}}(S_{n_0,n_1}, \mathbf{X}_1) \mid Y_1 = 0 \sim a_1\, \chi^2_d(\lambda_1) - a_2\, \chi^2_d(\lambda_2), \tag{7.41}$$

for positive constants

$$
\begin{aligned}
a_1 &= \frac{1}{4n_0n_1}\left(n_0 + n_1 + \sqrt{(n_0+n_1)(n_0 - 3n_1 + 4n_0n_1)}\right), \\
a_2 &= -\frac{1}{4n_0n_1}\left(n_0 + n_1 - \sqrt{(n_0+n_1)(n_0 - 3n_1 + 4n_0n_1)}\right),
\end{aligned} \tag{7.42}
$$

where $\chi^2_d(\lambda_1)$ and $\chi^2_d(\lambda_2)$ are independent noncentral chi-square random variables with d degrees of freedom and noncentrality parameters

$$
\begin{aligned}
\lambda_1 &= \frac{n_0n_1}{2(n_0+n_1)}\left(1 + \sqrt{\frac{n_0+n_1}{4n_0n_1 + n_0 - 3n_1}}\right)\delta^2, \\
\lambda_2 &= \frac{n_0n_1}{2(n_0+n_1)}\left(1 - \sqrt{\frac{n_0+n_1}{4n_0n_1 + n_0 - 3n_1}}\right)\delta^2.
\end{aligned} \tag{7.43}
$$

Proof: The discriminant $W_{\bar{L}}(S_{n_0,n_1}, \mathbf{X}_1) \mid Y_1 = 0$ can be represented by a symmetric quadratic form

$$
W_{\bar{L}}(S_{n_0,n_1}, \mathbf{X}_1) \mid Y_1 = 0 \sim \frac{1}{2}\mathbf{Z}^T C\, \mathbf{Z}, \tag{7.44}
$$

where vector $\mathbf{Z} = (\mathbf{X}_1, \ldots, \mathbf{X}_n)$ is a vector of size $(n_0 + n_1)p$ and

$$
C = H \otimes \Sigma^{-1}, \tag{7.45}
$$

where "$\otimes$" denotes the Kronecker product of two matrices, with

$$
H = \left(\begin{array}{c|c|c}
\frac{2n_0 - 1}{n_0^2} & \frac{n_0-1}{n_0^2}\mathbf{1}^T_{n_0-1} & -\frac{1}{n_1}\mathbf{1}^T_{n_1} \\
\hline
\frac{n_0-1}{n_0^2}\mathbf{1}_{n_0-1} & -\frac{1}{n_0^2}\mathbf{1}_{(n_0-1)\times(n_0-1)} & \mathbf{0}_{(n_0-1)\times n_1} \\
\hline
-\frac{1}{n_1}\mathbf{1}_{n_1} & \mathbf{0}_{n_1\times(n_0-1)} & \frac{1}{n_1^2}\mathbf{1}_{n_1\times n_1}
\end{array}\right)_{(n_0+n_1)\times(n_0+n_1)}. \tag{7.46}
$$

In addition, it can be shown that matrix H has two nonzero eigenvalues, given by

$$
\begin{aligned}
\lambda_1 &= \frac{1}{2n_0n_1}\left(n_0 + n_1 + \sqrt{(n_0+n_1)(4n_0n_1 + n_0 - 3n_1)}\right) > 0, \\
\lambda_2 &= \frac{1}{2n_0n_1}\left(n_0 + n_1 - \sqrt{(n_0+n_1)(4n_0n_1 + n_0 - 3n_1)}\right) < 0.
\end{aligned} \tag{7.47}
$$

The eigenvectors of matrix H corresponding to these eigenvalues can be likewise determined in closed form. By using the spectral decomposition theorem, one can write the quadratic form in (7.44) as a function of these eigenvalues and eigenvectors, and the result follows. ■

As expected, applying the representation of $W_{\bar{L}}(S_{n_0,n_1}, \mathbf{X}_1) \mid Y_1 = 0$ in the previous theorem to the computation of $E[\hat{\varepsilon}^{r,0}_{n_0,n_1}]$ leads to the same expression given by Moran.

Theorem 7.3 shows that the distribution of $W_{\bar{L}}(S_{n_0,n_1}, \mathbf{X}_1) \mid Y_1 = 0$ is a function only of the Mahalanobis distance δ between populations and the sample sizes n_0 and n_1. Hence, as was the case for the true error rate, flipping all class labels to obtain a representation for $W_{\bar{L}}(\overline{S_{n_0,n_1}}, \mathbf{X}_1) \mid Y_1 = 1$ requires only switching the sample sizes, since δ is invariant to the switch. Anti-symmetry of $W_{\bar{L}}$ implies that a representation for $W_{\bar{L}}(S_{n_0,n_1}, \mathbf{X}_1) \mid Y = 1$ can be obtained by multiplying this by -1. Therefore,

$$W_{\bar{L}}(S_{n_0,n_1}, \mathbf{X}_1) \mid Y_1 = 1 \sim \bar{a}_1\, \chi^2_d(\bar{\lambda}_1) - \bar{a}_2\, \chi^2_d(\bar{\lambda}_2), \tag{7.48}$$

where $\bar{a}_1$ and $\bar{a}_2$ are obtained from a_1 and a_2 by interchanging n_0 and n_1 and multiplying by -1, whereas $\bar{\lambda}_1$ and $\bar{\lambda}_2$ are obtained from λ_1 and λ_2 by interchanging n_0 and n_1. Both representations are only a function of δ, n_0, and n_1, and hence so are the resubstitution error rates $E[\hat{\varepsilon}^{r,0}_{n_0,n_1}]$, $E[\hat{\varepsilon}^{r,1}_{n_0,n_1}]$, and $E[\hat{\varepsilon}^{r}_{n_0,n_1}]$.

In addition, the exchangeability property, c.f. (5.37), is satisfied, since the only distributional parameter that appears in the representation is δ, which is invariant to the flip. As discussed in Section 5.5.2, this implies that $E[\hat{\varepsilon}^{r,0}_{n_0,n_1}] = E[\hat{\varepsilon}^{r,1}_{n_1,n_0}]$ and $E[\hat{\varepsilon}^{r,1}_{n_0,n_1}] = E[\hat{\varepsilon}^{r,0}_{n_1,n_0}]$. In addition, for a balanced design $n_0 = n_1 = n/2$, all the expressions become greatly simplified, with

$$a_1 = a_2 = -\bar{a}_1 = -\bar{a}_2 = \frac{1 + \sqrt{n-1}}{n} \tag{7.49}$$

and $\bar{\lambda}_1 = \lambda_1$, $\bar{\lambda}_2 = \lambda_2$, so that $E[\hat{\varepsilon}^{r}_{n/2,n/2}] = E[\hat{\varepsilon}^{r,0}_{n/2,n/2}] = E[\hat{\varepsilon}^{r,1}_{n/2,n/2}]$, regardless of the values of the prior probabilities c_0 and c_1.

7.2.1.4 Bootstrapped LDA Discriminant on a Test Point As in Section 6.3.4, let $S^{\mathbf{C}_0,\mathbf{C}_1}_{n_0,n_1}$ be a bootstrap sample corresponding to multinomial vectors $\mathbf{C}_0 = (C_{01}, \ldots, C_{0n_0})$ and $\mathbf{C}_1 = (C_{11}, \ldots, C_{1n_1})$. The vectors $\mathbf{C}_0$ and $\mathbf{C}_1$ do not necessarily add up to n_0 and n_1 separately, but together add up to the total sample size n. This allows one to apply the results in this section to the computation of the expectations needed in (5.63) to obtain $E[\hat{\varepsilon}^{\,\mathrm{zero}}_n]$ in the mixture-sample multivariate case.

In order to ease notation, we assume as before that the sample indices are sorted such that $Y_1 = \cdots = Y_{n_0} = 0$ and $Y_{n_0+1} = \cdots = Y_{n_0+n_1} = 1$, without loss of generality. John's LDA discriminant applied to $S_{n_0,n_1}^{\mathbf{C}_0,\mathbf{C}_1}$ can be written as

$$W_{\bar{L}}\left(S_{n_0,n_1}^{\mathbf{C}_0,\mathbf{C}_1},\mathbf{x}\right) = \left(\mathbf{x} - \bar{\mathbf{X}}^{C_0,C_1}\right)^T \Sigma^{-1}\left(\bar{\mathbf{X}}_0^{C_0} - \bar{\mathbf{X}}_1^{C_1}\right), \tag{7.50}$$

where

$$\bar{\mathbf{X}}_0^{\mathbf{C}_0} = \frac{\sum_{i=1}^{n_0} C_{0i}\mathbf{X}_i}{\sum_{i=1}^{n_0} C_{0i}} \quad \text{and} \quad \bar{\mathbf{X}}_1^{\mathbf{C}_1} = \frac{\sum_{i=1}^{n_1} C_{1i}\mathbf{X}_{n_0+i}}{\sum_{i=1}^{n_1} C_{1i}} \tag{7.51}$$

are bootstrap sample means, and $\bar{\mathbf{X}}^{\mathbf{C}_0,\mathbf{C}_1} = \frac{1}{2}(\bar{\mathbf{X}}_0^{\mathbf{C}_0} + \bar{\mathbf{X}}_1^{\mathbf{C}_1})$.

The next theorem extends Theorem 7.2 to the bootstrap case.

Theorem 7.4 *(Vu et al., 2014) John's LDA discriminant applied to a bootstrap sample and a test point can be represented in the separate-sampling homoskedastic Gaussian case as*

$$W_{\bar{L}}\left(S_{n_0,n_1}^{\mathbf{C}_0,\mathbf{C}_1},\mathbf{X}\right) \mid \mathbf{C}_0,\mathbf{C}_1, Y=0 \;\sim\; a_1\,\chi_d^2(\lambda_1) - a_2\,\chi_d^2(\lambda_2), \tag{7.52}$$

for positive constants

$$\begin{aligned} a_1 &= \frac{1}{4}\left(s_1 - s_0 + \sqrt{(s_0+s_1)(s_0+s_1+4)}\right),\\ a_2 &= \frac{1}{4}\left(s_0 - s_1 + \sqrt{(s_0+s_1)(s_0+s_1+4)}\right), \end{aligned} \tag{7.53}$$

where $\chi_d^2(\lambda_1)$ and $\chi_d^2(\lambda_2)$ are independent noncentral chi-square random variables with d degrees of freedom and noncentrality parameters

$$\begin{aligned} \lambda_1 &= \frac{1}{8a_1}\frac{\left(\sqrt{s_0+s_1} + \sqrt{s_0+s_1+4}\right)^2}{\sqrt{(s_0+s_1)(s_0+s_1+4)}}\,\delta^2,\\ \lambda_2 &= \frac{1}{8a_2}\frac{\left(\sqrt{s_0+s_1} - \sqrt{s_0+s_1+4}\right)^2}{\sqrt{(s_0+s_1)(s_0+s_1+4)}}\,\delta^2, \end{aligned} \tag{7.54}$$

with

$$s_0 = \frac{\sum_{i=1}^{n_0} C_{0i}^2}{(\sum_{i=1}^{n_0} C_{0i})^2} \quad \textit{and} \quad s_1 = \frac{\sum_{i=1}^{n_1} C_{1i}^2}{(\sum_{i=1}^{n_1} C_{1i})^2}. \tag{7.55}$$

Proof: The argument proceeds as in the proof of Theorem 7.2. First, the same transformation applied in the proof of that theorem is applied to reduce the populations to

$$\Pi_0 \sim N(\nu_0, I_d) \quad \text{and} \quad \Pi_1 \sim N(\nu_1, I_d), \tag{7.56}$$

where

$$\nu_0 = \left(\frac{\delta}{2}, 0, \dots, 0\right) \quad \text{and} \quad \nu_1 = \left(-\frac{\delta}{2}, 0, \dots, 0\right). \tag{7.57}$$

Next, we define the random vectors

$$\mathbf{U} = \frac{2}{\sqrt{s_0 + s_1 + 4}}\left(\mathbf{X} - \bar{\mathbf{X}}^{\mathbf{C}_0,\mathbf{C}_1}\right), \quad \mathbf{V} = \frac{1}{\sqrt{s_0 + s_1}}\left(\bar{\mathbf{X}}_0^{\mathbf{C}_0} - \bar{\mathbf{X}}_1^{\mathbf{C}_1}\right). \tag{7.58}$$

For $Y = 0$, $\mathbf{U}$ and $\mathbf{V}$ are Gaussian random vectors such that

$$\mathbf{U} \sim N\left(\frac{2\nu_0}{\sqrt{s_0 + s_1 + 4}}, I_d\right), \quad \mathbf{V} \sim N\left(\frac{\nu_0 - \nu_1}{\sqrt{s_0 + s_1}}, I_d\right). \tag{7.59}$$

In addition, the vector $\mathbf{Z} = (\mathbf{U}, \mathbf{V})$ is jointly Gaussian, with covariance matrix

$$\Sigma_{\mathbf{Z}} = \begin{bmatrix} I_d & \rho I_d \\ \rho I_d & I_d \end{bmatrix}, \tag{7.60}$$

where

$$\rho = \frac{s_1 - s_0}{\sqrt{(s_0 + s_1)(s_0 + s_1 + 4)}}. \tag{7.61}$$

Now we write

$$W_{\bar{L}}\left(S_{n_0,n_1}^{\mathbf{C}_0,\mathbf{C}_1}, \mathbf{X}\right) = \frac{\sqrt{(s_0 + s_1)(s_0 + s_1 + 4)}}{2}\, \mathbf{U}^T\mathbf{V}, \tag{7.62}$$

and the rest of the proof proceeds exactly as in the proof of Theorem 7.2. ■

It is easy to check that the result in Theorem 7.4 above reduces to the one in Theorem 7.2 when $\mathbf{C}_0 = \mathbf{1}_{n_0}$ and $\mathbf{C}_1 = \mathbf{1}_{n_1}$ (i.e., the bootstrap sample equals the original sample).

The previous representation is only a function of the Mahalanobis distance δ between the populations and the parameters s_0 and s_1. This implies that flipping all class labels to obtain a representation for $W_{\bar{L}}(S_{n_0,n_1}^{\overline{\mathbf{C}_0,\mathbf{C}_1}}, \mathbf{X}) \mid Y = 1$ requires only switching s_0 and s_1 (i.e., switching the sample sizes and switching $\mathbf{C}_0$ and $\mathbf{C}_1$). By anti-symmetry of the discriminant, we conclude that multiplying this by -1 produces a representation of $W_{\bar{L}}(S_{n_0,n_1}^{\mathbf{C}_0,\mathbf{C}_1}, \mathbf{X}) \mid Y = 1$. Therefore,

$$W_{\bar{L}}\left(S_{n_0,n_1}^{\mathbf{C}_0,\mathbf{C}_1}, \mathbf{X}\right) \mid Y = 1 \sim \bar{a}_1\, \chi_d^2(\bar{\lambda}_1) - \bar{a}_2\, \chi_d^2(\bar{\lambda}_2), \tag{7.63}$$

where $\bar{a}_1$ and $\bar{a}_2$ are obtained from a_1 and a_2 by exchanging s_0 and s_1 and multiplying by -1, while $\bar{\lambda}_1$ and $\bar{\lambda}_2$ are obtained from λ_1 and λ_2 by exchanging s_0 and s_1. Section 7.2.2.1 illustrates the application of this representation, as well as the previous two, in obtaining the weight for unbiased convex bootstrap error estimation for John's LDA discriminant.

7.2.1.5 QDA Discriminant on a Test Point In a remarkable set of two papers, McFarland and Richards provide a statistical representation for the QDA discriminant for the equal-means case (McFarland and Richards, 2001) and the general heteroskedastic case (McFarland and Richards, 2002). We give below the latter result, the proof of which is provided in detail in the cited reference.

Theorem 7.5 *(McFarland and Richards, 2002) Assume that $n_0 + n_1 > d$. Consider a vector $\xi = (\xi_1, \dots, \xi_d)^T$, a diagonal matrix $\Lambda = \mathrm{diag}(\lambda_1, \dots, \lambda_d)$, and an orthogonal $d \times d$ matrix U such that*

$$\begin{aligned} \Sigma_1^{-1/2}\Sigma_0\Sigma_1^{-1/2} &= U\Lambda U^T, \\ \xi &= U^T\Sigma_1^{-1/2}(\boldsymbol{\mu}_0 - \boldsymbol{\mu}_1). \end{aligned} \tag{7.64}$$

The QDA discriminant applied to a test point can be represented in the heteroskedastic Gaussian case as

$$\begin{aligned} &W_Q(S_{n_0,n_1}, \mathbf{X}) \,|\, Y = 0 \sim \\ &\frac{1}{2}\sum_{i=1}^{d}\left[\frac{(n_1-1)(n_1\lambda_i+1)}{n_1 T_1}\left(\omega_i Z_{1i} + \sqrt{1-\omega_i^2}\, Z_{2i} + \gamma_i\right)^2 - \frac{(n_0-1)(n_0+1)}{n_0 T_0} Z_{1i}^2\right] \\ &+\frac{1}{2}\left[\log\left(\frac{(n_0-1)^d\, T_1}{(n_1-1)^d\, T_0}\right) - \log|\Sigma_1^{-1}\Sigma_0| + \sum_{i=1}^{d-1}\log\left(\frac{n_1-i}{n_0-i}F_i\right)\right], \end{aligned} \tag{7.65}$$

where

$$\omega_i = \sqrt{\frac{n_0 n_1 \lambda_i}{(n_0+1)(\lambda_i n_1 + 1)}}, \qquad \gamma_i = \sqrt{\frac{n_1}{\lambda_i n_1 + 1}}\, \xi_i, \tag{7.66}$$

and $T_0, T_1, Z_{11}, \dots, Z_{1d}, Z_{21}, \dots, Z_{2d}, F_1, \dots, F_{d-1}$ are mutually independent random variables, such that T_i is chi-square distributed with $n_i - d$ degrees of freedom, for

$i = 0, 1$, $Z_{11}, \ldots, Z_{1d}, Z_{21}, \ldots, Z_{2d}$ are zero-mean, unit-variance Gaussian random variables, and F_i is F-distributed with $n_1 - i$ and $n_0 - i$ degrees of freedom, for $i = 1, \ldots, d-1$.

Notice that the distribution of $W_Q(S_{n_0,n_1}, \mathbf{X}) \mid Y = 0$ is a function of the sample sizes n_0 and n_1 and the distributional parameters Λ and ξ. The latter can be understood as follows. The QDA discriminant is invariant to a nonsingular affine transformation $X' = A\mathbf{X} + b$ to all sample vectors. That is, if $S' = \{(A\mathbf{X}_1 + b, Y_1), \ldots, (A\mathbf{X}_n + b, Y_n)\}$, then $W_Q(S', A\mathbf{X} + b) = W_Q(S, \mathbf{X})$, as can be easily seen from (7.5). Consider the nonsingular affine transformation

$$\mathbf{X}' = U^T \Sigma_1^{-\frac{1}{2}} \left(\mathbf{X} - \boldsymbol{\mu}_1 \right). \tag{7.67}$$

Clearly, the transformed Gaussian populations are given by

$$\Pi'_0 \sim N_d(\xi, \Lambda) \quad \text{and} \quad \Pi'_1 \sim N_d(\mathbf{0}, I_d). \tag{7.68}$$

Hence, the parameter ξ indicates the distance between the means, while Λ specifies how dissimilar the covariance matrices are. Due to the invariance of the QDA discriminant to affine transformations, one can proceed with these transformed populations, without loss of generality, and this is actually what is done in the proof of Theorem 7.5. However, unlike the Mahalanobis distance δ in the linear homoskedastic case, the parameters ξ and Λ are not invariant to a switch of the populations. In fact, it can be shown (see Hua et al. (2005a) for a detailed proof) that the parameters $\bar{\xi}$ and $\bar{\Lambda}$ after the switch of populations are related to the original parameters via

$$\bar{\Lambda} = \Lambda^{-1} \quad \text{and} \quad \bar{\xi} = -\Lambda^{-\frac{1}{2}} \xi. \tag{7.69}$$

Therefore, obtaining a representation for $W_Q(\overline{S_{n_0,n_1}}, \mathbf{X}) \mid Y = 1$ requires switching the sample sizes and replacing Λ and ξ by Λ^{-1} and $\Lambda^{-1}\xi$, respectively, in the expressions appearing in Theorem 7.5. Anti-symmetry of W_Q then implies that a representation for $W_Q(S_{n_0,n_1}, \mathbf{X}) \mid Y = 1$ can be obtained by multiplication by -1. Unfortunately, the exchangeability property (5.26) does not hold in general, since the distributional parameters are not invariant to the population shift. This means that the nice properties one is accustomed to in the homoskedastic LDA case are not true in the heteroskedastic QDA case. For example, in the balanced design case, $n_0 = n_1 = n/2$, it is not true for the QDA rule, in general, that $E[\varepsilon] = E[\varepsilon^0_{n/2,n/2}] = E[\varepsilon^1_{n/2,n/2}]$.

7.2.2 Computational Methods

In this section, we show how the statistical representations given in the previous subsections can be useful for Monte-Carlo computation of the error moments and error distribution based on the discriminant distribution. The basic idea is to simulate a large number of realizations of the various random variables in the representation to obtain an empirical distribution of the discriminant. Some of the random variables encountered in these representations, such as doubly-noncentral F variables, are not easy to simulate from and approximation methods are required. In addition, approximation is required to compute multivariate orthant probabilities necessary for high-order moments and sampling distributions.

7.2.2.1 Expected Error Rates Monte-Carlo computation of expected error rates can be accomplished by using the statistical representations given earlier, provided that it is possible to sample from the random variables appearing in the representations. For example, in order to use Theorem 7.1 to compute the expected true error rate $E[\varepsilon^0_{n_0,n_1}] = P(W_L(S_{n_0,n_1}, \mathbf{X}) \leq 0 \,|\, Y = 0)$ for Anderson's LDA discriminant, one needs to sample from Gaussian, *t*, and central chi-square distributions, which is fairly easy to do (for surveys on methods for sampling from such random variables, see Johnson et al. (1994, 1995)).

On the other hand, computing the expected true error rate, expected resubstitution error rate, and expected bootstrap error rate for John's LDA discriminant, using Theorems 7.2, 7.3, 7.4, respectively, involves simulating from a doubly noncentral F distribution, which is not a trivial task. For example, Moran proposes in Moran (1975) a complex procedure to address this computation, based on work by Price (1964), which only applies to even dimensionality d. We describe below a simpler approach to this problem, called the *Imhof-Pearson three-moment method*, which is applicable to even and odd dimensionality (Imhof, 1961). This consists of approximating a noncentral $\chi^2_d(\lambda)$ random variable with a central χ^2_h random variable by equating the first three moments of their distributions. This approach was employed in Zollanvari et al. (2009, 2010), where it was found to be very accurate.

To fix ideas, consider the expectation of the resubstitution error rate for John's LDA discriminant. We get from Theorem 7.3 that

$$\begin{aligned} E\left[\hat{\varepsilon}^{r,0}_{n_0,n_1}\right] &= P(W_{\bar{L}}(S_{n_0,n_1}, \mathbf{X}_1) \leq 0 \mid Y_1 = 0) \\ &= P\left(a_1\, \chi^2_d(\lambda_1) - a_2\, \chi^2_d(\lambda_2) \leq 0\right) = P\left(\frac{\chi^2_d(\lambda_1)}{\chi^2_d(\lambda_2)} < \frac{a_2}{a_1}\right). \end{aligned} \tag{7.70}$$

The ratio of two noncentral chi-square random variables in the previous equation is distributed as a doubly noncentral F variable. The Imhof-Pearson three-moment approximation to this distribution gives

$$E\left[\hat{\varepsilon}^{r,0}_{n_0,n_1}\right] = P\left(\frac{\chi^2_d(\lambda_1)}{\chi^2_d(\lambda_2)} < \frac{a_2}{a_1}\right) \approx P\left(\chi^2_{h_0} < y_0\right), \tag{7.71}$$

where $\chi^2_{h_0}$ is a central chi-square random variable with h_0 degrees of freedom, with

$$\begin{aligned} h_0 &= \frac{r_2^3}{r_3^2}, \\ y_0 &= h_0 - r_1\sqrt{\frac{h_0}{r_2}}, \end{aligned} \tag{7.72}$$

and

$$r_i = a_1^i (d + i\lambda_1) + a_2^i (d + i\lambda_2), \quad i = 1, 2, 3. \tag{7.73}$$

The approximation is valid only for $r_3 > 0$ (Imhof, 1961). However, it can be shown that in this case $r_3 = f_1(n_0, n_1)d + f_2(n_0, n_1)\delta^2$, where $f_1(n_0, n_1) > 0$ and $f_2(n_0, n_1) > 0$, so that it is always the case that $r_3 > 0$ (Zollanvari et al., 2009).

For $E[\hat{\varepsilon}^{r,1}_{n_0,n_1}]$, we use (7.48) and obtain the approximation

$$E\left[\hat{\varepsilon}^{r,1}_{n_0,n_1}\right] = P\left(\frac{\chi_d^2(\bar{\lambda}_1)}{\chi_d^2(\bar{\lambda}_2)} < \frac{\bar{a}_2}{\bar{a}_1}\right) \approx P\left(\chi^2_{h_1} < y_1\right), \tag{7.74}$$

with

$$h_1 = \frac{s_2^3}{s_3^2},$$

$$y_1 = h_1 + s_1\sqrt{\frac{h_1}{s_2}}, \tag{7.75}$$

and

$$s_i = \bar{a}_1^i (d + i\bar{\lambda}_1) + \bar{a}_2^i (d + i\bar{\lambda}_2), \quad i = 1, 2, 3. \tag{7.76}$$

However, similarly as before, it can be shown (Zollanvari et al., 2009) that $s_3 = g_1(n_0, n_1)d + g_2(n_0, n_1)\delta^2$, where $g_1(n_0, n_1) > 0$ and $g_2(n_0, n_1) > 0$, so that it is always the case that $s_3 > 0$, and the approximation applies.

As an illustration of results in this chapter so far, and their application to the mixture-sampling case, let us compute the weight for unbiased convex bootstrap error estimation, given by (5.67), for John's LDA discriminant. Given a multivariate homoskedastic Gaussian model, one can compute $E[\varepsilon_n]$ by using (5.23), (5.25), and Theorem 7.2; similarly, one can obtain $E[\hat{\varepsilon}_n^r]$ by using (5.30), (5.32), (5.36), and Theorem 7.3; finally, one can find $E[\hat{\varepsilon}_n^{\text{zero}}]$ by using (5.63), (5.64), and Theorem 7.4. In each case, the Imhof-Pearson three-moment approximation method is used to compute the expected error rates from the respective statistical distributions.

As in the univariate case, in the homoskedastic, balanced-sampling, multivariate Gaussian case all expected error rates become functions only of n and δ, and therefore so does the weight w^*. As the Bayes error is $\varepsilon_{\text{bay}} = \Phi(-\delta/2)$, there is a one-to-one correspondence between Bayes error and the Mahalanobis distance, and the weight w^* is a function only of the Bayes error ε_{bay} and the sample size n, being insensitive to the prior probabilities c_0 and c_1.

Figure 7.1 displays the value of w^* computed with the previous expressions in the homoskedastic, balanced-sampling, bivariate Gaussian case ($d = 2$) for several

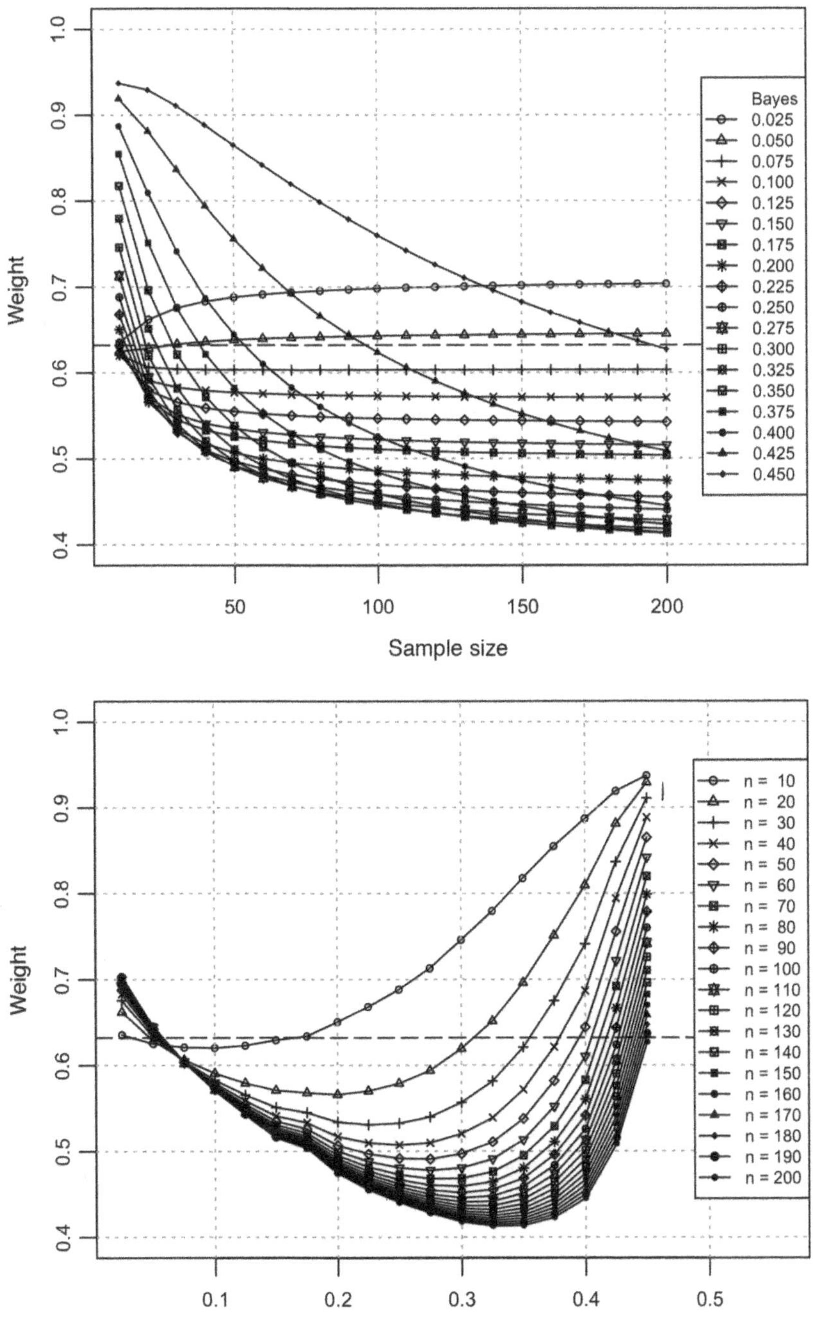

FIGURE 7.1 Required weight w^* for unbiased convex bootstrap estimation plotted against sample size and Bayes error in the homoskedastic, balanced-sampling, bivariate Gaussian case.

sample sizes and Bayes errors. The values are exact up to the (accurate) Imhof-Pearson approximation. We can see in Figure 7.1 that there is considerable variation in the value of w^* and it can be far from the heuristic 0.632 weight; however, as the sample size increases, w^* appears to settle around an asymptotic fixed value. In contrast to the univariate case, these asymptotic values here seem to be strongly dependent on the Bayes error and significantly smaller than the heuristic 0.632, except for very small Bayes errors. As in the univariate case, convergence to the asymptotic value is faster for smaller Bayes errors. These facts again help explain the good performance of the original convex .632 bootstrap error estimator for moderate sample sizes and small Bayes errors.

7.2.2.2 Higher-Order Moments and Sampling Distributions As mentioned earlier, computation of higher-order moments and sampling distributions require multivariate orthant probabilities involving the discriminant. However, computing an orthant probability such as $P(W(\mathbf{X}_1) \leq 0, \dots, W(\mathbf{X}_k) \leq 0)$ requires a statistical representation for the vector $(W(\mathbf{X}_1), \dots, W(\mathbf{X}_k))$, which is currently not available. The approach discussed in this section obtains simple and accurate approximations of these multivariate probabilities by using the marginal univariate ones.

For instance, consider the sampling distribution of the mixture-sampling resubstitution estimator of the overall error rate in Theorem 5.4. We propose to approximately compute the required multivariate orthant probabilities by using univariate ones. The idea is to assume that the random variables $W(\mathbf{X}_i)$ are only weakly dependent, so that the joint probability "factors," to a good degree of accuracy. Evidence is presented in Zollanvari et al. (2009) that this approximation can be accurate at small sample sizes. The approximation is given by

$$\begin{aligned}
P(W(\mathbf{X}_1) &\leq 0, \dots, W(\mathbf{X}_l) \leq 0, W(\mathbf{X}_{l+1}) > 0, \dots, W(\mathbf{X}_{n_0}) > 0,\\
&W(\mathbf{X}_{n_0+1}) > 0, \dots, W(\mathbf{X}_{n_0+k-l}) > 0, W(\mathbf{X}_{n_0+k-l+1}) \leq 0, \dots,\\
&W(\mathbf{X}_n) \leq 0 \mid Y_1 = \cdots = Y_l = 0, Y_{l+1} = \cdots = Y_{n_0+k-l} = 1,\\
&Y_{n_0+k-l+1} = \cdots = Y_n = 0)\\
&\approx p^l(1-p)^{(n_0-l)} q^{(k-l)} (1-q)^{n_1-(k-l)}, \qquad (7.77)
\end{aligned}$$

where $p = P(W(\mathbf{X}_1) \leq 0 \mid Y_1 = 0)$ and $q = P(W(\mathbf{X}_1) > 0 \mid Y_1 = 1)$ are computed by the methods described earlier. The approximate sampling distribution thus becomes

$$\begin{aligned}
P\left(\hat{\varepsilon}_n^r = \frac{k}{n}\right) &\approx \sum_{n_0=0}^{n} \binom{n}{n_0} c_0^{n_0} c_1^{n_1} \sum_{l=0}^{k} \binom{n_0}{l} \binom{n_1}{k-l}\\
&\times p^l(1-p)^{(n_0-l)} q^{(k-l)} (1-q)^{n_1-(k-l)}, \qquad (7.78)
\end{aligned}$$

for $k = 0, 1, \dots, n$.

This "quasi-binomial" distribution is a sum of two independent binomial random variables $U \sim B(n_0, p)$ and $V \sim B(n_1, q)$, where U and V are the number of errors in class 0 and 1, respectively. Butler and Stephens give approximations to the cumulative probabilities of such distributions (Butler and Stephens, 1993) that can be useful to compute the error rate cumulative distribution function $P(\hat{\varepsilon}_n^r \leq \frac{k}{n})$ for large n_0 and n_1.

It is easy to check that if $p = q$ (e.g., for an LDA discriminant under a homoskedastic Gaussian model and a balanced design), then the approximation to $P(\hat{\varepsilon}_n^r = \frac{k}{n})$ reduces to a plain binomial distribution:

$$P\left(\hat{\varepsilon}_n^r = \frac{k}{n}\right) \simeq \binom{n}{k} p^k (1-p)^{n-k}. \tag{7.79}$$

From the mean and variance of single binomial random variables, one directly obtains from this approximation that

$$\begin{aligned} E\left[\hat{\varepsilon}_n^r\right] &= p, \\ \operatorname{Var}\left[\hat{\varepsilon}_n^r\right] &\approx \frac{1}{n} p(1-p). \end{aligned} \tag{7.80}$$

The expression obtained for $E[\hat{\varepsilon}_n^r]$ is in fact exact, while the expression for $\operatorname{Var}[\hat{\varepsilon}_n^r]$ coincides with the approximation obtained by Foley (1972).

7.3 LARGE-SAMPLE METHODS

Classical asymptotic theory treats the behavior of a statistic as the sample size goes to infinity for fixed dimensionality and number of parameters. In the case of Pattern Recognition, there are a few such results concerning error rates for LDA under multivariate Gaussian populations, the most well-known of which is perhaps the classical work of Okamoto (1963). Another well-known example is the asymptotic bound on the bias of resubstitution given by McLachlan (1976). An extensive survey of these approaches is provided by McLachlan (1992) (see also Wyman et al. (1990)).

In this section, we focus instead on a less well-known large-sample approach: the *double-asymptotic method*, invented independently by S. Raudys and A. Kolmogorov (Raudys and Young, 2004; Serdobolskii, 2000) in 1967, which considers the behavior of a statistic as both sample size and dimensionality (or, more generally, number of parameters) increase to infinity in a controlled fashion, where the ratio between sample size and dimensionality converges to a finite constant (Deev, 1970; Meshalkin and Serdobolskii, 1978; Raudys, 1972, 1997; Raudys and Pikelis, 1980; Raudys and Young, 2004; Serdobolskii, 2000). This "increasing dimension limit" is also known, in the Machine Learning literature, as the "thermodynamic limit" (Haussler et al., 1994; Raudys and Young, 2004). The finite-sample approximations obtained via these asymptotic expressions have been shown to be remarkably accurate in small-sample cases, outperforming the approximations obtained by classical asymptotic methods (Pikelis, 1976; Wyman et al., 1990).

The double-asymptotic approach is formalized as follows. Under separate sampling, we assume that $n_0 \to \infty$, $n_1 \to \infty$, $p \to \infty$, $p/n_0 \to \lambda_0$, $p/n_1 \to \lambda_1$, where p is the dimensionality (for historical reasons, "p," rather than "d," is used to denote dimensionality throughout this section), and λ_0 and λ_1 are positive constants. These are called *Raudys-Kolmogorov asymptotic conditions* in Zollanvari et al. (2011). See Appendix C for the definitions of convergence of double ordinary and random sequences under the aforementioned double-asymptotic conditions, as well as associated results needed in the sequel.

The limiting linear ratio conditions $p/n_0 \to \lambda_0$ and $p/n_1 \to \lambda_1$ are quite natural, for several practical reasons. First, the ratio of dimensionality to sample size serves as a measure of complexity for the LDA classification rule (in fact, any linear classification rule): in this case, the VC dimension is $p + 1$ (c.f. Section B.3.1), so that p/n, where $n = n_0 + n_1$, gives the asymptotic relationship between the VC dimension and the sample size. In addition, for the special case of LDA with a known covariance matrix (John's Discriminant) and $n_0 = n_1$, it has long been known that if the features contribute equally to the Mahalanobis distance, then the optimal number of features is $p_{\mathrm{opt}} = n - 1$, so that choosing the optimal number of features leads to $p/n_k \to$ constant, for $k = 0, 1$ (Jain and Waller, 1978). Finally, a linear relation between the optimal number of features and the sample size has been observed empirically in the more general setting of LDA with unknown covariance matrix (Anderson's discriminant) assuming a blocked covariance matrix with identical off-diagonal correlation coefficient ρ; for instance, with $\rho = 0.125$, one has $p_{\mathrm{opt}} \approx n/3$, whereas with $\rho = 0.5$, one has $p_{\mathrm{opt}} \approx n/25$ (Hua et al., 2005b). Notice that the stronger constraint $p/n_0 = \lambda_0$ and $p/n_1 = \lambda_1$, for all p, n_0, n_1, is a special case of the more general assumption $p/n_0 \to \lambda_0$ and $p/n_1 \to \lambda_1$ made here.

The methods discussed in this section are used to derive "double-asymptotically exact" expressions for first- and second-order moments, including mixed moments, of the true, resubstitution, and leave-one-out error rates of LDA discriminants under multivariate Gaussian populations with a common covariance matrix, which appear in Deev (1970, 1972), Raudys (1972, 1978), Zollanvari et al. (2011). We focus on results for John's LDA discriminant. Expected error rates are studied in the next section by means of a double-asymptotic version of the Central Limit Theorem, proved in Appendix C — see Zollanvari et al. (2011) for an alternative approach. In Section 7.3.2, we will review results on second-order moments of error rates, which are derived in full detail in Zollanvari et al. (2011).

7.3.1 Expected Error Rates

7.3.1.1 True Error Consider a sequence of Gaussian discrimination problems of increasing dimensionality, such that $S^p_{n_0,n_1} = \{(\mathbf{X}^p_1, Y^p_1), \ldots, (\mathbf{X}^p_n, Y^p_n)\}$ is a sample consisting of n_0 points from a population $\Pi^p_0 \sim N_p(\boldsymbol{\mu}^p_0, \Sigma_p)$ and n_1 points from a population $\Pi^p_1 \sim N_p(\boldsymbol{\mu}^p_1, \Sigma_p)$, for $p = 1, 2, \ldots$

The distributional parameters $\boldsymbol{\mu}_0^p, \boldsymbol{\mu}_1^p, \Sigma_p$, for $p = 1, 2, \ldots$, are completely arbitrary, except for the requirement that the Mahalanobis distance between populations

$$\delta_p = \sqrt{(\boldsymbol{\mu}_1^p - \boldsymbol{\mu}_0^p)^T \Sigma_p^{-1} (\boldsymbol{\mu}_1^p - \boldsymbol{\mu}_0^p)} \tag{7.81}$$

must satisfy $\lim_{p\to\infty} \delta_p = \delta$, for a limiting Mahalanobis distance $\delta \geq 0$.

John's LDA discriminant is given by

$$W_{\bar{L}}\left(S_{n_0,n_1}^p, \mathbf{x}^p\right) = \left(\mathbf{x}^p - \bar{\mathbf{X}}^p\right)^T \Sigma_p^{-1}\left(\bar{\mathbf{X}}_0^p - \bar{\mathbf{X}}_1^p\right), \tag{7.82}$$

where $\mathbf{x}^p \in R^p$ (the superscript "p" will be used to make the dimensionality explicit), $\bar{\mathbf{X}}_0^p = \frac{1}{n_0}\sum_{i=1}^n \mathbf{X}_i^p I_{Y_i=0}$ and $\bar{\mathbf{X}}_1^p = \frac{1}{n_1}\sum_{i=1}^n \mathbf{X}_i^p I_{Y_i=1}$ are the sample means, and $\bar{\mathbf{X}}^p = \frac{1}{2}(\bar{\mathbf{X}}_0^p + \bar{\mathbf{X}}_1^p)$.

The exact distributions of $W_{\bar{L}}(S_{n_0,n_1}^p, \mathbf{X}^p) \mid Y^p = 0$ and $W_{\bar{L}}(S_{n_0,n_1}^p, \mathbf{X}^p) \mid Y^p = 1$ for finite n_0, n_1, and p are given by the difference of two noncentral chi-square random variables, as shown in Section 7.2.1.2. However, that still leaves one with a difficult computation to obtain $E[\varepsilon_{n_0,n_1}^0] = P(W_{\bar{L}}(S_{n_0,n_1}^p, \mathbf{X}^p) \leq 0 \mid Y^p = 0)$ and $E[\varepsilon_{n_0,n_1}^1] = P(W_{\bar{L}}(S_{n_0,n_1}^p, \mathbf{X}^p) > 0 \mid Y^p = 1)$, as these involve the cumulative distribution function of a doubly noncentral F random variable.

An alternative is to consider asymptotic approximations. The asymptotic distributions of $W_{\bar{L}}(S_{n_0,n_1}^p, \mathbf{X}^p) \mid Y^p = 0$ and $W_{\bar{L}}(S_{n_0,n_1}^p, \mathbf{X}^p) \mid Y^p = 1$ as $n_0, n_1 \to \infty$, with p fixed, are the univariate Gaussian distributions $N\left(\frac{1}{2}\delta_p^2, \delta_p^2\right)$ and $N\left(-\frac{1}{2}\delta_p^2, \delta_p^2\right)$, respectively (this is true also in the case of Anderson's LDA discriminant (Anderson, 1984)). These turn out to be the distributions of $D_L(\mathbf{X}^p) \mid Y^p = 0$ and $D_L(\mathbf{X}^p) \mid Y^p = 1$, respectively, where $D_L(\mathbf{X}^p)$ is the population-based discriminant in feature space R^p, defined in Section 1.2; this result was first established in Wald (1944). However, this does not yield useful approximations for $E[\varepsilon_{n_0,n_1}^0]$ and $E[\varepsilon_{n_0,n_1}^1]$ for finite n_0 and n_1.

The key to obtaining accurate approximations is to assume that both the sample sizes and the dimensionality go to infinity simultaneously, that is, to consider the double-asymptotic distribution of the discriminant, which is Gaussian, as given by the next theorem. This result is an application of a double-asymptotic version of the Central Limit Theorem (c.f. Theorem C.5 in Appendix C).

Theorem 7.6 *As $n_0, n_1, p \to \infty$, at rates $p/n_0 \to \lambda_0$ and $p/n_1 \to \lambda_1$,*

$$W_{\bar{L}}\left(S_{n_0,n_1}^p, \mathbf{X}^p\right) \mid Y^p = i \xrightarrow{D} N\left((-1)^i \frac{\delta^2}{2} + \frac{1}{2}\left(\lambda_1 - \lambda_0\right),\ \delta^2 + \lambda_0 + \lambda_1\right), \tag{7.83}$$

for $i = 0, 1$.

Proof: First we apply the affine transformation used in the proof of Theorem 7.2, for each $p = 1, 2, \ldots$, to obtain the transformed populations

$$\Pi_0^p \sim N\left(\nu_0^p, I_p\right) \quad \text{and} \quad \Pi_1^p \sim N\left(\nu_1^p, I_p\right), \quad p = 1, 2, \ldots, \tag{7.84}$$

where

$$\nu_0^p = \left(\frac{\delta_p}{2}, 0, \ldots, 0\right) \quad \text{and} \quad \nu_1^p = \left(-\frac{\delta_p}{2}, 0, \ldots, 0\right). \tag{7.85}$$

Due to the invariance of John's LDA discriminant to affine transformations, one can proceed with these transformed populations without loss of generality.

To establish (7.83), consider the case $Y^p = 0$ (the other case being entirely analogous). We can write

$$W_{\bar{L}}\left(S_{n_0,n_1}^p, \mathbf{X}^p\right) = U_{n_0,n_1,p} V_{n_0,n_1,p} + \sum_{k=2}^{p} T_{n_0,n_1,k}, \tag{7.86}$$

where, with $\mathbf{X}^p = (X_1^p, \ldots, X_p^p)$, $\bar{\mathbf{X}}^p = (\bar{X}_1^p, \ldots, \bar{X}_p^p)$, $\bar{\mathbf{X}}_0^p = (\bar{X}_{01}^p, \ldots, \bar{X}_{0p}^p)$, and $\bar{\mathbf{X}}_1^p = (\bar{X}_{11}^p, \ldots, \bar{X}_{1p}^p)$,

$$\begin{aligned}
U_{n_0,n_1,p} &= \left(X_1^p - \bar{X}_1^p\right) \sim N\left(\frac{\delta_p}{2}, 1 + \frac{1}{4}\left(\frac{1}{n_0} + \frac{1}{n_1}\right)\right), \\
V_{n_0,n_1,p} &= \left(\bar{X}_{01}^p - \bar{X}_{11}^p\right) \sim N\left(\delta_p, \frac{1}{n_0} + \frac{1}{n_1}\right), \\
T_{n_0,n_1,k} &= \left(X_k^p - \bar{X}_k^p\right)\left(\bar{X}_{0k}^p - \bar{X}_{1k}^p\right), \quad k = 2, \ldots, p.
\end{aligned} \tag{7.87}$$

Note that $U_{n_0,n_1,p} V_{n_0,n_1,p}$ is independent of $T_{n_0,n_1,k}$ for each $k = 2, \ldots, p$, and that $T_{n_0,n_1,k}$ is independent of $T_{n_0,n_1,l}$ for $k \neq l$. In addition, as $n_0, n_1, p \to \infty$,

$$\begin{aligned}
U_{n_0,n_1,p} &\xrightarrow{D} N\left(\frac{\delta}{2}, 1\right), \\
V_{n_0,n_1,p} &\xrightarrow{D} \delta.
\end{aligned} \tag{7.88}$$

In particular, Corollary C.1 in Appendix C implies that convergence also happens as $n_0, n_1, p \to \infty$ at rates $p/n_0 \to \lambda_0$ and $p/n_1 \to \lambda_1$, for any given $\lambda_0 > 0$ and $\lambda_1 > 0$. By combining this with the double-asymptotic Slutsky's Theorem (c.f. Theorem C.4 in Appendix C), we may conclude that, for $n_0, n_1, p \to \infty$ at rates $p/n_0 \to \lambda_0$ and $p/n_1 \to \lambda_1$,

$$U_{n_0,n_1,p} V_{n_0,n_1,p} \xrightarrow{D} N\left(\frac{\delta^2}{2}, \delta^2\right). \tag{7.89}$$

Now, by exploiting the independence among sample points, one obtains after some algebra that

$$E[T_{n_0,n_1,k}] = \frac{1}{2}\left(\frac{1}{n_1} - \frac{1}{n_0}\right),$$

$$\mathrm{Var}(T_{n_0,n_1,k}) = \left(\frac{1}{n_0} + \frac{1}{n_1}\right) + \frac{1}{2}\left(\frac{1}{n_0^2} + \frac{1}{n_1^2}\right), \tag{7.90}$$

so that we can apply the double-asymptotic Central Limit Theorem (c.f. Theorem C.5 in Appendix C), to conclude that, as $n_0, n_1, p \to \infty$ at rates $p/n_0 \to \lambda_0$ and $p/n_1 \to \lambda_1$,

$$\sum_{k=2}^{p} T_{n_0,n_1,k} \xrightarrow{D} N\left(\frac{1}{2}(\lambda_1 - \lambda_0),\ \lambda_0 + \lambda_1\right). \tag{7.91}$$

The result now follows from (7.86), (7.89), (7.91), and Theorem C.2 in Appendix C. ■

The next result is an immediate consequence of the previous theorem; a similar result was proven independently using a different method in Fujikoshi (2000). Here, as elsewhere in the book, Φ denotes the cumulative distribution function of a zero-mean, unit-variance Gaussian random variable.

Corollary 7.1 *As $n_0, n_1, p \to \infty$ at rates $p/n_0 \to \lambda_0$ and $p/n_1 \to \lambda_1$,*

$$E\left[\varepsilon^0_{n_0,n_1}\right] \longrightarrow \Phi\left(-\frac{\delta}{2}\frac{1 + \frac{1}{\delta^2}(\lambda_1 - \lambda_0)}{\sqrt{1 + \frac{1}{\delta^2}(\lambda_0 + \lambda_1)}}\right). \tag{7.92}$$

The corresponding result for $E[\varepsilon^1_{n_0,n_1}]$ can be obtained by exchanging λ_0 and λ_1.

Notice that, as $\lambda_0, \lambda_1 \to 0$, the right-hand side in (7.92) tends to $\varepsilon_{\mathrm{bay}} = \Phi(-\frac{\delta}{2})$, the optimal classification error, as $p \to \infty$, and this is true for both $E[\varepsilon^0_{n_0,n_1}]$ and $E[\varepsilon^1_{n_0,n_1}]$. Therefore, the overall expected classification error $E[\varepsilon_{n_0,n_1}]$ converges to $\varepsilon_{\mathrm{bay}}$ as both sample sizes grow infinitely faster than the dimensionality, regardless of the class prior probabilities and the Mahalanobis distance.

A double-asymptotically exact approximation for $E[\varepsilon^0_{n_0,n_1}]$ can be obtained from the previous corollary by letting $\lambda_0 \approx p/n_0$, $\lambda_1 \approx p/n_1$, and $\delta \approx \delta_p$:

$$E\left[\varepsilon^0_{n_0,n_1}\right] \approx \Phi\left(-\frac{\delta_p}{2}\,\frac{1+\frac{p}{\delta_p^2}\left(\frac{1}{n_1}-\frac{1}{n_0}\right)}{\sqrt{1+\frac{p}{\delta_p^2}\left(\frac{1}{n_0}+\frac{1}{n_1}\right)}}\right), \tag{7.93}$$

with the corresponding approximation for $E[\varepsilon^1_{n_0,n_1}]$ obtained by simply exchanging n_0 and n_1. This approximation is double-asymptotically exact, as shown by Corollary 7.1. The expression (7.93) appears in Zollanvari et al. (2011), where asymptotic exactness is proved using a different method.

In Raudys (1972), Raudys used a Gaussian approximation to the distribution of the discriminant to propose a different approximation to the expected true classification error:

$$E\left[\varepsilon^0_{n_0,n_1}\right] \approx \Phi\left(-\frac{E[W_{\bar{L}}(S^p_{n_0,n_1},\mathbf{X}^p) \mid Y^p=0]}{\sqrt{\mathrm{Var}(W_{\bar{L}}(S^p_{n_0,n_1},\mathbf{X}^p) \mid Y^p=0)}}\right). \tag{7.94}$$

The corresponding approximation to $E[\varepsilon^1_{n_0,n_1}]$ is obtained by replacing $Y^p=0$ by $Y^p=1$ and removing the negative sign. Raudys evaluated this expression only in the case $n_0=n_1=n/2$, in which case, as we showed before, $E[\varepsilon_{n/2,n/2}]=E[\varepsilon^0_{n/2,n/2}]=E[\varepsilon^1_{n/2,n/2}]$. In Zollanvari et al. (2011), this expression is evaluated in the general case $n_0 \neq n_1$, yielding

$$E\left[\varepsilon^0_{n_0,n_1}\right] \approx \Phi\left(-\frac{\delta_p}{2}\,\frac{1+\frac{p}{\delta_p^2}\left(\frac{1}{n_1}-\frac{1}{n_0}\right)}{\sqrt{1+\frac{1}{n_1}+\frac{p}{\delta_p^2}\left(\frac{1}{n_0}+\frac{1}{n_1}\right)+\frac{p}{2\delta_p^2}\left(\frac{1}{n_0^2}+\frac{1}{n_1^2}\right)}}\right), \tag{7.95}$$

with the corresponding approximation for $E[\varepsilon^1_{n_0,n_1}]$ obtained by simply exchanging n_0 and n_1. Comparing (7.93) and (7.95), we notice that they differ by a term $\frac{1}{n_1}+\frac{p}{2\delta_p^2}\left(\frac{1}{n_0^2}+\frac{1}{n_1^2}\right)$ inside the radical. But this term clearly goes to zero as $n_0, n_1, p \to \infty$, at rates $p/n_0 \to \lambda_0$ and $p/n_1 \to \lambda_1$. Therefore, the two approximations (7.93) and (7.95) are asymptotically equivalent. As a result, the double-asymptotic exactness of (7.95) is established by virtue of Corollary 7.1.[1]

[1] This was claimed in Raudys and Young (2004) without proof in the case $n_0=n_1=n/2$.

7.3.1.2 Resubstitution Error Rate The counterpart of Theorem 7.6 when the discriminant is evaluated on a training point is given next.

Theorem 7.7 *As $n_0, n_1, p \to \infty$ at rates $p/n_0 \to \lambda_0$ and $p/n_1 \to \lambda_1$,*

$$W_{\bar{L}}\left(S^p_{n_0,n_1}, \mathbf{X}^p_1\right) \mid Y^p = i \xrightarrow{D} N\left(\frac{\delta^2}{2} + \frac{1}{2}\left(\lambda_0 + \lambda_1\right),\ \delta^2 + \lambda_0 + \lambda_1\right). \tag{7.96}$$

for $i = 0, 1$.

Proof: The proof of (7.96) proceeds analogously to the proof of (7.83) in Theorem 7.6, the key difference being that now

$$E[T_{n_0,n_1,k}] = \frac{1}{2}\left(\frac{1}{n_0} + \frac{1}{n_1}\right),$$

$$\mathrm{Var}(T_{n_0,n_1,k}) = \left(\frac{1}{n_0} + \frac{1}{n_1}\right) + \frac{1}{2}\left(\frac{1}{n_0^2} + \frac{1}{n_1^2}\right) + \frac{1}{n_0^2}\left(2 - \frac{3}{n_0}\right). \tag{7.97}$$

■

The next result is an immediate consequence of the previous theorem.

Corollary 7.2 *As $n_0, n_1, p \to \infty$ at rates $p/n_0 \to \lambda_0$ and $p/n_1 \to \lambda_1$,*

$$E\left[\hat{\varepsilon}^{r,01}_{n_0,n_1}\right], E\left[\hat{\varepsilon}^{r,0}_{n_0,n_1}\right], E\left[\hat{\varepsilon}^{r,1}_{n_0,n_1}\right] \longrightarrow \Phi\left(-\frac{\delta}{2}\sqrt{1 + \frac{1}{\delta^2}\left(\lambda_0 + \lambda_1\right)}\right). \tag{7.98}$$

Notice that, as $\lambda_0, \lambda_1 \to 0$, the right-hand side in (7.98) tends to $\varepsilon_{\text{bay}} = \Phi(-\frac{\delta}{2})$ as $p \to \infty$. From the previous corresponding result for the true error, we conclude that the bias of the resubstitution estimator disappears as both sample sizes grow infinitely faster than the dimensionality, regardless of the class prior probabilities and the Mahalanobis distance.

A double-asymptotically exact approximation for the resubstitution error rates can be obtained from the previous corollary by letting $\lambda_0 \approx p/n_0$, $\lambda_1 \approx p/n_1$, and $\delta \approx \delta_p$:

$$E\left[\hat{\varepsilon}^{r,01}_{n_0,n_1}\right] \approx E\left[\hat{\varepsilon}^{r,0}_{n_0,n_1}\right] \approx E\left[\hat{\varepsilon}^{r,1}_{n_0,n_1}\right] \approx \Phi\left(-\frac{\delta_p}{2}\sqrt{1 + \frac{p}{\delta_p^2}\left(\frac{1}{n_0} + \frac{1}{n_1}\right)}\right). \tag{7.99}$$

This approximation is double-asymptotically exact, as shown by Corollary 7.2, but it is only expected to be accurate at large sample sizes, when the difference between the three expected error rates in (7.99) becomes very small. This approximation appears in Zollanvari et al. (2011), where asymptotic exactness is proved using a different method. If (7.99) is evaluated for $n_0 = n_1 = n/2$, in which case, as we showed before, $E[\hat{\varepsilon}^{r,01}_{n/2,n/2}] = E[\hat{\varepsilon}^{r,0}_{n/2,n/2}] = E[\hat{\varepsilon}^{r,1}_{n/2,n/2}]$, an expression published in Raudys and Young (2004) is produced.

The counterpart of approximation (7.94) can be obtained by replacing $\mathbf{X}^p$ and Y^p by $\mathbf{X}_1^p$ and Y_1^p, respectively:

$$E\left[\varepsilon^{r,0}_{n_0,n_1}\right] \approx \Phi\left(- \frac{E\left[W_{\bar{L}}\left(S^p_{n_0,n_1}, \mathbf{X}_1^p\right) \mid Y_1^p = 0\right]}{\sqrt{\mathrm{Var}\left(W_{\bar{L}}\left(S^p_{n_0,n_1}, \mathbf{X}_1\right) \mid Y_1^p = 0\right)}} \right). \tag{7.100}$$

The corresponding approximation to $E[\varepsilon^{r,1}_{n_0,n_1}]$ is obtained by replacing $Y_1^p = 0$ by $Y_1^p = 1$ and removing the negative sign. In Zollanvari et al. (2011), (7.100) is evaluated to yield

$$E\left[\varepsilon^{r,0}_{n_0,n_1}\right] \approx \Phi\left(-\frac{\delta_p}{2} \frac{1 + \frac{p}{\delta_p^2}\left(\frac{1}{n_0} + \frac{1}{n_1}\right)}{\sqrt{1 + \frac{1}{n_1} + \frac{p}{\delta_p^2}\left(\frac{1}{n_0} + \frac{1}{n_1}\right) + \frac{p}{2\delta_p^2}\left(\frac{1}{n_1^2} - \frac{1}{n_0^2}\right)}} \right), \tag{7.101}$$

with the corresponding approximation for $E[\varepsilon^{r,1}_{n_0,n_1}]$ obtained by simply exchanging n_0 and n_1. Comparing (7.99) and (7.101), we notice that they differ by a term $\frac{1}{n_1} + \frac{p}{2\delta_p^2}\left(\frac{1}{n_1^2} - \frac{1}{n_0^2}\right)$ inside the radical. But this term clearly goes to zero as $n_0, n_1, p \to \infty$, at rates $p/n_0 \to \lambda_0$ and $p/n_1 \to \lambda_1$. Therefore, the two approximations (7.99) and (7.101) are asymptotically equivalent. As a result the double-asymptotic exactness of (7.101) is established by virtue of Corollary 7.2.

7.3.1.3 Leave-One-Out Error Rate

From (5.52), $E[\hat{\varepsilon}^{l,0}_{n_0,n_1}] = E[\varepsilon^0_{n_0-1,n_1}]$ and $E[\hat{\varepsilon}^{l,1}_{n_0,n_1}] = E[\varepsilon^1_{n_0,n_1-1}]$; hence, (7.93) and (7.95) can be used here, by replacing n_0 by $n_0 - 1$ and n_1 by $n_1 - 1$ at the appropriate places. For $E[\hat{\varepsilon}^{l,0}_{n_0,n_1}]$, this yields the following asymptotic approximations:

$$E\left[\hat{\varepsilon}^{l,0}_{n_0,n_1}\right] \approx \Phi\left(-\frac{\delta_p}{2} \frac{1 + \frac{p}{\delta_p^2}\left(\frac{1}{n_1} - \frac{1}{n_0 - 1}\right)}{\sqrt{1 + \frac{p}{\delta_p^2}\left(\frac{1}{n_0 - 1} + \frac{1}{n_1}\right)}} \right), \tag{7.102}$$

and

$$E\left[\hat{\varepsilon}^{\,l,0}_{n_0,n_1}\right] \approx \Phi\left(-\frac{\delta_p}{2}\frac{1+\frac{p}{\delta_p^2}\left(\frac{1}{n_1}-\frac{1}{n_0-1}\right)}{\sqrt{1+\frac{1}{n_1}+\frac{p}{\delta_p^2}\left(\frac{1}{n_0-1}+\frac{1}{n_1}\right)+\frac{p}{2\delta_p^2}\left(\frac{1}{(n_0-1)^2}+\frac{1}{n_1^2}\right)}}\right), \tag{7.103}$$

with the corresponding approximations for $E[\hat{\varepsilon}^{\,l,1}_{n_0,n_1}]$ obtained by simply exchanging n_0 and n_1. As before, (7.102) and (7.103) differ only by terms that go to zero as $n_0, n_1, p \to \infty$ at rates $p/n_0 \to \lambda_0$ and $p/n_1 \to \lambda_1$, and are thus asymptotically equivalent.

7.3.2 Second-Order Moments of Error Rates

In this section, we develop double-asymptotically exact approximations to the second-order moments, including mixed moments, of the true, resubstitution, and leave-one-out error rates.

7.3.2.1 *True Error Rate* We start by considering the extension of (7.94) to second moments. The standard bivariate Gaussian distribution function is given by (c.f. Appendix A.11)

$$\Phi(a,b;\rho) = \int_{-\infty}^{a}\int_{-\infty}^{b} \frac{1}{2\pi\sqrt{1-\rho^2}} \exp\left\{-\frac{1}{2(1-\rho)^2)}\left(x^2+y^2-2\rho xy\right)\right\} dx\,dy. \tag{7.104}$$

This corresponds to the distribution function of a joint bivariate Gaussian vector with zero means, unit variances, and correlation coefficient ρ. Note that $\Phi(a,\infty;\rho) = \Phi(a)$ and $\Phi(a,b;0) = \Phi(a)\Phi(b)$. For simplicity of notation, we write $\Phi(a,a;\rho)$ as $\Phi(a;\rho)$.

The rectangular-area probabilities involving any jointly Gaussian pair of variables (U, V) can be written in terms of Φ:

$$P(U \le u, V \le v) = \Phi\left(\frac{u-E[U]}{\sqrt{\mathrm{Var}(U)}}, \frac{v-E[V]}{\sqrt{\mathrm{Var}(V)}}; \rho_{UV}\right), \tag{7.105}$$

where ρ_{UV} is the correlation coefficient between U and V. From (5.73) and (7.105), we obtain the second-order extension of Raudys' formula (7.94):

$$E\left[\left(\varepsilon^0_{n_0,n_1}\right)^2\right] = P\left(W_{\bar{L}}\left(S^p_{n_0,n_1}, \mathbf{X}^{p,1}\right) \le 0, W_{\bar{L}}\left(S^p_{n_0,n_1}, \mathbf{X}^{p,2}\right) \le 0 \mid Y^{p,1} = 0, Y^{p,2} = 0\right)$$

$$\approx \Phi\left(-\frac{E\left[W_{\bar{L}}\left(S^p_{n_0,n_1}, \mathbf{X}^p\right) \mid Y^p = 0\right]}{\sqrt{\mathrm{Var}\left(W_{\bar{L}}\left(S^p_{n_0,n_1}, \mathbf{X}^p\right) \mid Y^p = 0\right)}}; \right.$$
$$\left. \frac{\mathrm{Cov}\left(W_{\bar{L}}\left(S^p_{n_0,n_1}, \mathbf{X}^{p,1}\right), W_{\bar{L}}\left(S^p_{n_0,n_1}, \mathbf{X}^{p,2}\right) \mid Y^{p,1} = 0, Y^{p,2} = 0\right)}{\mathrm{Var}\left(W_{\bar{L}}\left(S^p_{n_0,n_1}, \mathbf{X}^p\right) \mid Y^p = 0\right)} \right), \tag{7.106}$$

where $(\mathbf{X}^p, Y^p)$, $(\mathbf{X}^{p,1}, Y^{p,1})$, and $(\mathbf{X}^{p,2}, Y^{p,2})$ are independent test points.

In the general case $n_0 \neq n_1$, evaluating the terms in (7.106) yields (see Exercise 7.7)

$$E\left[\left(\varepsilon^0_{n_0,n_1}\right)^2\right] \approx \Phi\left(-\frac{\delta_p}{2} \frac{1 + \frac{p}{\delta_p^2}\left(\frac{1}{n_1} - \frac{1}{n_0}\right)}{f_0\left(n_0, n_1, p, \delta_p^2\right)}; \frac{\frac{1}{n_1} + \frac{p}{2\delta_p^2}\left(\frac{1}{n_0^2} + \frac{1}{n_1^2}\right)}{f_0\left(n_0, n_1, p, \delta_p^2\right)^2} \right), \tag{7.107}$$

where

$$f_0\left(n_0, n_1, p, \delta_p^2\right) = \sqrt{1 + \frac{1}{n_1} + \frac{p}{\delta_p^2}\left(\frac{1}{n_0} + \frac{1}{n_1}\right) + \frac{p}{2\delta_p^2}\left(\frac{1}{n_0^2} + \frac{1}{n_1^2}\right)}. \tag{7.108}$$

Note that (7.107) is the second-order extension of (7.95). The corresponding approximation for $E[(\varepsilon^1_{n_0,n_1})^2]$ is obtained by exchanging n_0 and n_1. An analogous approximation for $E[\varepsilon^0_{n_0,n_1}\varepsilon^1_{n_0,n_1}]$ can be similarly developed (see Exercise 7.7).

These approximations are double-asymptotically exact. This fact can be proved using the following theorem.

Theorem 7.8 *(Zollanvari et al., 2011) As $n_0, n_1, p \to \infty$ at rates $p/n_0 \to \lambda_0$ and $p/n_1 \to \lambda_1$,*

$$\varepsilon^0_{n_0,n_1} \xrightarrow{P} \Phi\left(-\frac{\delta}{2} \frac{1 + \frac{1}{\delta^2}\left(\lambda_1 - \lambda_0\right)}{\sqrt{1 + \frac{1}{\delta^2}\left(\lambda_0 + \lambda_1\right)}} \right). \tag{7.109}$$

The corresponding result for $\varepsilon^1_{n_0,n_1}$ can be obtained by exchanging λ_0 and λ_1.

Proof: Note that

$$\varepsilon^0_{n_0,n_1} = P\left(W_{\bar{L}}\left(S^p_{n_0,n_1},\mathbf{X}^p\right) \le 0 \mid Y^p = 0, S^p_{n_0,n_1}\right)$$
$$= \Phi\left(-\frac{E\left[W_{\bar{L}}\left(S^p_{n_0,n_1},\mathbf{X}^p\right) \mid Y^p = 0, S^p_{n_0,n_1}\right]}{\sqrt{\mathrm{Var}\left(W_{\bar{L}}\left(S^p_{n_0,n_1},\mathbf{X}^p\right) \mid Y^p = 0, S^p_{n_0,n_1}\right)}}\right), \qquad (7.110)$$

where we used the fact that $W_{\bar{L}}(S^p_{n_0,n_1},\mathbf{X}^p)$, with $S^p_{n_0,n_1}$ fixed, is a linear function of $\mathbf{X}^p$ and thus has a Gaussian distribution. Now, apply the affine transformation used in the proofs of Theorems 7.2 and 7.6 to obtain the transformed populations in (7.84) and (7.85). Due to the invariance of John's LDA discriminant to affine transformations, one can proceed with these transformed populations without loss of generality.

The numerator of the fraction in (7.110) is given by

$$\hat{G}_0 \overset{\Delta}{=} E\left[W_{\bar{L}}\left(S^p_{n_0,n_1},\mathbf{X}^p\right) \mid Y^p = 0, S^p_{n_0,n_1}\right] = \left(v^p_0 - \bar{\mathbf{X}}^p\right)^T\left(\bar{\mathbf{X}}^p_0 - \bar{\mathbf{X}}^p_1\right). \quad (7.111)$$

Note that $\mathbf{U} = v^p_0 - \bar{\mathbf{X}}^p$ and $\mathbf{V} = \bar{\mathbf{X}}^p_0 - \bar{\mathbf{X}}^p_1$ are Gaussian vectors, with mean vectors and covariance matrices

$$E[\mathbf{U}] = \left(\frac{\delta_p}{2}, 0, \ldots, 0\right), \quad E[\mathbf{V}] = \left(\delta_p, 0, \ldots, 0\right),$$
$$\Sigma_{\mathbf{U}} = \frac{1}{4}\left(\frac{1}{n_0} + \frac{1}{n_1}\right)I_p, \quad \Sigma_{\mathbf{V}} = \left(\frac{1}{n_0} + \frac{1}{n_1}\right)I_p. \qquad (7.112)$$

Using these facts and Isserlis' formula for the expectation and variance of a product of Gaussian random variables (Kan, 2008) one obtains, after some algebraic manipulation,

$$E[\hat{G}_0] = \frac{\delta_p^2}{2} + \frac{p}{2}\left(\frac{1}{n_1} - \frac{1}{n_0}\right) \to \frac{1}{2}(\delta^2 + \lambda_1 - \lambda_0) \overset{\Delta}{=} G_0,$$
$$\mathrm{Var}(\hat{G}_0) = \frac{\delta_p^2}{n_1} + \frac{p}{2}\left(\frac{1}{n_0^2} + \frac{1}{n_1^2}\right) \to 0, \qquad (7.113)$$

as $n_0, n_1, p \to \infty$ at rates $p/n_0 \to \lambda_0$ and $p/n_1 \to \lambda_1$. By a simple application of Chebyshev's inequality, it follows that

$$\hat{G}_0 \xrightarrow{P} G_0 = \frac{1}{2}(\delta^2 + \lambda_1 - \lambda_0), \qquad (7.114)$$

as $n_0, n_1, p \to \infty$ at rates $p/n_0 \to \lambda_0$ and $p/n_1 \to \lambda_1$.

The square of the denominator of the fraction in (7.110) can be written as

$$\hat{D}_0 \overset{\Delta}{=} \mathrm{Var}\left(W_{\bar{L}}\left(S^p_{n_0,n_1}, \mathbf{X}^p\right) \mid Y^p = 0, S^p_{n_0,n_1}\right)$$
$$= \left(\bar{\mathbf{X}}^p_0 - \bar{\mathbf{X}}^p_1\right)^T \left(\bar{\mathbf{X}}^p_0 - \bar{\mathbf{X}}^p_1\right) \overset{\Delta}{=} \hat{\delta}^2. \tag{7.115}$$

Notice that

$$\frac{\hat{\delta}^2}{\dfrac{1}{n_0} + \dfrac{1}{n_1}} \sim \chi^2_1\left(\frac{\delta^2}{\dfrac{1}{n_0} + \dfrac{1}{n_1}}\right) + \chi^2_{p-1}, \tag{7.116}$$

that is, the sum of a noncentral and a central independent chi-square random variable, with the noncentrality parameter and degrees of freedom indicated. It follows that

$$E[\hat{D}_0] = \delta^2_p + p\left(\frac{1}{n_0} + \frac{1}{n_1}\right) \to \delta^2 + \lambda_0 + \lambda_1 \overset{\Delta}{=} D_0,$$

$$\mathrm{Var}(\hat{D}_0) = \left(\frac{1}{n_0} + \frac{1}{n_1}\right)^2 \left[\mathrm{Var}\left(\chi^2_1\left(\frac{\delta^2}{\dfrac{1}{n_0} + \dfrac{1}{n_1}}\right)\right) + \mathrm{Var}\left(\chi^2_{p-1}\right)\right]$$
$$= 4\delta^2\left(\frac{1}{n_0} + \frac{1}{n_1}\right) + 2p\left(\frac{1}{n_0} + \frac{1}{n_1}\right)^2 \to 0, \tag{7.117}$$

as $n_0, n_1, p \to \infty$ at rates $p/n_0 \to \lambda_0$ and $p/n_1 \to \lambda_1$. By a simple application of Chebyshev's inequality, it follows that

$$\hat{D}_0 \xrightarrow{P} D_0 = \delta^2 + \lambda_0 + \lambda_1, \tag{7.118}$$

as $n_0, n_1, p \to \infty$ at rates $p/n_0 \to \lambda_0$ and $p/n_1 \to \lambda_1$.

Finally, by an application of Theorem C.7 in Appendix C, it follows that

$$\varepsilon^0_{n_0,n_1} = \Phi\left(-\frac{\hat{G}_0}{\sqrt{\hat{D}_0}}\right) \xrightarrow{P} \Phi\left(-\frac{G_0}{\sqrt{D_0}}\right), \tag{7.119}$$

as $n_0, n_1, p \to \infty$ at rates $p/n_0 \to \lambda_0$ and $p/n_1 \to \lambda_1$. By following the same steps, it is easy to see that the corresponding result for $\varepsilon^1_{n_0,n_1}$ can be obtained by exchanging λ_0 and λ_1. ■

Note that the previous theorem provides an independent proof of Corollary 7.1, by virtue of Theorem C.6 in Appendix C.

Let $h_e(\lambda_0, \lambda_1)$ denote the right-hand side of (7.109). Theorem 7.8, via Theorems C.3 and C.7 in Appendix C, implies that $(\varepsilon^0_{n_0,n_1})^2$, $\varepsilon^0_{n_0,n_1}\varepsilon^1_{n_0,n_1}$, and $(\varepsilon^1_{n_0,n_1})^2$ converge in probability to $h_e^2(\lambda_0, \lambda_1)$, $h_e(\lambda_0, \lambda_1)h_e(\lambda_1, \lambda_0)$, and $h_e^2(\lambda_1, \lambda_0)$, respectively, as $n_0, n_1, p \to \infty$ at rates $p/n_0 \to \lambda_0$ and $p/n_1 \to \lambda_1$. Application of Theorem C.6 in Appendix C then proves the next corollary.

Corollary 7.3 *As $n_0, n_1, p \to \infty$ at rates $p/n_0 \to \lambda_0$ and $p/n_1 \to \lambda_1$,*

$$E\left[(\varepsilon^0_{n_0,n_1})^2\right] \longrightarrow \left[\Phi\left(-\frac{\delta}{2}\frac{1+\frac{1}{\delta^2}(\lambda_1-\lambda_0)}{\sqrt{1+\frac{1}{\delta^2}(\lambda_0+\lambda_1)}}\right)\right]^2,$$

$$E\left[\varepsilon^0_{n_0,n_1}\varepsilon^1_{n_0,n_1}\right] \longrightarrow \Phi\left(-\frac{\delta}{2}\frac{1+\frac{1}{\delta^2}(\lambda_0-\lambda_1)}{\sqrt{1+\frac{1}{\delta^2}(\lambda_0+\lambda_1)}}\right)\Phi\left(-\frac{\delta}{2}\frac{1+\frac{1}{\delta^2}(\lambda_1-\lambda_0)}{\sqrt{1+\frac{1}{\delta^2}(\lambda_0+\lambda_1)}}\right),$$

$$E\left[\left(\varepsilon^1_{n_0,n_1}\right)^2\right] \longrightarrow \left[\Phi\left(-\frac{\delta}{2}\frac{1+\frac{1}{\delta^2}(\lambda_0-\lambda_1)}{\sqrt{1+\frac{1}{\delta^2}(\lambda_0+\lambda_1)}}\right)\right]^2. \tag{7.120}$$

The following double-asymptotically exact approximations can be obtained from the previous corollary by letting $\lambda_0 \approx p/n_0$, $\lambda_1 \approx p/n_1$, and $\delta \approx \delta_p$:

$$E\left[\left(\varepsilon^0_{n_0,n_1}\right)^2\right] \approx \left[\Phi\left(-\frac{\delta_p}{2}\frac{1+\frac{p}{\delta_p^2}\left(\frac{1}{n_1}-\frac{1}{n_0}\right)}{\sqrt{1+\frac{p}{\delta_p^2}\left(\frac{1}{n_0}+\frac{1}{n_1}\right)}}\right)\right]^2, \tag{7.121}$$

$$E\left[\varepsilon^0_{n_0,n_1}\varepsilon^1_{n_0,n_1}\right] \approx \Phi\left(-\frac{\delta_p}{2}\frac{1+\frac{p}{\delta_p^2}\left(\frac{1}{n_1}-\frac{1}{n_0}\right)}{\sqrt{1+\frac{p}{\delta_p^2}\left(\frac{1}{n_0}+\frac{1}{n_1}\right)}}\right)\Phi\left(-\frac{\delta_p}{2}\frac{1+\frac{p}{\delta_p^2}\left(\frac{1}{n_0}-\frac{1}{n_1}\right)}{\sqrt{1+\frac{p}{\delta_p^2}\left(\frac{1}{n_0}+\frac{1}{n_1}\right)}}\right), \tag{7.122}$$

$$E\left[\left(\varepsilon^{1}_{n_0,n_1}\right)^2\right] \approx \left[\Phi\left(-\frac{\delta_p}{2}\,\frac{1+\frac{p}{\delta_p^2}\left(\frac{1}{n_0}-\frac{1}{n_1}\right)}{\sqrt{1+\frac{p}{\delta_p^2}\left(\frac{1}{n_0}+\frac{1}{n_1}\right)}}\right)\right]^2. \tag{7.123}$$

Note that (7.121) is simply the square of the approximation (7.93).

A key fact is that by removing terms that tend to zero as $n_0, n_1, p \to \infty$ at rates $p/n_0 \to \lambda_0$ and $p/n_1 \to \lambda_1$ in (7.107) produces (7.121), revealing that these two approximations are asymptotically equivalent, and establishing the double-asymptotic exactness of (7.107). Analogous arguments regarding the approximations of $E[\varepsilon^0_{n_0,n_1}\varepsilon^1_{n_0,n_1}]$ and $E[(\varepsilon^1_{n_0,n_1})^2]$ can be readily made (see Exercise 7.7).

7.3.2.2 Resubstitution Error Rate All the approximations described in this section can be shown to be double-asymptotically exact by using results proved in Zollanvari et al. (2011). We refer the reader to that reference for the details.[2] We start by considering the extension of (7.100) to second moments. From (5.77),

$$\begin{aligned} E\left[\left(\hat{\varepsilon}^{r,0}_{n_0,n_1}\right)^2\right] &= \frac{1}{n_0}P\left(W_{\bar{L}}\left(S^p_{n_0,n_1},\mathbf{X}^p_1\right) \le 0 \mid Y^p_1 = 0\right) \\ &+ \frac{n_0-1}{n_0}P\left(W_{\bar{L}}\left(S^p_{n_0,n_1},\mathbf{X}^p_1\right) \le 0, W_{\bar{L}}\left(S^p_{n_0,n_1},\mathbf{X}^p_2\right) \le 0 \mid Y^p_1 = 0, Y^p_2 = 0\right). \end{aligned} \tag{7.124}$$

A Gaussian approximation to the bivariate probability in the previous equation is obtained by applying (7.105):

$$\begin{aligned} &P\left(W_{\bar{L}}\left(S^p_{n_0,n_1},\mathbf{X}^p_1\right) \le 0, W_{\bar{L}}\left(S^p_{n_0,n_1},\mathbf{X}^p_2\right) \le 0 \mid Y^p_1 = 0, Y^p_2 = 0\right) \\ &\approx \Phi\left(-\frac{E\left[W_{\bar{L}}\left(S^p_{n_0,n_1},\mathbf{X}^p_1\right) \mid Y^p_1 = 0\right]}{\sqrt{\mathrm{Var}\left(W_{\bar{L}}\left(S^p_{n_0,n_1},\mathbf{X}^p_1\right) \mid Y^p_1 = 0\right)}};\right. \\ &\left.\frac{\mathrm{Cov}\left(W_{\bar{L}}\left(S^p_{n_0,n_1},\mathbf{X}^p_1\right), W_{\bar{L}}\left(S^p_{n_0,n_1},\mathbf{X}^p_2\right) \mid Y^p_1 = 0, Y^p_2 = 0\right)}{\mathrm{Var}\left(W_{\bar{L}}\left(S^p_{n_0,n_1},\mathbf{X}^p_1\right) \mid Y^p_1 = 0\right)}\right), \end{aligned} \tag{7.125}$$

[2] We point out that it is incorrectly stated in Zollanvari et al. (2011) that (7.126) and (7.136) are approximations of $E[(\hat{\varepsilon}^{r,0}_{n_0,n_1})^2]$ and $E[(\hat{\varepsilon}^{l,0}_{n_0,n_1})^2]$, respectively; in fact, the correct expressions are given in (7.128) and (7.137), respectively. This is a minor mistake that does not change the validity of any of the results in Zollanvari et al. (2011).

In the general case $n_0 \neq n_1$, evaluating (7.125) yields (see Exercise 7.8)

$$P\left(W_{\bar{L}}\left(S^p_{n_0,n_1}, \mathbf{X}^p_1\right) \leq 0, W_{\bar{L}}\left(S^p_{n_0,n_1}, \mathbf{X}^p_2\right) \leq 0 \mid Y^p_1 = 0, Y^p_2 = 0\right)$$
$$\approx \Phi\left(-\frac{\delta_p}{2}\frac{1+\frac{p}{\delta_p^2}\left(\frac{1}{n_0}+\frac{1}{n_1}\right)}{g_0\left(n_0,n_1,p,\delta_p^2\right)};\frac{\frac{1}{n_1}+\frac{p}{2\delta_p^2}\left(\frac{1}{n_1^2}-\frac{1}{n_0^2}\right)}{g_0^2\left(n_0,n_1,p,\delta_p^2\right)}\right), \quad (7.126)$$

where

$$g_0\left(n_0,n_1,p,\delta^2\right) = \sqrt{1+\frac{1}{n_1}+\frac{p}{\delta^2}\left(\frac{1}{n_0}+\frac{1}{n_1}\right)+\frac{p}{2\delta^2}\left(\frac{1}{n_1^2}-\frac{1}{n_0^2}\right)}. \quad (7.127)$$

Combining (7.101), (7.124), and (7.126) produces the following asymptotically exact approximation:

$$E\left[\left(\hat{\varepsilon}^{r,0}_{n_0,n_1}\right)^2\right] \approx \frac{1}{n_0}\Phi\left(-\frac{\delta_p}{2}\frac{1+\frac{p}{\delta_p^2}\left(\frac{1}{n_0}+\frac{1}{n_1}\right)}{\sqrt{1+\frac{1}{n_1}+\frac{p}{\delta_p^2}\left(\frac{1}{n_0}+\frac{1}{n_1}\right)+\frac{p}{2\delta_p^2}\left(\frac{1}{n_1^2}-\frac{1}{n_0^2}\right)}}\right)$$
$$+\frac{n_0-1}{n_0}\Phi\left(-\frac{\delta_p}{2}\frac{1+\frac{p}{\delta_p^2}\left(\frac{1}{n_0}+\frac{1}{n_1}\right)}{g_0\left(n_0,n_1,p,\delta_p^2\right)};\frac{\frac{1}{n_1}+\frac{p}{2\delta_p^2}\left(\frac{1}{n_1^2}-\frac{1}{n_0^2}\right)}{g_0^2\left(n_0,n_1,p,\delta_p^2\right)}\right). \quad (7.128)$$

The corresponding approximation for $E[(\hat{\varepsilon}^{r,1}_{n_0,n_1})^2]$ is obtained by exchanging n_0 and n_1 in (7.128), and an approximation for $E[\hat{\varepsilon}^{r,0}_{n_0,n_1}\hat{\varepsilon}^{r,1}_{n_0,n_1}]$ can be similarly developed (see Exercise 7.8).

Removing terms that tend to zero as $n_0, n_1, p \to \infty$ at rates $p/n_0 \to \lambda_0$ and $p/n_1 \to \lambda_1$ in (7.128) leads to the asymptotically equivalent approximation

$$E\left[\left(\hat{\varepsilon}^{r,0}_{n_0,n_1}\right)^2\right] \approx \Phi\left(-\frac{\delta_p}{2}\sqrt{1+\frac{p}{\delta^2}\left(\frac{1}{n_0}+\frac{1}{n_1}\right)}\right)$$
$$\times\left[\frac{1}{n_0}+\frac{n_0-1}{n_0}\Phi\left(-\frac{\delta_p}{2}\sqrt{1+\frac{p}{\delta_p^2}\left(\frac{1}{n_0}+\frac{1}{n_1}\right)}\right)\right]. \quad (7.129)$$

Corresponding approximations for $E[\hat{\varepsilon}^{r,0}_{n_0,n_1}\hat{\varepsilon}^{r,1}_{n_0,n_1}]$ and $E[(\hat{\varepsilon}^{r,1}_{n_0,n_1})^2]$ are obtained similarly (see Exercise 7.8).

The approximation for the cross-moment between the true and resubstitution error rates can be written by using (5.84) and (7.105):

$$E\left[\varepsilon^0_{n_0,n_1}\hat{\varepsilon}^{r,0}_{n_0,n_1}\right] = P\Big(W_{\bar{L}}\left(S^p_{n_0,n_1},\mathbf{X}^p\right) \leq 0, W_{\bar{L}}\left(S^p_{n_0,n_1},\mathbf{X}^p_1\right) \leq 0 \mid Y^p = 0, Y^p_1 = 0\Big)$$

$$\approx \Phi\left(-\frac{E\left[W_{\bar{L}}\left(S^p_{n_0,n_1},\mathbf{X}^p \mid Y^p = 0\right)\right]}{\sqrt{\mathrm{Var}\left(W_{\bar{L}}\left(S^p_{n_0,n_1},\mathbf{X}^p \mid Y^p = 0\right)\right)}}, -\frac{E\left[W_{\bar{L}}\left(S^p_{n_0,n_1},\mathbf{X}^p_1\right) \mid Y^p_1 = 0\right]}{\sqrt{\mathrm{Var}\left(W_{\bar{L}}\left(S^p_{n_0,n_1},\mathbf{X}^p_1\right) \mid Y^p_1 = 0\right)}};\right.$$

$$\left.\frac{\mathrm{Cov}\left(W_{\bar{L}}\left(S^p_{n_0,n_1},\mathbf{X}^p\right), W_{\bar{L}}\left(S^p_{n_0,n_1},\mathbf{X}^p_1\right) \mid Y^p = 0, Y_1 = 0\right)}{\sqrt{\mathrm{Var}\left(W_{\bar{L}}\left(S^p_{n_0,n_1},\mathbf{X}^p\right) \mid Y^p = 0\right)}\sqrt{\mathrm{Var}\left(W_{\bar{L}}\left(S^p_{n_0,n_1},\mathbf{X}^p_1\right) \mid Y^p_1 = 0\right)}}\right). \tag{7.130}$$

In the general case $n_0 \neq n_1$, evaluating the terms in (7.130) yields (see Exercise 7.9)

$$E\left[\varepsilon^0_{n_0,n_1}\hat{\varepsilon}^{r,0}_{n_0,n_1}\right] \approx \Phi\left(-\frac{\delta_p}{2}\frac{1+\frac{p}{\delta_p^2}\left(\frac{1}{n_1}-\frac{1}{n_0}\right)}{f_0\left(n_0,n_1,p,\delta_p^2\right)}, -\frac{\delta_p}{2}\frac{1+\frac{p}{\delta_p^2}\left(\frac{1}{n_0}+\frac{1}{n_1}\right)}{g_0\left(n_0,n_1,p,\delta_p^2\right)};\right.$$

$$\left.\frac{\frac{1}{n_1}+\frac{p}{2\delta_p^2}\left(\frac{1}{n_1^2}-\frac{1}{n_0^2}\right)}{f_0\left(n_0,n_1,p,\delta_p^2\right)g_0\left(n_0,n_1,p,\delta_p^2\right)}\right). \tag{7.131}$$

One obtains an approximation for $E[\varepsilon^0_{n_0,n_1}\hat{\varepsilon}^{r,1}_{n_0,n_1}]$ from (7.131) by replacing $g_0(n_0,n_1,p,\delta_p^2)$ by $g_1(n_0,n_1,p,\delta_p^2) = g_0(n_1,n_0,p,\delta_p^2)$ (see Exercise 7.9):

$$E\left[\varepsilon^0_{n_0,n_1}\hat{\varepsilon}^{r,1}_{n_0,n_1}\right] \approx \Phi\left(-\frac{\delta_p}{2}\frac{1+\frac{p}{\delta_p^2}\left(\frac{1}{n_1}-\frac{1}{n_0}\right)}{f_0\left(n_0,n_1,p,\delta_p^2\right)}, -\frac{\delta_p}{2}\frac{1+\frac{p}{\delta_p^2}\left(\frac{1}{n_0}+\frac{1}{n_1}\right)}{g_1\left(n_0,n_1,p,\delta_p^2\right)};\right.$$

$$\left.\frac{\frac{1}{n_1}+\frac{p}{2\delta_p^2}\left(\frac{1}{n_1^2}-\frac{1}{n_0^2}\right)}{f_0\left(n_0,n_1,p,\delta_p^2\right)g_1\left(n_0,n_1,p,\delta_p^2\right)}\right). \tag{7.132}$$

The corresponding approximations for $E[\varepsilon^1_{n_0,n_1}\hat{\varepsilon}^{r,0}_{n_0,n_1}]$ and $E[\varepsilon^1_{n_0,n_1}\hat{\varepsilon}^{r,1}_{n_0,n_1}]$ are obtained from the approximations for $E[\varepsilon^0_{n_0,n_1}\hat{\varepsilon}^{r,1}_{n_0,n_1}]$ and $E[\varepsilon^0_{n_0,n_1}\hat{\varepsilon}^{r,0}_{n_0,n_1}]$, respectively, by exchanging n_0 and n_1 (see Exercise 7.9).

Removing terms tending to zero as $n_0, n_1, p \to \infty$ at rates $p/n_0 \to \lambda_0$ and $p/n_1 \to \lambda_1$ in (7.131) gives the approximation

$$E\left[\varepsilon^0_{n_0,n_1}\hat{\varepsilon}^{r,0}_{n_0,n_1}\right] \approx \Phi\left(-\frac{\delta_p}{2}\frac{1+\frac{p}{\delta_p^2}\left(\frac{1}{n_1}-\frac{1}{n_0}\right)}{\sqrt{1+\frac{p}{\delta_p^2}\left(\frac{1}{n_0}+\frac{1}{n_1}\right)}}\right)\Phi\left(-\frac{\delta_p}{2}\sqrt{1+\frac{p}{\delta_p^2}\left(\frac{1}{n_0}+\frac{1}{n_1}\right)}\right). \tag{7.133}$$

Note that the right-hand side of this approximation is simply the product of the right-hand sides of (7.93) and (7.99). Corresponding approximations for $E[\varepsilon^0_{n_0,n_1}\hat{\varepsilon}^{r,1}_{n_0,n_1}]$, $E[\varepsilon^1_{n_0,n_1}\hat{\varepsilon}^{r,0}_{n_0,n_1}]$, and $E[\varepsilon^1_{n_0,n_1}\hat{\varepsilon}^{r,1}_{n_0,n_1}]$ are obtained similarly (see Exercise 7.9).

7.3.2.3 Leave-One-Out Error Rate

As in the previous section, all the approximations described in this section can be shown to be double-asymptotically exact by using results proved in Zollanvari et al. (2011). The results for leave-one-out are obtained from those for resubstitution by replacing $W_{\bar{L}}(S^p_{n_0,n_1},\mathbf{X}^p_1)$ and $W_{\bar{L}}(S^p_{n_0,n_1},\mathbf{X}^p_2)$ by the deleted discriminants $W^{(1)}_{\bar{L}}(S^p_{n_0,n_1},\mathbf{X}^p_1)$ and $W^{(2)}_{\bar{L}}(S^p_{n_0,n_1},\mathbf{X}^p_2)$, respectively (c.f. Section 5.5.3). Hence, as in (7.124),

$$\begin{aligned}E\left[\left(\hat{\varepsilon}^{l,0}_{n_0,n_1}\right)^2\right] &= \frac{1}{n_0}P\left(W^{(1)}_{\bar{L}}\left(S^p_{n_0,n_1},\mathbf{X}^p_1\right)\le 0 \mid Y^p_1=0\right)\\ &+\frac{n_0-1}{n_0}P\left(W^{(1)}_{\bar{L}}\left(S^p_{n_0,n_1},\mathbf{X}^p_1\right)\le 0, W^{(2)}_{\bar{L}}\left(S^p_{n_0,n_1},\mathbf{X}^p_2\right)\le 0 \mid Y^p_1=0, Y^p_2=0\right)\end{aligned} \tag{7.134}$$

and, as in (7.125), a Gaussian approximation to the bivariate probability in the previous equation is obtained by applying (7.105):

$$\begin{aligned}&P\left(W^{(1)}_{\bar{L}}\left(S^p_{n_0,n_1},\mathbf{X}^p_1\right)\le 0, W^{(2)}_{\bar{L}}\left(S^p_{n_0,n_1},\mathbf{X}^p_2\right)\le 0 \mid Y^p_1=0, Y^p_2=0\right)\\ &\quad\approx \Phi\left(-\frac{E\left[W^{(1)}_{\bar{L}}\left(S^p_{n_0,n_1},\mathbf{X}^p_1\right)\mid Y^p_1=0\right]}{\sqrt{\mathrm{Var}(W^{(1)}_{\bar{L}}(S^p_{n_0,n_1},\mathbf{X}^p_1)\mid Y^p_1=0)}};\right.\\ &\qquad\left.\frac{\mathrm{Cov}\left(W^{(1)}_{\bar{L}}\left(S^p_{n_0,n_1},\mathbf{X}^p_1\right), W^{(2)}_{\bar{L}}\left(S^p_{n_0,n_1},\mathbf{X}^p_2\right)\mid Y^p_1=0, Y^p_2=0\right)}{\mathrm{Var}(W^{(1)}_{\bar{L}}\left(S^p_{n_0,n_1},\mathbf{X}^p_1\right)\mid Y^p_1=0)}\right).\end{aligned} \tag{7.135}$$

In the general case $n_0 \neq n_1$, evaluating (7.135) yields (see Exercise 7.10)

$$P\left(W_{\bar{L}}^{(1)}\left(S_{n_0,n_1}^p, \mathbf{X}_1^p\right) \leq 0, W_{\bar{L}}^{(2)}\left(S_{n_0,n_1}^p, \mathbf{X}_2^p\right) \leq 0 \mid Y_1^p = 0, Y_2^p = 0\right)$$

$$\approx \Phi\left(-\frac{\delta_p}{2}\frac{1+\frac{p}{\delta_p^2}\left(\frac{1}{n_1}-\frac{1}{n_0-1}\right)}{f_0\left(n_0-1,n_1,p,\delta_p^2\right)};\right.$$

$$\left.\frac{\frac{1}{n_1}+\frac{p}{2\delta_p^2}\left(\frac{1}{n_1^2}+\frac{2}{(n_0-1)^4}-\frac{(n_0-2)^2}{(n_0-1)^4}\right)}{f_0^2\left(n_0-1,n_1,p,\delta_p^2\right)}\right). \quad (7.136)$$

Combining (7.103), (7.134), and (7.136) yields the following asymptotically exact approximation:

$$E\left[\left(\hat{\varepsilon}_{n_0,n_1}^{l,0}\right)^2\right] \approx \frac{1}{n_0}\Phi\left(-\frac{\delta_p}{2}\frac{1+\frac{p}{\delta_p^2}\left(\frac{1}{n_1}-\frac{1}{n_0-1}\right)}{\sqrt{1+\frac{1}{n_1}+\frac{p}{\delta_p^2}\left(\frac{1}{n_0-1}+\frac{1}{n_1}\right)+\frac{p}{2\delta_p^2}\left(\frac{1}{(n_0-1)^2}+\frac{1}{n_1^2}\right)}}\right)$$

$$+\frac{n_0-1}{n_0}\Phi\left(-\frac{\delta_p}{2}\frac{1+\frac{p}{\delta_p^2}\left(\frac{1}{n_1}-\frac{1}{n_0-1}\right)}{f_0\left(n_0-1,n_1,p,\delta_p^2\right)};\frac{\frac{1}{n_1}+\frac{p}{2\delta_p^2}\left(\frac{1}{n_1^2}+\frac{2}{(n_0-1)^4}-\frac{(n_0-2)^2}{(n_0-1)^4}\right)}{f_0^2\left(n_0-1,n_1,p,\delta_p^2\right)}\right).$$

(7.137)

The corresponding approximation for $E[(\hat{\varepsilon}_{n_0,n_1}^{l,1})^2]$ is obtained by exchanging n_0 and n_1 in (7.137), and an approximation for $E[\hat{\varepsilon}_{n_0,n_1}^{l,0}\hat{\varepsilon}_{n_0,n_1}^{l,1}]$ can be similarly developed (see Exercise 7.10).

Removing terms that tend to zero as $n_0, n_1, p \to \infty$ at rates $p/n_0 \to \lambda_0$ and $p/n_1 \to \lambda_1$ in (7.137) leads to the asymptotically equivalent approximation

$$E\left[\left(\hat{\varepsilon}^{l,0}_{n_0,n_1}\right)^2\right] \approx \Phi\left(-\frac{\delta_p}{2}\frac{1+\frac{p}{\delta_p^2}\left(\frac{1}{n_1}-\frac{1}{n_0-1}\right)}{\sqrt{1+\frac{p}{\delta_p^2}\left(\frac{1}{n_0-1}+\frac{1}{n_1}\right)}}\right)$$
$$\times\left[\frac{1}{n_0}+\frac{n_0-1}{n_0}\Phi\left(-\frac{\delta_p}{2}\frac{1+\frac{p}{\delta_p^2}\left(\frac{1}{n_1}-\frac{1}{n_0-1}\right)}{\sqrt{1+\frac{p}{\delta_p^2}\left(\frac{1}{n_0-1}+\frac{1}{n_1}\right)}}\right)\right]. \quad (7.138)$$

Corresponding approximations for $E[\hat{\varepsilon}^{l,0}_{n_0,n_1}\hat{\varepsilon}^{l,1}_{n_0,n_1}]$ and $E[(\hat{\varepsilon}^{l,0}_{n_0,n_1})^2]$ are obtained similarly (see Exercise 7.10).

The approximation for the cross-moment between the true and leave-one-out error rates is obtained by replacing $W_{\bar{L}}(S^p_{n_0,n_1},\mathbf{X}^p_1)$ by $W^{(1)}_{\bar{L}}(S^p_{n_0,n_1},\mathbf{X}^p_1)$ in (7.130). In the general case $n_0 \neq n_1$, this gives (see Exercise 7.11)

$$E[\varepsilon^0_{n_0,n_1}\hat{\varepsilon}^{l,0}_{n_0,n_1}] \approx \Phi\left(-\frac{\delta_p}{2}\frac{1+\frac{p}{\delta_p^2}\left(\frac{1}{n_1}-\frac{1}{n_0}\right)}{f_0(n_0,n_1,p,\delta_p^2)}, -\frac{\delta_p}{2}\frac{1+\frac{p}{\delta_p^2}\left(\frac{1}{n_1}-\frac{1}{n_0-1}\right)}{f_0(n_0-1,n_1,p,\delta_p^2)};\right.$$
$$\left.\frac{\frac{1}{n_1}+\frac{p}{2\delta_p^2}\left(\frac{1}{n_1^2}-\frac{1}{n_0^2}\right)}{f_0(n_0,n_1,p,\delta_p^2)f_0(n_0-1,n_1,p,\delta_p^2)}\right). \quad (7.139)$$

One obtains an approximation for $E[\varepsilon^0_{n_0,n_1}\hat{\varepsilon}^{l,1}_{n_0,n_1}]$ from (7.139) by replacing $f_0(n_0-1,n_1,p,\delta_p^2)$ by $f_0(n_0,n_1-1,p,\delta_p^2)$ (see Exercise 7.11):

$$E\left[\varepsilon^0_{n_0,n_1}\hat{\varepsilon}^{l,1}_{n_0,n_1}\right] \approx \Phi\left(-\frac{\delta_p}{2}\frac{1+\frac{p}{\delta_p^2}\left(\frac{1}{n_1}-\frac{1}{n_0}\right)}{f_0(n_0,n_1,p,\delta_p^2)}, -\frac{\delta_p}{2}\frac{1+\frac{p}{\delta_p^2}\left(\frac{1}{n_0}-\frac{1}{n_1-1}\right)}{f_0(n_0,n_1-1,p,\delta_p^2)};\right.$$
$$\left.\frac{\frac{1}{n_1}+\frac{p}{2\delta_p^2}\left(\frac{1}{n_1^2}-\frac{1}{n_0^2}\right)}{f_0(n_0,n_1,p,\delta_p^2)f_0(n_0,n_1-1,p,\delta_p^2)}\right). \quad (7.140)$$

The corresponding approximations for $E[\varepsilon^1_{n_0,n_1}\hat{\varepsilon}^{l,0}_{n_0,n_1}]$ and $E[\varepsilon^1_{n_0,n_1}\hat{\varepsilon}^{l,1}_{n_0,n_1}]$ are obtained from the approximations for $E[\varepsilon^0_{n_0,n_1}\hat{\varepsilon}^{l,1}_{n_0,n_1}]$ and $E[\varepsilon^0_{n_0,n_1}\hat{\varepsilon}^{l,0}_{n_0,n_1}]$, respectively, by exchanging n_0 and n_1 (see Exercise 7.11).

Removing terms that tend to zero as $n_0, n_1, p \to \infty$ at rates $p/n_0 \to \lambda_0$ and $p/n_1 \to \lambda_1$ in (7.139) gives the approximation

$$E\left[\varepsilon^0_{n_0,n_1}\hat{\varepsilon}^{l,0}_{n_0,n_1}\right] \approx \Phi\left(-\frac{\delta_p}{2}\frac{1+\frac{p}{\delta_p^2}\left(\frac{1}{n_1}-\frac{1}{n_0}\right)}{\sqrt{1+\frac{p}{\delta_p^2}\left(\frac{1}{n_0}+\frac{1}{n_1}\right)}}\right) \times\Phi\left(-\frac{\delta_p}{2}\frac{1+\frac{p}{\delta_p^2}\left(\frac{1}{n_1}-\frac{1}{n_0-1}\right)}{\sqrt{1+\frac{p}{\delta_p^2}\left(\frac{1}{n_0-1}+\frac{1}{n_1}\right)}}\right). \tag{7.141}$$

Note that the right-hand side of this approximation is simply the product of the right-hand sides of (7.93) and (7.102). The corresponding approximations for $E[\varepsilon^0_{n_0,n_1}\hat{\varepsilon}^{l,1}_{n_0,n_1}]$, $E[\varepsilon^1_{n_0,n_1}\hat{\varepsilon}^{l,0}_{n_0,n_1}]$, and $E[\varepsilon^1_{n_0,n_1}\hat{\varepsilon}^{l,1}_{n_0,n_1}]$ are obtained similarly (see Exercise 7.11).

EXERCISES

7.1 Use Theorem 7.1 to write out the expressions for $E[\varepsilon^0_{n_0,n_1}]$, $E[\varepsilon^1_{n_0,n_1}]$, and $E[\varepsilon_{n_0,n_1}]$, simplifying the equations as much as possible. Repeat using the alternative expression (7.14).

7.2 Write the expressions in Theorems 7.1, 7.2, 7.3, and 7.5 for the balanced design case $n_0 = n_1 = n/2$. Describe the behavior of the resulting expressions as a function of increasing n and p.

7.3 Find the eigenvectors of matrix H corresponding to the eigenvalues in (7.47). Then use the spectral decomposition theorem to write the quadratic form in (7.44) as a function of these eigenvalues and eigenvectors, thus completing the proof of Theorem 7.3.

7.4 Generate Gaussian synthetic data from two populations $\Pi_0 \sim N_d(\mathbf{0}, I)$ and $\Pi_1 \sim N_d(\boldsymbol{\mu}, I)$. Note that $\delta = ||\boldsymbol{\mu}||$. Compute Monte-Carlo approximations for $E[\varepsilon^0_{n_0,n_1}]$ and $E[\varepsilon^{r,0}_{n_0,n_1}]$ for John's LDA classification rule with $n_0 = 10, 20, \ldots, 50$, $n_1 = 10, 20, \ldots, 50$, and $\delta = 1, 2, 3, 4$ (the latter correspond to

Bayes errors $\varepsilon_{\text{bay}} = 30.9\%$, 15.9%, 6.7%, and 2.3%, respectively). Compare to the values obtained via Theorems 7.2 and 7.3 and the Imhof-Pearson three-moment method of Section 7.2.2.1.

7.5 Derive the approximation (7.95) for $E[\varepsilon^0_{n_0,n_1}]$ by evaluating the terms in (7.94).

7.6 Derive the approximation (7.101) for $E[\hat{\varepsilon}^{r,0}_{n_0,n_1}]$ by evaluating the terms in (7.100).

7.7 Regarding the approximation (7.107) for $E[(\varepsilon^0_{n_0,n_1})^2]$:

(a) Derive this expression by evaluating the terms in (7.106).

(b) Obtain analogous approximations for $E[\varepsilon^0_{n_0,n_1}\varepsilon^1_{n_0,n_1}]$ and $E[(\varepsilon^1_{n_0,n_1})^2]$ in terms of n_0, n_1, and δ_p.

(c) Remove terms that tend to zero as $n_0, n_1, p \to \infty$ at rates $p/n_0 \to \lambda_0$ and $p/n_1 \to \lambda_1$ in the result of part (b) in order to obtain (7.122) and (7.123), thus establishing the double-asymptotic exactness of the approximations in part (b).

7.8 Regarding the approximation (7.128) for $E[(\hat{\varepsilon}^{r,0}_{n_0,n_1})^2]$:

(a) Derive this expression by evaluating the terms in (7.125).

(b) Obtain analogous approximations for $E[\hat{\varepsilon}^{r,0}_{n_0,n_1}\hat{\varepsilon}^{r,1}_{n_0,n_1}]$ and $E[(\hat{\varepsilon}^{r,1}_{n_0,n_1})^2]$ in terms of n_0, n_1, and δ_p.

(c) Remove terms that tend to zero as $n_0, n_1, p \to \infty$ at rates $p/n_0 \to \lambda_0$ and $p/n_1 \to \lambda_1$ in the result of part (b) in order to obtain asymptotically equivalent approximations similar to (7.129).

7.9 Regarding the approximation (7.131) for $E[\varepsilon^0_{n_0,n_1}\hat{\varepsilon}^{r,0}_{n_0,n_1}]$:

(a) Derive this expression by evaluating the terms in (7.130).

(b) Show that the approximation (7.132) for $E[\varepsilon^0_{n_0,n_1}\hat{\varepsilon}^{r,1}_{n_0,n_1}]$ can be obtained by replacing $g_0(n_0, n_1, p, \delta_p^2)$ by $g_1(n_0, n_1, p, \delta_p^2)$ in (7.131).

(c) Verify that corresponding approximations for $E[\varepsilon^1_{n_0,n_1}\hat{\varepsilon}^{r,1}_{n_0,n_1}]$ and $E[\varepsilon^1_{n_0,n_1}\hat{\varepsilon}^{r,0}_{n_0,n_1}]$ result from exchanging n_0 and n_1 in (7.131) and (7.132), respectively.

(d) Remove terms that tend to zero as $n_0, n_1, p \to \infty$ at rates $p/n_0 \to \lambda_0$ and $p/n_1 \to \lambda_1$ in the expressions obtained in parts (b) and (c) in order to obtain approximations for $E[\varepsilon^0_{n_0,n_1}\hat{\varepsilon}^{r,1}_{n_0,n_1}]$, $E[\varepsilon^1_{n_0,n_1}\hat{\varepsilon}^{r,0}_{n_0,n_1}]$, and $E[\varepsilon^1_{n_0,n_1}\hat{\varepsilon}^{r,1}_{n_0,n_1}]$ analogous to the one for $E[\varepsilon^0_{n_0,n_1}\hat{\varepsilon}^{r,0}_{n_0,n_1}]$ in (7.133).

7.10 Regarding the approximation (7.137) for $E[(\hat{\varepsilon}^{l,0}_{n_0,n_1})^2]$:

(a) Derive this expression by evaluating the terms in (7.135).

(b) Obtain analogous approximations for $E[\hat{\varepsilon}^{l,0}_{n_0,n_1}\hat{\varepsilon}^{l,1}_{n_0,n_1}]$ and $E[(\hat{\varepsilon}^{l,1}_{n_0,n_1})^2]$ in terms of n_0, n_1, and δ_p.

(c) Remove terms that tend to zero as $n_0, n_1, p \to \infty$ at rates $p/n_0 \to \lambda_0$ and $p/n_1 \to \lambda_1$ in the result of part (b) in order to obtain asymptotically equivalent approximations similar to (7.138).

7.11 Regarding the approximation (7.139) for $E[\varepsilon^0_{n_0,n_1} \hat{\varepsilon}^{l,0}_{n_0,n_1}]$:

(a) Derive this expression by evaluating the terms in (7.130) with $W_{\bar{L}}(S^p_{n_0,n_1}, \mathbf{X}^p_1)$ replaced by $W^{(1)}_{\bar{L}}(S^p_{n_0,n_1}, \mathbf{X}^p_1)$.

(b) Show that the approximation (7.140) for $E[\varepsilon^0_{n_0,n_1} \hat{\varepsilon}^{l,1}_{n_0,n_1}]$ can be obtained by replacing $f_0(n_0 - 1, n_1, p, \delta^2_p)$ by $f_0(n_0, n_1 - 1, p, \delta^2_p)$ in (7.139).

(c) Verify that corresponding approximations for $E[\varepsilon^1_{n_0,n_1} \hat{\varepsilon}^{l,1}_{n_0,n_1}]$ and $E[\varepsilon^1_{n_0,n_1} \hat{\varepsilon}^{l,0}_{n_0,n_1}]$ result from exchanging n_0 and n_1 in (7.139) and (7.140), respectively.

(d) Remove terms that tend to zero as $n_0, n_1, p \to \infty$ at rates $p/n_0 \to \lambda_0$ and $p/n_1 \to \lambda_1$ in the expressions obtained in parts (b) and (c) in order to obtain approximations for $E[\varepsilon^0_{n_0,n_1} \hat{\varepsilon}^{l,1}_{n_0,n_1}]$, $E[\varepsilon^1_{n_0,n_1} \hat{\varepsilon}^{l,0}_{n_0,n_1}]$, and $E[\varepsilon^1_{n_0,n_1} \hat{\varepsilon}^{l,1}_{n_0,n_1}]$ analogous to the one for $E[\varepsilon^0_{n_0,n_1} \hat{\varepsilon}^{l,0}_{n_0,n_1}]$ in (7.141).

7.12 Rewrite all approximations derived in Section 7.3 and Exercises 7.6–7.10 for the balanced design case $n_0 = n_1 = n/2$, simplifying the equations as much as possible. Describe the behavior of the resulting expressions as functions of increasing n and p.

7.13 Using the model in Exercise 7.4, compute Monte-Carlo approximations for $E[\varepsilon^0_{n_0,n_1}]$, $E[\hat{\varepsilon}^{r,0}_{n_0,n_1}]$, $E[(\varepsilon^0_{n_0,n_1})^2]$, $E[(\hat{\varepsilon}^{r,0}_{n_0,n_1})^2]$, and $E[(\hat{\varepsilon}^{l,0}_{n_0,n_1})^2]$ for John's LDA classification rule with n_0, n_1, and δ as in Exercise 7.4, and $p = 5, 10, 15, 20$. Use these results to verify the accuracy of the double-asymptotic approximations (7.93) and (7.95), (7.99) and (7.101), (7.107) and (7.121), (7.128) and (7.129), and (7.137) and (7.138), respectively. Each approximation in these pairs is asymptotically equivalent to the other, but they yield distinct finite-sample approximations. Based on the simulation results, how do you compare their relative accuracy?

CHAPTER 8

BAYESIAN MMSE ERROR ESTIMATION

There are three possibilities regarding bounding the RMS of an error estimator: (1) the feature–label distribution is known and the classification problem collapses to finding the Bayes classifier and Bayes error, so there is no classifier design or error estimation issue; (2) there is no prior knowledge concerning the feature–label distribution and (typically) we have no bound or one that is too loose for practice; (3) the actual feature–label distribution is contained in an uncertainty class of feature–label distributions and this uncertainty class must be sufficiently constrained (equivalently, the prior knowledge must be sufficiently great) that an acceptable bound can be achieved with an acceptable sample size. In the latter case, if we posit a prior distribution governing the uncertainty class of feature–label distributions, then it would behoove us to utilize the prior knowledge in obtaining an error estimate, rather than employing some *ad hoc* rule such as bootstrap or cross-validation, even if we must assume a non-informative prior distribution. This can be accomplished by finding a minimum mean-square error (MMSE) estimator of the error, in which case the uncertainty will manifest itself in a Bayesian framework relative to a space of feature–label distributions and random samples (Dalton and Dougherty, 2011b).

Error Estimation for Pattern Recognition, First Edition. Ulisses M. Braga-Neto and Edward R. Dougherty.

8.1 THE BAYESIAN MMSE ERROR ESTIMATOR

Let S_n be a sample[1] from a feature–label distribution that is a member of an uncertainty class of distributions parameterized by a random variable[2] θ, governed by a *prior distribution* $\pi(\theta)$, such that each θ corresponds to a feature–label distribution, denoted in this chapter by $f_\theta(\mathbf{x}, y)$. Letting $\varepsilon_n(\theta, S_n)$ denote the true error on f_θ of the designed classifier ψ_n, we desire a sample-dependent MMSE estimator $\hat{\varepsilon}(S_n)$ of $\varepsilon_n(\theta, S_n)$; that is, we wish to find the minimum of $E_{\theta,S_n}[|\varepsilon_n(\theta, S_n) - \xi(S_n)|^2]$ over all Borel measurable functions $\xi(S_n)$ (c.f. Appendix A.2). It is well-known that the solution is the conditional expectation

$$\hat{\varepsilon}(S_n) = E_\theta[\varepsilon_n(\theta, S_n) \mid S_n]\,, \tag{8.1}$$

which is called the *Bayesian MMSE error estimator*. Besides minimizing $E_{\theta,S_n}[|\varepsilon_n(\theta, S_n) - \xi(S_n)|^2]$, $\hat{\varepsilon}(S_n)$ is also an unbiased estimator over the distribution $f(\theta, S_n)$ of θ and S_n, namely,

$$E_{S_n}[\hat{\varepsilon}(S_n)] = E_{\theta,S_n}[\varepsilon_n(\theta, S_n)]\,. \tag{8.2}$$

The fact that $\hat{\varepsilon}(S_n)$ is an unbiased MMSE estimator of $\varepsilon_n(\theta, S_n)$ over $f(\theta, S_n)$ does not tell us how well $\hat{\varepsilon}(S_n)$ estimates $\varepsilon_n(\bar{\theta}, S_n)$ for some specific $\bar{\theta}$. This has to do with the expected squared difference $E_{S_n}[|\varepsilon_n(\bar{\theta}, S_n) - \hat{\varepsilon}(S_n)|^2]$, which is bounded according to

$$\begin{aligned} E_{S_n}[|\varepsilon_n(\bar{\theta}, S_n) - \hat{\varepsilon}(S_n)|^2] &= E_{S_n}\left[\left|\varepsilon_n(\bar{\theta}, S_n) - \int \varepsilon_n(\theta, S_n) f(\theta \mid S_n) d\theta\right|^2\right] \\ &= E_{S_n}\left[\left|\int \varepsilon_n(\theta, S_n)[f(\theta \mid S_n) - \delta(\theta - \bar{\theta})] d\theta\right|^2\right] \\ &\le E_{S_n}\left[\left(\int \varepsilon_n(\theta, S_n)|f(\theta \mid S_n) - \delta(\theta - \bar{\theta})| d\theta\right)^2\right], \end{aligned} \tag{8.3}$$

where $\delta(\theta)$ is the delta generalized function. This distribution inequality reveals that the accuracy of the estimate for the actual feature–label distribution $f_{\bar{\theta}}(\mathbf{x}, y)$ depends upon the degree to which the mass of the conditional distribution for θ given S_n is concentrated at $\bar{\theta}$ on average across the sampling distribution. Although the Bayesian MMSE error estimate is not guaranteed to be the optimal error estimate for any particular feature–label distribution, for a given sample, and as long as the classifier is a deterministic function of the sample only, it is both optimal on average with respect to MSE and unbiased when averaged over all parameters and samples. We

[1] In this chapter, we return our focus to mixture sampling.

[2] In this chapter, we employ small letters to denote the random parameter vector θ, which is a common convention in Bayesian statistics.

would like $f(\theta \mid S_n)$ to be close to $\delta(\theta - \bar{\theta})$; however, concentrating the conditional mass at a particular value of θ can have detrimental effects if that value is not close to $\bar{\theta}$.

The conditional distribution $f(\theta \mid S_n)$ characterizes uncertainty with regard to the actual value $\bar{\theta}$ of θ. This conditional distribution is called the *posterior distribution* $\pi^*(\theta) = f(\theta \mid S_n)$, where we tacitly keep in mind conditioning on the sample. Based on these conventions, we write the Bayesian MMSE error estimator (8.1) as

$$\hat{\varepsilon} = E_{\pi^*}[\varepsilon_n]. \tag{8.4}$$

The posterior distribution characterizes uncertainty with regard to $\bar{\theta}$. We desire $\hat{\varepsilon}(S_n)$ to estimate $\varepsilon_n(\bar{\theta}, S_n)$ but are uncertain of $\bar{\theta}$. Recognizing this uncertainty, we can obtain an unbiased MMSE estimator for $\varepsilon_n(\theta, S_n)$, which means good performance across all possible values of θ relative to the distribution of θ and S_n, with the performance of $\hat{\varepsilon}(S_n)$ for a particular value $\theta = \bar{\theta}$ depending on the concentration of $\pi^*(\theta)$ relative to $\bar{\theta}$.

In binary classification, θ is a random vector composed of three parts: the parameters of the class-0 conditional distribution θ_0, the parameters of the class-1 conditional distribution θ_1, and the class-0 probability $c = c_0$ (with $c_1 = 1 - c$ for class 1). We let Θ denote the parameter space for θ, let Θ_y be the parameter space containing all permitted values for θ_y, and write the class-conditional distributions as $f_{\theta_y}(\mathbf{x}|y)$. The marginal prior densities of the class-conditional distributions are denoted $\pi(\theta_y)$, for $y = 0, 1$, and that of the class probabilities is denoted $\pi(c)$. We also refer to these prior distributions as "prior probabilities."

To facilitate analytic representations, we assume that c, θ_0, and θ_1 are all independent prior to observing the data. This assumption imposes limitations; for instance, we cannot assume Gaussian distributions with the same unknown covariance for both classes, nor can we use the same parameter in both classes; however, it allows us to separate the posterior joint density $\pi^*(\theta)$ and ultimately to separate the Bayesian error estimator into components representing the error contributed by each class.

Once the priors $\pi(\theta_y)$ and $\pi(c)$ have been established, we have

$$\pi^*(\theta) = f(c, \theta_0, \theta_1 \mid S_n) = f(c \mid S_n, \theta_0, \theta_1) f(\theta_0 \mid S_n, \theta_1) f(\theta_1 \mid S_n). \tag{8.5}$$

We assume that, given n_0, c is independent from S_n and the distribution parameters for each class. Hence,

$$f(c \mid S_n, \theta_0, \theta_1) = f(c \mid n_0, S_n, \theta_0, \theta_1) = f(c \mid n_0). \tag{8.6}$$

In addition, we assume that, given n_0, the sample S_{n_0} and distribution parameters for class 0 are independent from the sample S_{n_1} and distribution parameters for class 1.

Hence,

$$\begin{aligned} f(\boldsymbol{\theta}_0 \mid S_n, \boldsymbol{\theta}_1) &= f(\boldsymbol{\theta}_0 \mid n_0, S_{n_0}, S_{n_1}, \boldsymbol{\theta}_1) \\ &= \frac{f(\boldsymbol{\theta}_0, S_{n_0} \mid n_0, S_{n_1}, \boldsymbol{\theta}_1)}{f(S_{n_0} \mid n_0, S_{n_1}, \boldsymbol{\theta}_1)} \\ &= \frac{f(\boldsymbol{\theta}_0, S_{n_0} \mid n_0)}{f(S_{n_0} \mid n_0)} \\ &= f(\boldsymbol{\theta}_0 \mid n_0, S_{n_0}) \\ &= f(\boldsymbol{\theta}_0 \mid S_{n_0}), \end{aligned} \tag{8.7}$$

where in the last line, as elsewhere in this chapter, we assume that n_0 is given when S_{n_0} is given. Given n_1, analogous statements apply and

$$f(\boldsymbol{\theta}_1 \mid S_n) = f(\boldsymbol{\theta}_1 \mid n_1, S_{n_0}, S_{n_1}) = f(\boldsymbol{\theta}_1 \mid S_{n_1}). \tag{8.8}$$

Combining these results, we have that c, $\boldsymbol{\theta}_0$, and $\boldsymbol{\theta}_1$ remain independent posterior to observing the data:

$$\pi^*(\boldsymbol{\theta}) = f(c \mid n_0) f(\boldsymbol{\theta}_0 \mid S_{n_0}) f(\boldsymbol{\theta}_1 \mid S_{n_1}) = \pi^*(c)\pi^*(\boldsymbol{\theta}_0)\pi^*(\boldsymbol{\theta}_1), \tag{8.9}$$

where $\pi^*(\boldsymbol{\theta}_0)$, $\pi^*(\boldsymbol{\theta}_1)$, and $\pi^*(c)$ are the marginal posterior densities for the parameters $\boldsymbol{\theta}_0$, $\boldsymbol{\theta}_1$, and c, respectively.

When finding the posterior probabilities for the parameters, we need to only consider the sample points from the corresponding class. Using Bayes' rule,

$$\begin{aligned} \pi^*(\boldsymbol{\theta}_y) &= f(\boldsymbol{\theta}_y \mid S_{n_y}) \\ &\propto \pi(\boldsymbol{\theta}_y) f(S_{n_y} \mid \boldsymbol{\theta}_y) \\ &= \pi(\boldsymbol{\theta}_y) \prod_{i:y_i=y} f_{\boldsymbol{\theta}_y}(\mathbf{x}_i \mid y), \end{aligned} \tag{8.10}$$

where the constant of proportionality can be found by normalizing the integral of $\pi^*(\boldsymbol{\theta}_y)$ to 1. The term $f(S_{n_y} \mid \boldsymbol{\theta}_y)$ is called the *likelihood function.*

Although we call $\pi(\boldsymbol{\theta}_y)$ the "prior probabilities," they are not required to be valid density functions. The priors are called "improper" if the integral of $\pi(\boldsymbol{\theta}_y)$ is infinite, that is, if $\pi(\boldsymbol{\theta}_y)$ induces a σ-finite measure but not a finite probability measure. When improper priors are used, Bayes' rule does not apply. Hence, assuming the posterior is integrable, we take (8.10) as the definition of the posterior distribution, normalizing it so that its integral is equal to 1.

As was mentioned in Section 1.3, $n_0 \sim \text{Binomial}(n, c)$ given c:

$$\pi^*(c) = f(c|n_0) \propto \pi(c) f(n_0|c) \propto \pi(c) c^{n_0} (1-c)^{n_1}. \tag{8.11}$$

In Dalton and Dougherty (2011b,c), three models for the prior distributions of c are considered: beta, uniform, and known.

If the prior distribution for c is beta(α, β) distributed, then the posterior distribution for c can be simplified from (8.11). From this beta-binomial model, the form of $\pi^*(c)$ is still a beta distribution and

$$\pi^*(c) = \frac{c^{n_0+\alpha-1}(1-c)^{n_1+\beta-1}}{B(n_0+\alpha, n_1+\beta)}, \tag{8.12}$$

where B is the beta function. The expectation of this distribution is given by

$$E_{\pi^*}[c] = \frac{n_0+\alpha}{n+\alpha+\beta}. \tag{8.13}$$

In the case of uniform priors, $\alpha = 1$, $\beta = 1$, and

$$\pi^*(c) = \frac{(n+1)!}{n_0!n_1!} c^{n_0}(1-c)^{n_1}, \tag{8.14}$$

$$E_{\pi^*}[c] = \frac{n_0+1}{n+2}. \tag{8.15}$$

Finally, if c is known, then, $E_{\pi^*}[c] = c$.

Owing to the posterior independence between c, $\boldsymbol{\theta}_0$, and $\boldsymbol{\theta}_1$, and since ε_n^y is a function of $\boldsymbol{\theta}_y$ only, the Bayesian MMSE error estimator can be expressed as

$$\begin{aligned}\hat{\varepsilon} &= E_{\pi^*}[\varepsilon_n] \\ &= E_{\pi^*}\left[c\varepsilon_n^0 + (1-c)\varepsilon_n^1\right] \\ &= E_{\pi^*}[c]E_{\pi^*}\left[\varepsilon_n^0\right] + (1-E_{\pi^*}[c])E_{\pi^*}\left[\varepsilon_n^1\right].\end{aligned} \tag{8.16}$$

Note that $E_{\pi^*}[\varepsilon_n^y]$ may be viewed as the posterior expectation for the error contributed by class y. With a fixed classifier and given $\boldsymbol{\theta}_y$, the true error, $\varepsilon_n^y(\boldsymbol{\theta}_y)$, is deterministic and

$$E_{\pi^*}\left[\varepsilon_n^y\right] = \int_{\boldsymbol{\Theta}_y} \varepsilon_n^y(\boldsymbol{\theta}_y)\pi^*(\boldsymbol{\theta}_y)d\boldsymbol{\theta}_y. \tag{8.17}$$

Letting

$$\hat{\varepsilon}^y = E_{\pi^*}\left[\varepsilon_n^y\right], \tag{8.18}$$

(8.16) takes the form

$$\hat{\varepsilon} = E_{\pi^*}[c]\hat{\varepsilon}^0 + (1-E_{\pi^*}[c])\hat{\varepsilon}^1. \tag{8.19}$$

8.2 SAMPLE-CONDITIONED MSE

As seen in earlier chapters, the standard approach to evaluating the performance of an error estimator is to find its MSE (RMS) for a given feature–label distribution and classification rule. Expectations are taken with respect to the sampling distribution, so that performance is across the set of samples, which is where randomness is manifested. When considering Bayesian MMSE error estimation, there are two sources of randomness: (1) the sampling distribution and (2) uncertainty in the underlying feature–label distribution. Thus, for a fixed sample, one can examine performance across the uncertainty class. This leads to the *sample-conditioned MSE,* $\text{MSE}(\hat{\varepsilon}|S_n)$ (Dalton and Dougherty, 2012a). This is of great practical value because one can update the RMS during the sampling process to have a real-time performance measure. In particular, one can employ *censored sampling*, where sampling is performed until $\text{MSE}(\hat{\varepsilon}|S_n)$ is satisfactory or until both $\text{MSE}(\hat{\varepsilon}|S_n)$ and the error estimate are satisfactory (Dalton and Dougherty, 2012b).

Applying basic properties of expectation yields

$$\begin{aligned}
\text{MSE}(\hat{\varepsilon}|S_n) &= E_\theta\left[(\varepsilon_n(\theta) - \hat{\varepsilon})^2|S_n\right] \\
&= E_\theta\left[(\varepsilon_n(\theta) - \hat{\varepsilon})\varepsilon_n(\theta)|S_n\right] + E_\theta\left[(\varepsilon_n(\theta) - \hat{\varepsilon})\hat{\varepsilon}|S_n\right] \\
&= E_\theta\left[(\varepsilon_n(\theta) - \hat{\varepsilon})\varepsilon_n(\theta)|S_n\right] \\
&= E_\theta\left[(\varepsilon_n(\theta))^2|S_n\right] - (\hat{\varepsilon})^2 \\
&= \text{Var}_\theta\left(\varepsilon_n(\theta)|S_n\right). \qquad (8.20)
\end{aligned}$$

Owing to the posterior independence between θ_0, θ_1, and c, this can be expanded via the Conditional Variance Formula (c.f. Appendix A.9) to obtain

$$\begin{aligned}
\text{MSE}(\hat{\varepsilon}|S_n) &= \text{Var}_{c,\theta_0,\theta_1}\left(c\varepsilon_n^0(\theta_0) + (1-c)\varepsilon_n^1(\theta_1)|S_n\right) \\
&= \text{Var}_c\left(E_{\theta_0,\theta_1}\left[c\varepsilon_n^0(\theta_0) + (1-c)\varepsilon_n^1(\theta_1)|c,S_n\right]|S_n\right) \\
&\quad + E_c\left[\text{Var}_{\theta_0,\theta_1}\left(c\varepsilon_n^0(\theta_0) + (1-c)\varepsilon_n^1(\theta_1)|c,S_n\right)|S_n\right]. \qquad (8.21)
\end{aligned}$$

Further decomposing the inner expectation and variance yields

$$\begin{aligned}
\text{MSE}(\hat{\varepsilon}|S_n) &= \text{Var}_c\left(c\hat{\varepsilon}^0 + (1-c)\hat{\varepsilon}^1|S_n\right) \\
&\quad + E_{\pi^*}\left[c^2\right]\text{Var}_{\theta_0}\left(\varepsilon_n^0(\theta_0)|S_n\right) + E_{\pi^*}\left[(1-c)^2\right]\text{Var}_{\theta_1}\left(\varepsilon_n^1(\theta_1)|S_n\right) \\
&= \text{Var}_{\pi^*}(c)\left(\hat{\varepsilon}^0 - \hat{\varepsilon}^1\right)^2 \\
&\quad + E_{\pi^*}\left[c^2\right]\text{Var}_{\pi^*}\left(\varepsilon_n^0(\theta_0)\right) + E_{\pi^*}\left[(1-c)^2\right]\text{Var}_{\pi^*}\left(\varepsilon_n^1(\theta_1)\right). \qquad (8.22)
\end{aligned}$$

Using the identity

$$\text{Var}_{\pi^*}\left(\varepsilon_n^y(\theta_y)\right) = E_{\pi^*}[\left(\varepsilon_n^y(\theta_y)\right)^2] - \left(\hat{\varepsilon}^y\right)^2, \qquad (8.23)$$

we obtain the representation

$$\begin{aligned}\mathrm{MSE}(\hat{\varepsilon}|S_n) = &-\left(E_{\pi^*}[c]\right)^2(\hat{\varepsilon}^0)^2 - 2\mathrm{Var}_{\pi^*}(c)\hat{\varepsilon}^0\hat{\varepsilon}^1 - \left(E_{\pi^*}[1-c]\right)^2(\hat{\varepsilon}^1)^2 \\ &+ E_{\pi^*}\left[c^2\right]E_{\pi^*}[\left(\varepsilon_n^0(\theta_0)\right)^2] + E_{\pi^*}[(1-c)^2]E_{\pi^*}[\left(\varepsilon_n^1(\theta_1)\right)^2]. \end{aligned} \tag{8.24}$$

Since $\hat{\varepsilon}^y = E_{\pi^*}[\varepsilon_n^y(\theta_y)]$, excluding the expectations involving c, we need to find the first and second moments $E_{\pi^*}[\varepsilon_n^y]$ and $E_{\pi^*}[(\varepsilon_n^y)^2]$, for $y = 0, 1$, in order to obtain $\mathrm{MSE}(\hat{\varepsilon}|S_n)$.

Having evaluated the conditional MSE of the Bayesian MMSE error estimate, one can obtain the conditional MSE for an arbitrary error estimate $\hat{\varepsilon}_\bullet$ (Dalton and Dougherty, 2012a). Since $E_\theta[(\varepsilon_n(\theta)|S_n] = \hat{\varepsilon}$,

$$\begin{aligned}\mathrm{MSE}(\hat{\varepsilon}_\bullet \mid S_n) &= E_\theta\left[(\varepsilon_n(\theta) - \hat{\varepsilon}_\bullet)^2|S_n\right] \\ &= E_\theta\left[(\varepsilon_n(\theta) - \hat{\varepsilon} + \hat{\varepsilon} - \hat{\varepsilon}_\bullet)^2|S_n\right] \\ &= E_\theta\left[(\varepsilon_n(\theta) - \hat{\varepsilon})^2|S_n\right] \\ &\quad + 2\left(\hat{\varepsilon} - \hat{\varepsilon}_\bullet\right)E_\theta\left[\varepsilon_n(\theta) - \hat{\varepsilon}|S_n\right] + (\hat{\varepsilon} - \hat{\varepsilon}_\bullet)^2 \\ &= \mathrm{MSE}(\hat{\varepsilon}|S_n) + (\hat{\varepsilon} - \hat{\varepsilon}_\bullet)^2. \end{aligned} \tag{8.25}$$

Note that (8.25) shows that the conditional MSE of the Bayesian MMSE error estimator lower bounds the conditional MSE of any other error estimator.

According to (8.24), ignoring the *a priori* class probabilities, to find both the Bayesian MMSE error estimate and the sample-conditioned MSE, it suffices to evaluate the first-order posterior moments, $E_{\pi^*}[\varepsilon_n^0]$ and $E_{\pi^*}[\varepsilon_n^1]$, and the second-order posterior moments, $E_{\pi^*}[(\varepsilon_n^0)^2]$ and $E_{\pi^*}[(\varepsilon_n^1)^2]$. We will now do this for discrete classification and linear classification for Gaussian class-conditional densities. The moment analysis depends on evaluating a number of complex integrals and some of the purely integration equalities will be stated as lemmas, with reference to the literature for their proofs.

8.3 DISCRETE CLASSIFICATION

For discrete classification we consider an arbitrary number of bins with generalized beta priors. Using the same notation of Section 1.5, define the parameters for each class to contain all but one bin probability, that is, $\theta_0 = \{p_1, p_2, \ldots, p_{b-1}\}$ and $\theta_1 = \{q_1, q_2, \ldots, q_{b-1}\}$. With this model, each parameter space is defined as the set of all valid bin probabilities, for example $\{p_1, p_2, \ldots, p_{b-1}\} \in \Theta_0$ if and only if $0 \le p_i \le 1$ for $i = 1, \ldots, b-1$ and $\sum_{i=1}^{b-1} p_i \le 1$. Given θ_0, the last bin probability is defined by $p_b = 1 - \sum_{i=1}^{b-1} p_i$. We use the Dirichlet priors

$$\pi(\theta_0) \propto \prod_{i=1}^{b} p_i^{\alpha_i^0 - 1} \quad \text{and} \quad \pi(\theta_1) \propto \prod_{i=1}^{b} q_i^{\alpha_i^1 - 1}, \tag{8.26}$$

where $\alpha_i^y > 0$. As we will shortly see, these are conjugate priors taking the same form as the posteriors. If $\alpha_i^y = 1$ for all i and y, we obtain uniform priors. As we increase a specific α_i^y, it is as if we bias the corresponding bin with α_i^y samples from the corresponding class before ever observing the data.

For discrete classifiers, it is convenient to work with ordered bin dividers in the integrals rather than the bin probabilities directly. Define the linear one-to-one change of variables

$$a_{(i)}^0 = \begin{cases} 0\,, & i = 0\,, \\ \sum_{k=1}^{i} p_k & i = 1, \ldots, b-1\,, \\ 1\,, & i = b\,. \end{cases} \tag{8.27}$$

Define $a_{(i)}^1$ similarly using q_k. We use the subscript (i) instead of just i to emphasize that the $a_{(i)}^y$ are ordered so that $0 \le a_{(1)}^y \le \ldots \le a_{(b-1)}^y \le 1$. The bin probabilities are determined by the partitions the $a_{(i)}^y$ make in the interval [0, 1], that is, $p_i = a_{(i)}^0 - a_{(i-1)}^0$, for $i = 1, \ldots, b$. The Jacobean determinant of this transformation is 1, so integrals over the p_i may be converted to integrals over the $a_{(i)}^0$ by simply replacing p_i with $a_{(i)}^0 - a_{(i-1)}^0$ and defining the new integration region characterized by $0 \le a_{(1)}^y \le \ldots \le a_{(b-1)}^y \le 1$. To obtain the posterior distributions $\pi^*(\boldsymbol{\theta}_0)$ and $\pi^*(\boldsymbol{\theta}_1)$ using these changes of variable, we will employ the following lemma. As in Section 1.5, let U_i and V_i, for $i = 1, \ldots, b$ denote the number of sample points observed in each bin for classes 0 and 1, respectively. We begin by stating a technical lemma, deferring to the literature for the proof.

Lemma 8.1 *(Dalton and Dougherty, 2011b) Let $b \ge 2$ be an integer and let $U_i > -1$ for $i = 1, \ldots, b$. If $0 = a_{(0)} \le a_{(1)} \le \ldots \le a_{(b-1)} \le a_{(b)} = 1$, then*

$$\int_0^1 \int_0^{a_{(b-1)}} \cdots \int_0^{a_{(3)}} \int_0^{a_{(2)}} \prod_{i=1}^{b} \left(a_{(i)} - a_{(i-1)}\right)^{U_i} da_{(1)} da_{(2)} \ldots da_{(b-2)} da_{(b-1)}$$
$$= \frac{\prod_{k=1}^{b} \Gamma\left(U_k + 1\right)}{\Gamma\left(\sum_{i=1}^{b} U_i + b\right)}\,. \tag{8.28}$$

Theorem 8.1 *(Dalton and Dougherty, 2011b) In the discrete model with Dirichlet priors (8.26), the posterior distributions for $\boldsymbol{\theta}_0$ and $\boldsymbol{\theta}_1$ are given by*

$$\pi^*(\boldsymbol{\theta}_0) = \frac{\Gamma\left(n_0 + \sum_{i=1}^{b} \alpha_i^0\right)}{\prod_{k=1}^{b} \Gamma\left(U_k + \alpha_k^0\right)} \prod_{i=1}^{b} p_i^{U_i + \alpha_i^0 - 1}\,, \tag{8.29}$$

$$\pi^*(\boldsymbol{\theta}_1) = \frac{\Gamma\left(n_1 + \sum_{i=1}^{b} \alpha_i^1\right)}{\prod_{k=1}^{b} \Gamma\left(V_k + \alpha_k^1\right)} \prod_{i=1}^{b} q_i^{V_i + \alpha_i^1 - 1}\,. \tag{8.30}$$

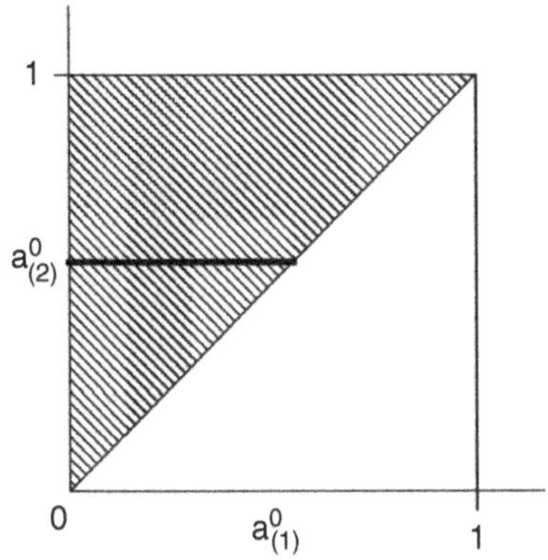

FIGURE 8.1 Region of integration in $E_{\pi^*}[\varepsilon_n^0]$ for $b = 3$. Reproduced from Dalton and Dougherty (2011b) with permission from IEEE.

Proof: Focusing on class 0 and utilizing (8.27), note that $f_{\theta_0}(x_i|0)$ is equal to the bin probability corresponding to bin x_i, so that we have the likelihood function

$$\prod_{i=1}^{n_0} f_{\theta_0}(x_i|0) = \prod_{i=1}^{b} p_i^{U_i} = \prod_{i=1}^{b} \left(a^0_{(i)} - a^0_{(i-1)}\right)^{U_i}. \tag{8.31}$$

The posterior parameter density is still proportional to the product of the bin probabilities:

$$\pi^*(\theta_0) \propto \prod_{i=1}^{b} \left(a^0_{(i)} - a^0_{(i-1)}\right)^{U_i+\alpha_i^0-1}. \tag{8.32}$$

To find the proportionality constant in $\pi^*(\theta_0)$, partition the region of integration so that the bin dividers are ordered. An example of this region is shown in Figure 8.1 for $b = 3$. Let the last bin divider, $a^0_{(b-1)}$, vary freely between 0 and 1. Once this divider is fixed, the next smallest bin divider can vary from 0 to $a^0_{(b-1)}$. This continues until we reach the first bin divider, $a^0_{(1)}$, which can vary from 0 to $a^0_{(2)}$. The proportionality constant for $\pi^*(\theta_0)$ can be found by applying Lemma 8.1 to (8.32) to obtain

$$\pi^*(\theta_0) = \frac{\Gamma\left(n_0 + \sum_{i=1}^{b} \alpha_i^0\right)}{\prod_{k=1}^{b} \Gamma\left(U_k + \alpha_k^0\right)} \prod_{i=1}^{b} \left(a^0_{(i)} - a^0_{(i-1)}\right)^{U_i+\alpha_i^0-1}. \tag{8.33}$$

■

Theorem 8.2 *(Dalton and Dougherty, 2011b, 2012a) In the discrete model with Dirichlet priors (8.26), for any fixed discrete classifier, ψ_n, the first-order posterior* *moments are given by*

$$E_{\pi^*}\left[\varepsilon_n^0\right] = \sum_{j=1}^{b} \frac{U_j + \alpha_j^0}{n_0 + \sum_{i=1}^{b} \alpha_i^0} I_{\psi_n(j)=1}, \tag{8.34}$$

$$E_{\pi^*}\left[\varepsilon_n^1\right] = \sum_{j=1}^{b} \frac{V_j + \alpha_j^1}{n_1 + \sum_{i=1}^{b} \alpha_i^1} I_{\psi_n(j)=0}, \tag{8.35}$$

while the second-order posterior moments are given by

$$E_{\pi^*}\left[\left(\varepsilon_n^0\right)^2\right] = \frac{1 + \sum_{j=1}^{b} I_{\psi_n(j)=1}\left(U_j + \alpha_j^0\right)}{1 + n_0 + \sum_{i=1}^{b} \alpha_i^0} \frac{\sum_{j=1}^{b} I_{\psi_n(j)=1}\left(U_j + \alpha_j^0\right)}{n_0 + \sum_{i=1}^{b} \alpha_i^0}, \tag{8.36}$$

$$E_{\pi^*}\left[\left(\varepsilon_n^1\right)^2\right] = \frac{1 + \sum_{j=1}^{b} I_{\psi_n(j)=0}\left(V_j + \alpha_j^1\right)}{1 + n_1 + \sum_{i=1}^{b} \alpha_i^1} \frac{\sum_{j=1}^{b} I_{\psi_n(j)=0}\left(V_j + \alpha_j^1\right)}{n_1 + \sum_{i=1}^{b} \alpha_i^1}. \tag{8.37}$$

Proof: We prove the case $y = 0$, the case $y = 1$ being completely analogous. Utilizing the representation $p_i = a_{(i)}^0 - a_{(i-1)}^0$ employed previously in Theorem 8.1, $E_{\pi^*}[\varepsilon_n^y]$ is found from (8.17):

$$E_{\pi^*}\left[\varepsilon_n^0\right] = \int_0^1 \cdots \int_0^{a_{(2)}} \varepsilon_n^0(\boldsymbol{\theta}_0)\pi^*(\boldsymbol{\theta}_0)da_{(1)}^0 \ldots da_{(b-1)}^0, \tag{8.38}$$

where $\varepsilon_n^0(\boldsymbol{\theta}_0)$ is the true error for class 0, and $\pi^*(\boldsymbol{\theta}_0)$ is given by Theorem 8.1. Using the region of integration illustrated in Figure 8.1, the expected true error for class 0 is

$$\begin{aligned}
E_{\pi^*}\left[\varepsilon_n^0\right] &= \int_0^1 \int_0^{a_{(b-1)}^0} \cdots \int_0^{a_{(3)}^0} \int_0^{a_{(2)}^0} \varepsilon_n^0(\boldsymbol{\theta}_0)\pi^*(\boldsymbol{\theta}_0)da_{(1)}^0 da_{(2)}^0 \ldots da_{(b-2)}^0 da_{(b-1)}^0 \\
&= \int_0^1 \cdots \int_0^{a_{(2)}^0} \left(\sum_{j=1}^{b} \left(a_{(j)}^0 - a_{(j-1)}^0\right) I_{\psi_n(j)=1}\right) \\
&\quad \times \left(\frac{\Gamma\left(n_0 + \sum_{i=1}^{b} \alpha_i^0\right)}{\prod_{k=1}^{b} \Gamma\left(U_k + \alpha_k^0\right)} \prod_{i=1}^{b} \left(a_{(i)}^0 - a_{(i-1)}^0\right)^{U_i + \alpha_i^0 - 1}\right) da_{(1)}^0 \ldots da_{(b-1)}^0 \\
&= \sum_{j=1}^{b} \frac{\Gamma\left(n_0 + \sum_{i=1}^{b} \alpha_i^0\right)}{\prod_{k=1}^{b} \Gamma\left(U_k + \alpha_k^0\right)} I_{\psi_n(j)=1} \\
&\quad \times \int_0^1 \cdots \int_0^{a_{(2)}^0} \prod_{i=1}^{b} \left(a_{(i)}^0 - a_{(i-1)}^0\right)^{U_i + \alpha_i^0 - 1 + \delta_{i-j}} da_{(1)}^0 \ldots da_{(b-1)}^0,
\end{aligned} \tag{8.39}$$

where δ_i is the Kronecker delta function, equal to 1 if $i = 0$ and 0 otherwise. The inner integrals can also be solved using Lemma 8.1:

$$
\begin{aligned}
E_{\pi^*}\left[\varepsilon_n^0\right] &= \sum_{j=1}^{b} \frac{\Gamma\left(n_0 + \sum_{i=1}^{b} \alpha_i^0\right)}{\prod_{k=1}^{b} \Gamma\left(U_k + \alpha_k^0\right)} I_{\psi_n(j)=1} \frac{\prod_{k=1}^{b} \Gamma\left(U_k + \alpha_k^0 + \delta_{k-j}\right)}{\Gamma\left(\sum_{i=1}^{b}(U_i + \alpha_i^0 - 1 + \delta_{i-j}) + b\right)} \\
&= \sum_{j=1}^{b} \frac{\Gamma\left(n_0 + \sum_{i=1}^{b} \alpha_i^0\right)}{\prod_{k=1}^{b} \Gamma\left(U_k + \alpha_k^0\right)} I_{\psi_n(j)=1} \frac{\left(U_j + \alpha_j^0\right) \prod_{k=1}^{b} \Gamma\left(U_k + \alpha_k^0\right)}{\left(n_0 + \sum_{i=1}^{b} \alpha_i^0\right) \Gamma\left(n_0 + \sum_{i=1}^{b} \alpha_i^0\right)} \\
&= \sum_{j=1}^{b} \frac{U_j + \alpha_j^0}{n_0 + \sum_{i=1}^{b} \alpha_i^0} I_{\psi_n(j)=1} \, .
\end{aligned}
\tag{8.40}
$$

For the second moment,

$$
\begin{aligned}
E_{\pi^*}\left[\left(\varepsilon_n^0(\theta_0)\right)^2\right] &= \int_{\Theta_0} \left(\varepsilon_n^0(\theta_0)\right)^2 \pi^*(\theta_0) d\theta_0 \\
&= \int_0^1 \cdots \int_0^{a_{(2)}^0} \left(\sum_{j=1}^{b} \left(a_{(j)}^0 - a_{(j-1)}^0\right) I_{\psi_n(j)=1} \right)^2 \\
&\quad \times \left(\frac{\Gamma\left(n_0 + \sum_{i=1}^{b} \alpha_i^0\right)}{\prod_{i=1}^{b} \Gamma\left(U_i + \alpha_i^0\right)} \prod_{i=1}^{b} \left(a_{(i)}^0 - a_{(i-1)}^0\right)^{U_i + \alpha_i^0 - 1} \right) da_{(1)}^0 \ldots da_{(b-1)}^0 \\
&= \frac{\Gamma\left(n_0 + \sum_{i=1}^{b} \alpha_i^0\right)}{\prod_{i=1}^{b} \Gamma\left(U_i + \alpha_i^0\right)} \sum_{j=1}^{b} \sum_{k=1}^{b} I_{\psi_n(j)=1} I_{\psi_n(k)=1} \\
&\quad \times \int_0^1 \cdots \int_0^{a_{(2)}^0} \prod_{i=1}^{b} \left(a_{(i)}^0 - a_{(i-1)}^0\right)^{U_i + \alpha_i^0 - 1 + \delta_{i-j} + \delta_{i-k}} da_{(1)}^0 \ldots da_{(b-1)}^0 \\
&= \frac{\Gamma\left(n_0 + \sum_{i=1}^{b} \alpha_i^0\right)}{\prod_{i=1}^{b} \Gamma\left(U_i + \alpha_i^0\right)} \sum_{j=1}^{b} \sum_{k=1}^{b} I_{\psi_n(j)=1} I_{\psi_n(k)=1} \\
&\quad \times \frac{\prod_{i=1}^{b} \Gamma\left(U_i + \alpha_i^0 + \delta_{i-j} + \delta_{i-k}\right)}{\Gamma\left(b + \sum_{i=1}^{b} \left(U_i + \alpha_i^0 - 1 + \delta_{i-j} + \delta_{i-k}\right)\right)} \\
&= \frac{\Gamma\left(n_0 + \sum_{i=1}^{b} \alpha_i^0\right)}{\prod_{i=1}^{b} \Gamma\left(U_i + \alpha_i^0\right)} \sum_{j=1}^{b} \sum_{k=1}^{b} I_{\psi_n(j)=1} I_{\psi_n(k)=1} \\
&\quad \times \frac{\left(U_k + \alpha_k^0 + \delta_{k-j}\right) \left(U_j + \alpha_j^0\right) \prod_{i=1}^{b} \Gamma\left(U_i + \alpha_i^0\right)}{\left(1 + n_0 + \sum_{i=1}^{b} \alpha_i^0\right) \left(n_0 + \sum_{i=1}^{b} \alpha_i^0\right) \Gamma\left(n_0 + \sum_{i=1}^{b} \alpha_i^0\right)}
\end{aligned}
\tag{8.41}
$$

$$= \sum_{j=1}^{b} \sum_{k=1}^{b} I_{\psi_n(j)=1} I_{\psi_n(k)=1} \frac{\left(U_k + \alpha_k^0 + \delta_{k-j}\right)\left(U_j + \alpha_j^0\right)}{\left(1 + n_0 + \sum_{i=1}^{b} \alpha_i^0\right)\left(n_0 + \sum_{i=1}^{b} \alpha_i^0\right)}$$

$$= \frac{\sum_{j=1}^{b} I_{\psi_n(j)=1} \left(U_j + \alpha_j^0\right) \sum_{k=1}^{b} I_{\psi_n(k)=1} \left(U_k + \alpha_k^0 + \delta_{k-j}\right)}{\left(1 + n_0 + \sum_{i=1}^{b} \alpha_i^0\right)\left(n_0 + \sum_{i=1}^{b} \alpha_i^0\right)}$$

$$= \frac{1 + \sum_{j=1}^{b} I_{\psi_n(j)=1} \left(U_j + \alpha_j^0\right)}{1 + n_0 + \sum_{i=1}^{b} \alpha_i^0} \frac{\sum_{j=1}^{b} I_{\psi_n(j)=1} \left(U_j + \alpha_j^0\right)}{n_0 + \sum_{i=1}^{b} \alpha_i^0}, \tag{8.42}$$

where the fourth equality follows from Lemma 8.1 and the fifth equality follows from properties of the gamma function. ∎

Inserting (8.34) and (8.35) into (8.16) yields the Bayesian MMSE error estimate. Inserting (8.36) and (8.37) into (8.24) yields $\text{MSE}(\hat{\varepsilon}|S_n)$. In the special case where we have uniform priors for the bin probabilities ($\alpha_i^y = 1$ for all i and y) and uniform prior for c, the Bayesian MMSE error estimate is

$$\hat{\varepsilon} = \frac{n_0 + 1}{n + 2} \left(\sum_{i=1}^{b} \frac{U_i + 1}{n_0 + b} I_{\psi_n(i)=1} \right) + \frac{n_1 + 1}{n + 2} \left(\sum_{i=1}^{b} \frac{V_i + 1}{n_1 + b} I_{\psi_n(i)=0} \right). \tag{8.43}$$

To illustrate the RMS performance of Bayesian error estimators, we utilize noninformative uniform priors for an arbitrary number of bins using the histogram classification rule and random sampling throughout. Figure 8.2 gives the average RMS deviation from the true error, as a function of sample size, over random distributions and random samples in the model with uniform priors for the bin probabilities and c, and for bin sizes 2, 4, 8, and 16. As must be the case, it is optimal with respect to $E_{\theta,S_n}[|\varepsilon_n(\theta, S_n) - \xi(S_n)|^2]$. Even with uniform priors the Bayesian MMSE error estimator shows great improvement over resubstitution and leave-one-out, especially for small samples or a large number of bins.

In the next set of experiments, the distributions are fixed with $c = 0.5$ and we use the Zipf model, where $p_i \propto i^{-\alpha}$ and $q_i = p_{b-i+1}$, $i = 1, \ldots, b$, the parameter $\alpha \geq 0$ being a free parameter used to target a specific Bayes error, with larger α corresponding to smaller Bayes error. Figure 8.3 shows RMS as a function of Bayes error for sample sizes 5 and 20 for fixed Zipf distributions, again for bin sizes 2, 4, 8, and 16. These graphs show that performance for Bayesian error estimators is superior to resubstitution and leave-one-out for most distributions and tends to be especially favorable with moderate-to-high Bayes errors and smaller sample sizes. Performance of the Bayesian MMSE error estimators tends to be more uniform across all distributions, while the other error estimators, especially resubstitution, favor a small Bayes error. Bayesian MMSE error estimators are guaranteed to be optimal on average over

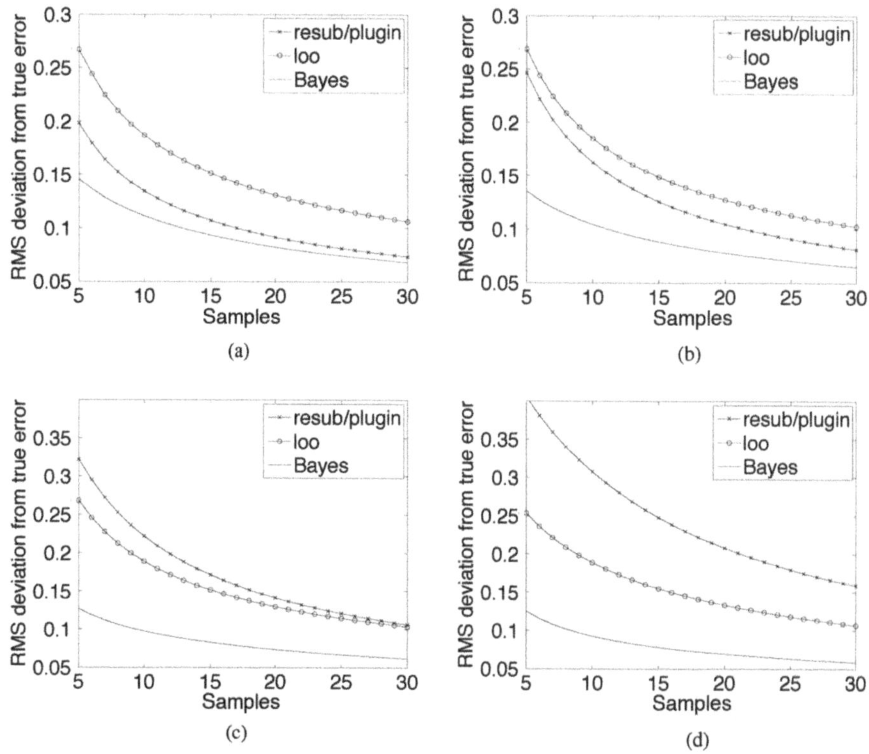

FIGURE 8.2 RMS deviation from true error for discrete classification and uniform priors with respect to sample size (bin probabilities and c uniform). (a) $b = 2$, (b) $b = 4$, (c) $b = 8$, (d) $b = 16$. Reproduced from Dalton and Dougherty (2011b) with permission from IEEE.

the ensemble of parameterized distributions modeled with respect to the given priors; however, they are not guaranteed to be optimal for a specific distribution. Performance of the Bayesian MMSE error estimator suffers for small Bayes errors under the assumption of uniform prior distributions. Should, however, one be confident that the Bayes error is small, then it would be prudent to utilize prior distributions that favor small Bayes errors, thereby correcting the problem (see (Dalton and Dougherty, 2011b) for the effect of using different priors). Recall that from previous analyses, both simulation and theoretical, one must assume a very small Bayes error to have any hope of a reasonable RMS for cross-validation, in which case the Bayesian MMSE error estimator would still retain superior performance across the uncertainty class.

Using Monte-Carlo techniques, we now consider the conditional MSE assuming Dirichlet priors with hyperparameters $\alpha_i^0 \propto 2b - 2i + 1$ and $\alpha_i^1 \propto 2i - 1$, where the α_i^y are normalized such that $\sum_{i=1}^{b} \alpha_i^y = b$ for $y = 0, 1$ and fixed $c = 0.5$ (see Dalton and Dougherty (2012b) for simulation details). The assumption on c means that the moments of c are given by $E_{\pi^*}[c] = E_{\pi^*}[1-c] = 0.5$, $E_{\pi^*}\left[c^2\right] = E_{\pi^*}\left[(1-c)^2\right] = 0.25$, and $\mathrm{Var}_{\pi^*}(c) = 0$. Figures 8.4a and 8.4b show the probability densities of

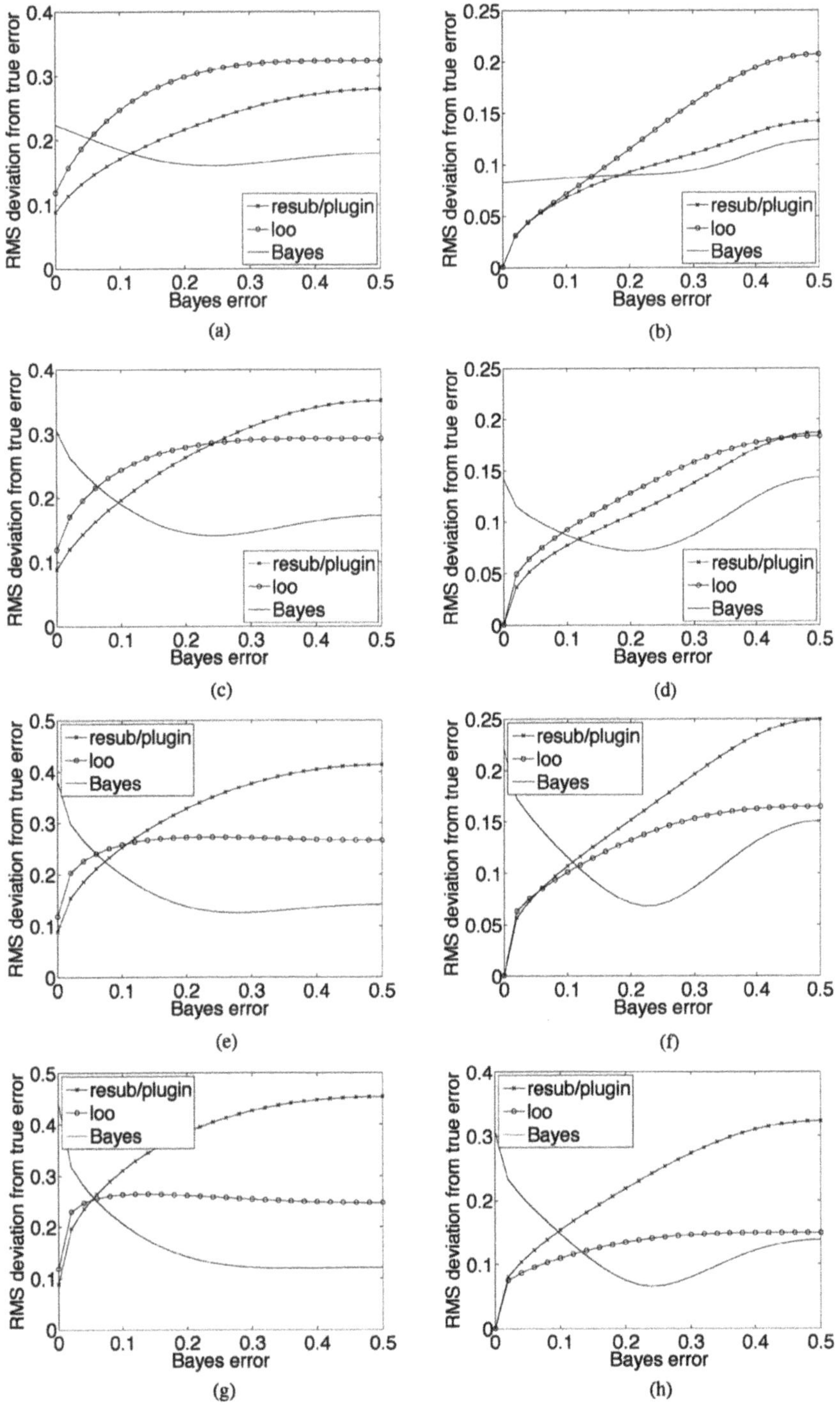

FIGURE 8.3 RMS deviation from true error for discrete classification and uniform priors with respect to Bayes error ($c = 0.5$). (a) $b = 2$, $n = 5$, (b) $b = 2$, $n = 20$, (c) $b = 4$, $n = 5$, (d) $b = 4$, $n = 20$, (e) $b = 8$, $n = 5$, (f) $b = 8$, $n = 20$, (g) $b = 16$, $n = 5$, (h) $b = 16$, $n = 20$. Reproduced from Dalton and Dougherty (2011b) with permission from IEEE.

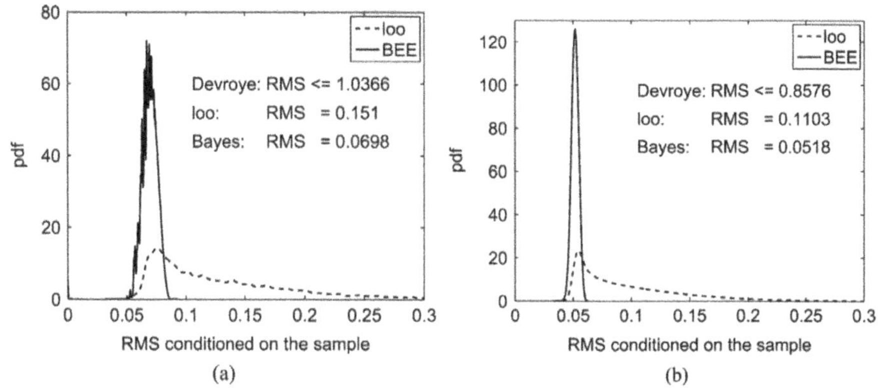

FIGURE 8.4 Probability densities for the conditional RMS of the leave-one-out and Bayesian error estimators. The sample sizes for each experiment were chosen so that the expected true error is 0.25. The unconditional RMS for both error estimators is also shown, as well as Devroye's distribution-free bound. (a) $b = 8$, $n = 16$. (b) $b = 16$, $n = 30$. Reproduced from Dalton and Dougherty (2012b) with permission from IEEE.

the conditional RMS for both the leave-one-out and Bayesian error estimators with settings $b = 8, n = 16$ and $b = 16, n = 30$, respectively, the sample sizes being chosen so that the expected true error is close to 0.25. Within each plot, we also show the unconditional RMS of both the leave-one-out and Bayesian error estimators, as well as the distribution-free RMS bound on the leave-one-out error estimator for the discrete histogram rule. In both parts of Figure 8.4 the density of the conditional RMS for the Bayesian error estimator is much tighter than that of leave-one-out. Moreover, the conditional RMS for the Bayesian error estimator is concentrated on lower values of RMS, so much so that in all cases the unconditional RMS of the Bayesian error estimator is less than half that of leave-one-out. Without any kind of modeling assumptions, distribution-free bounds on the unconditional RMS are too loose to be useful, with the bound from (2.44) exceeding 0.85 in both parts of the figure, whereas the Bayesian framework facilitates exact expressions for the sample-conditioned RMS.

In Section 8.5, we will show that $\text{MSE}(\hat{\varepsilon}|S_n) \to 0$ as $n \to \infty$ (almost surely relative to the sampling process) for the discrete model. Here, we show that there exists an upper bound on the conditional MSE as a function of the sample size under fairly general assumptions. This bound, provided in the next theorem, facilitates an interesting comparison between the performance of Bayesian MMSE error estimation and holdout.

Theorem 8.3 *(Dalton and Dougherty, 2012b) In the discrete model with the beta prior for c, if* $\alpha^0 \leq \sum_{i=1}^{b} \alpha_i^0$ *and* $\beta^0 \leq \sum_{i=1}^{b} \alpha_i^1$*, then*

$$\text{RMS}(\hat{\varepsilon}) \leq \sqrt{\frac{1}{4n}}. \tag{8.44}$$

Proof: Using the variance decomposition $\mathrm{Var}_{\pi^*}(\varepsilon_n^0(\theta_0)) = E_{\pi^*}[(\varepsilon_n^0(\theta_0))^2] - (\hat{\varepsilon}^0)^2$, applying (8.36), and simplifying yields

$$\begin{aligned}\mathrm{Var}_{\pi^*}\left(\varepsilon_n^0(\theta_0)\right) &= \left(\frac{\left(n_0 + \sum_{i=1}^b \alpha_i^0\right)\hat{\varepsilon}^0 + 1}{n_0 + \sum_{i=1}^b \alpha_i^0 + 1}\right)\hat{\varepsilon}^0 - \left(\hat{\varepsilon}^0\right)^2 \\ &= \frac{\hat{\varepsilon}^0(1-\hat{\varepsilon}^0)}{n_0 + \sum_{i=1}^b \alpha_i^0 + 1}. \end{aligned} \tag{8.45}$$

Similarly, for class 1,

$$\mathrm{Var}_{\pi^*}\left(\varepsilon_n^1(\theta_1)\right) = \frac{\hat{\varepsilon}^1(1-\hat{\varepsilon}^1)}{n_1 + \sum_{i=1}^b \alpha_i^1 + 1}. \tag{8.46}$$

Plugging these in (8.22) and applying the beta prior/posterior model for c yields

$$\begin{aligned}\mathrm{MSE}(\hat{\varepsilon}|S_n) = {} & \frac{\left(n_0+\alpha^0\right)\left(n_1+\beta^0\right)}{(n+\alpha^0+\beta^0)^2(n+\alpha^0+\beta^0+1)}(\hat{\varepsilon}^0-\hat{\varepsilon}^1)^2 \\ & + \frac{\left(n_0+\alpha^0\right)\left(n_0+\alpha^0+1\right)}{(n+\alpha^0+\beta^0)(n+\alpha^0+\beta^0+1)} \times \frac{\hat{\varepsilon}^0(1-\hat{\varepsilon}^0)}{n_0 + \sum_{i=1}^b \alpha_i^0 + 1} \\ & + \frac{\left(n_1+\beta^0\right)\left(n_1+\beta^0+1\right)}{(n+\alpha^0+\beta^0)(n+\alpha^0+\beta^0+1)} \times \frac{\hat{\varepsilon}^1(1-\hat{\varepsilon}^1)}{n_1 + \sum_{i=1}^b \alpha_i^1 + 1}. \end{aligned} \tag{8.47}$$

Applying the bounds on α^0 and β^0 in the theorem yields

$$\begin{aligned}\mathrm{MSE}(\hat{\varepsilon}|S_n) \le {} & \frac{1}{n+\alpha^0+\beta^0+1} \\ & \times \left(\frac{n_0+\alpha^0}{n+\alpha^0+\beta^0}\frac{n_1+\beta^0}{n+\alpha^0+\beta^0}(\hat{\varepsilon}^0-\hat{\varepsilon}^1)^2\right. \\ & \left. + \frac{n_0+\alpha^0}{n+\alpha^0+\beta^0}\hat{\varepsilon}^0(1-\hat{\varepsilon}^0) + \frac{n_1+\beta^0}{n+\alpha^0+\beta^0}\hat{\varepsilon}^1(1-\hat{\varepsilon}^1)\right) \\ = {} & \frac{1}{n+\alpha^0+\beta^0+1}\Big(E_{\pi^*}[c]\,E_{\pi^*}[1-c]\,(\hat{\varepsilon}^0-\hat{\varepsilon}^1)^2 \\ & + E_{\pi^*}[c]\,\hat{\varepsilon}^0(1-\hat{\varepsilon}^0) + E_{\pi^*}[1-c]\,\hat{\varepsilon}^1(1-\hat{\varepsilon}^1)\Big), \end{aligned} \tag{8.48}$$

where we have used $E_{\pi^*}[c] = (n_0 + \alpha^0)/(n + \alpha^0 + \beta^0)$ and $E_{\pi^*}[1-c] = (n_1 + \beta^0)/(n + \alpha^0 + \beta^0)$. To help simplify this equation further, define $x = \hat{\varepsilon}^0$, $y = \hat{\varepsilon}^1$, and $z = E_{\pi^*}[c]$. Then

$$\begin{aligned}\text{MSE}(\hat{\varepsilon}|S_n) &\le \frac{z(1-z)(x-y)^2 + zx(1-x) + (1-z)y(1-y)}{n + \alpha^0 + \beta^0 + 1} \\ &= \frac{zx + (1-z)y - (zx + (1-z)y)^2}{n + \alpha^0 + \beta^0 + 1}.\end{aligned} \tag{8.49}$$

From (8.19), $\hat{\varepsilon} = zx + (1-z)y$. Hence, since $0 \le \hat{\varepsilon} \le 1$,

$$\text{MSE}(\hat{\varepsilon}|S_n) \le \frac{\hat{\varepsilon} - (\hat{\varepsilon})^2}{n + \alpha^0 + \beta^0 + 1} \le \frac{1}{4(n + \alpha^0 + \beta^0 + 1)} \le \frac{1}{4n}, \tag{8.50}$$

where in the middle inequality we have used the fact that $w - w^2 = w(1-w) \le 1/4$ whenever $0 \le w \le 1$. Since this bound is only a function of the sample size, it holds if we remove the conditioning on S_n. ∎

Comparing (8.44) to the holdout bound in (2.28) , we observe that the RMS bound on the Bayesian MMSE error estimator is always lower than that of the holdout bound, since the holdout sample size m cannot exceed the full sample size n. Moreover, as $m \to n$ for full holdout, the holdout bound converges down to the Bayes estimate bound.

Using Monte-Carlo techniques, we illustrate the two bounds using a discrete model with bin size b, $\alpha^0 = \beta^0 = 1$, and the bin probabilities of class 0 and 1 having Dirichlet priors given by the hyperparameters $\alpha_i^0 \propto 2b - 2i + 1$ and $\alpha_i^1 \propto 2i - 1$, where the α_i^y are normalized such that $\sum_{i=1}^{b} \alpha_i^y = b$ for $y = 0, 1$ (see Dalton and Dougherty (2012b) for full Monte-Carlo details). Class 0 tends to assign more weight to bins with a low index and class 1 assigns a higher weight to bins with a high index. These priors satisfy the bounds on $\alpha^0$and β^0 in the theorem. The sample sizes for each experiment are chosen so that the expected true error of the classifier trained on the full sample is 0.25 when $c = 0.5$ is fixed (the true error being somewhat smaller than 0.25 because in the experiments c is uniform). Classification is done via the histogram rule. Results are shown in Figure 8.5 for $b = 8$ and $n = 16$: part (a) shows the expected true error and part (b) shows the RMS between the true and estimated errors, both as a function of the holdout sample size. As expected, the average true error of the classifier in the holdout experiment decreases and converges to the average true error of the classifier trained from the full sample as the holdout sample size decreases. In addition, the RMS performance of the Bayesian error estimator consistently surpasses that of the holdout error estimator, as suggested by the RMS bounds given in (8.44).

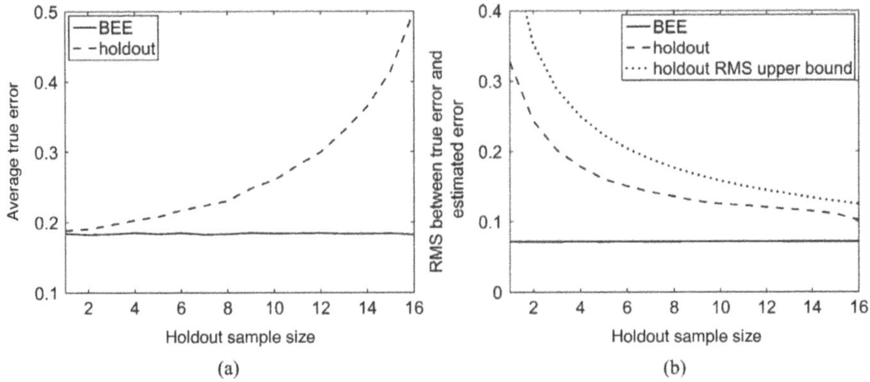

FIGURE 8.5 Comparison of the holdout error estimator and Bayesian error estimator with respect to the holdout sample size for a discrete model with $b = 8$ bins and fixed sample size $n = 16$. (a) Average true error. (b) RMS performance. Reproduced from Dalton and Dougherty (2012b) with permission from IEEE.

8.4 LINEAR CLASSIFICATION OF GAUSSIAN DISTRIBUTIONS

This section considers a Gaussian distribution with parameters $\boldsymbol{\theta}_y = \{\boldsymbol{\mu}_y, \Lambda_y\}$, where $\boldsymbol{\mu}_y$ is the mean with a parameter space comprised of the whole sample space R^d, and Λ_y is a collection of parameters that determine the covariance matrix Σ_y of the class. The parameter space of Λ_y, denoted $\boldsymbol{\Lambda}_y$, must permit only valid covariance matrices. Λ_y can be the covariance matrix itself, but we make a distinction to enable imposition of a structure on the covariance. Here we mainly focus on the general setting where $\boldsymbol{\Lambda}_y$ is the class of all positive definite matrices, in which case we simply write $\boldsymbol{\theta}_y = \{\boldsymbol{\mu}_y, \Sigma_y\}$. In Dalton and Dougherty (2011c, 2012a) constraints are placed on the covariance matrix and $\boldsymbol{\Lambda}_y$ conforms to those constraints. This model permits the use of improper priors; however, whether improper priors or not, the posterior must be a valid probability density.

Considering one class at a time, we assume that the sample covariance matrix $\hat{\Sigma}_y$ is nonsingular and priors are of the form

$$\pi(\boldsymbol{\theta}_y) \propto |\Sigma_y|^{-(\kappa+d+1)/2} \exp\left(-\frac{1}{2}\text{trace}\left(S\Sigma_y^{-1}\right)\right) \times |\Sigma_y|^{-1/2} \exp\left(-\frac{\nu}{2}(\boldsymbol{\mu}_y - \boldsymbol{m})^T \Sigma_y^{-1}(\boldsymbol{\mu}_y - \boldsymbol{m})\right), \tag{8.51}$$

where κ is a real number, S is a nonnegative definite, symmetric $d \times d$ matrix, $\nu \geq 0$ is a real number, and $\boldsymbol{m} \in R^d$. The hyperparameters $\boldsymbol{m}$ and S can be viewed as targets for the mean and the shape of the covariance, respectively. Also, the larger ν is, the more localized this distribution is about $\boldsymbol{m}$ and the larger κ is, the less the shape of Σ_y is allowed to vary. Increasing κ while fixing the other hyperparameters also defines a prior favoring smaller $|\Sigma_y|$. This prior may or may not be proper, depending on the definition of Λ_y and the parameters κ, S, ν, and $\boldsymbol{m}$. For instance, if $\Lambda_y = \Sigma_y$, then this prior is essentially the normal-inverse-Wishart distribution, which is the conjugate prior for the mean and covariance when sampling from normal distributions (DeGroot, 1970; Raiffa and Schlaifer, 1961). In this case, to insist on a proper prior we restrict $\kappa > d - 1$, S positive definite and symmetric, and $\nu > 0$.

Theorem 8.4 *(DeGroot, 1970) For fixed κ, S, ν, and m, the posterior density for the prior in (8.51) has the form*

$$\pi^*(\boldsymbol{\theta}_y) \propto |\Sigma_y|^{-(\kappa^*+d+1)/2} \exp\left(-\frac{1}{2}\mathrm{trace}\left(S^*\Sigma_y^{-1}\right)\right) \times |\Sigma_y|^{-\frac{1}{2}} \exp\left(-\frac{\nu^*}{2}\left(\boldsymbol{\mu}_y - \boldsymbol{m}^*\right)^T \Sigma_y^{-1}\left(\boldsymbol{\mu}_y - \boldsymbol{m}^*\right)\right), \tag{8.52}$$

where

$$\kappa^* = \kappa + n_y, \tag{8.53}$$

$$S^* = (n_y - 1)\hat{\Sigma}_y + S + \frac{n_y\nu}{n_y + \nu}(\hat{\boldsymbol{\mu}}_y - \boldsymbol{m})(\hat{\boldsymbol{\mu}}_y - \boldsymbol{m})^T, \tag{8.54}$$

$$\nu^* = \nu + n_y, \tag{8.55}$$

$$\boldsymbol{m}^* = \frac{n_y\hat{\boldsymbol{\mu}}_y + \nu\boldsymbol{m}}{n_y + \nu}. \tag{8.56}$$

Proof: From (8.10), following some algebraic manipulation,

$$\pi^*(\boldsymbol{\theta}_y) \propto \pi(\boldsymbol{\theta}_y)|\Sigma_y|^{-n_y/2} \exp\left(-\frac{1}{2}\mathrm{trace}\left((n_y - 1)\hat{\Sigma}_y\Sigma_y^{-1}\right) - \frac{n_y}{2}(\boldsymbol{\mu}_y - \hat{\boldsymbol{\mu}}_y)^T\Sigma_y^{-1}(\boldsymbol{\mu}_y - \hat{\boldsymbol{\mu}}_y)\right) \tag{8.57}$$

$$= |\Sigma_y|^{-\frac{\kappa+n_y+d+1}{2}} \exp\left(-\frac{1}{2}\mathrm{trace}\left(\left((n_y - 1)\hat{\Sigma}_y + S\right)\Sigma_y^{-1}\right)\right) \times |\Sigma_y|^{-\frac{1}{2}} \exp\left(-\frac{1}{2}\left(n_y(\boldsymbol{\mu}_y - \hat{\boldsymbol{\mu}}_y)^T\Sigma_y^{-1}(\boldsymbol{\mu}_y - \hat{\boldsymbol{\mu}}_y) + \nu(\boldsymbol{\mu}_y - \boldsymbol{m})^T\Sigma_y^{-1}(\boldsymbol{\mu}_y - \boldsymbol{m})\right)\right). \tag{8.58}$$

The theorem follows because

$$
\begin{aligned}
&n_y(\boldsymbol{\mu}_y - \hat{\boldsymbol{\mu}}_y)^T \Sigma_y^{-1} (\boldsymbol{\mu}_y - \hat{\boldsymbol{\mu}}_y) + \nu(\boldsymbol{\mu}_y - \boldsymbol{m})^T \Sigma_y^{-1} (\boldsymbol{\mu}_y - \boldsymbol{m}) \\
&= (n_y + \nu) \left(\boldsymbol{\mu}_y - \frac{n_y \hat{\boldsymbol{\mu}}_y + \nu \boldsymbol{m}}{n_y + \nu} \right)^T \Sigma_y^{-1} \left(\boldsymbol{\mu}_y - \frac{n_y \hat{\boldsymbol{\mu}}_y + \nu \boldsymbol{m}}{n_y + \nu} \right) \\
&\quad + \frac{n_y \nu}{n_y + \nu} (\hat{\boldsymbol{\mu}}_y - \boldsymbol{m})^T \Sigma_y^{-1} (\hat{\boldsymbol{\mu}}_y - \boldsymbol{m}) .
\end{aligned} \tag{8.59}
$$

■

Keeping in mind that the choice of Λ_y will affect the proportionality constant in $\pi^*(\boldsymbol{\theta}_y)$ and letting $f_{\{\boldsymbol{\mu},\Sigma\}}$ denote a Gaussian density with mean vector $\boldsymbol{\mu}$ and covariance Σ, we can write the posterior probability in (8.52) as

$$
\pi^*(\boldsymbol{\theta}_y) = \pi^*(\boldsymbol{\mu}_y | \Lambda_y) \pi^*(\Lambda_y) , \tag{8.60}
$$

where

$$
\pi^*(\boldsymbol{\mu}_y | \Lambda_y) = f_{\{m^*, \Sigma_y / \nu^*\}}(\boldsymbol{\mu}_y), \tag{8.61}
$$

$$
\pi^*(\Lambda_y) \propto |\Sigma_y|^{-(\kappa^* + d + 1)/2} \exp\left(-\frac{1}{2} \mathrm{trace}\left(S^* \Sigma_y^{-1} \right) \right) . \tag{8.62}
$$

Thus, for a fixed covariance matrix the posterior density for the mean, $\pi^*(\boldsymbol{\mu}_y | \Lambda_y)$, is Gaussian.

The posterior moments for ε_n^y follow from (8.60):

$$
E_{\pi^*}\left[\left(\varepsilon_n^y \right)^k \right] = \int_{\Lambda_y} \int_{R^d} \left(\varepsilon_n^y(\boldsymbol{\mu}_y, \Lambda_y) \right)^k \pi^*(\boldsymbol{\mu}_y | \Lambda_y) d\boldsymbol{\mu}_y \pi^*(\Lambda_y) d\Lambda_y , \tag{8.63}
$$

for $k = 1, 2, \ldots$ Let us now assume that the classifier discriminant is linear in form,

$$
\psi_n(\mathbf{x}) = \begin{cases} 0, & g(\mathbf{x}) \leq 0 , \\ 1 , & \text{otherwise}, \end{cases} \tag{8.64}
$$

where

$$
g(\mathbf{x}) = \mathbf{a}^T \mathbf{x} + b , \tag{8.65}
$$

with some constant vector $\mathbf{a}$ and constant scalar b, where $\mathbf{a}$ and b are not dependent on the parameter $\boldsymbol{\theta}$. With fixed distribution parameters and nonzero $\mathbf{a}$, the true error for this classifier applied to a class-y Gaussian distribution $f_{\{\boldsymbol{\mu}_y, \Sigma_y\}}$ is given by

$$
\varepsilon_n^y = \Phi\left(\frac{(-1)^y g(\boldsymbol{\mu}_y)}{\sqrt{\mathbf{a}^T \Sigma_y \mathbf{a}}} \right) , \tag{8.66}
$$

where $\Phi(x)$ is the cumulative distribution function of a standard $N(0, 1)$ Gaussian random variable. The following technical lemma, whose proof we omit, evaluates the inner integrals in (8.63), for $k = 1, 2$.

Lemma 8.2 *(Dalton and Dougherty, 2011c, 2012a) Suppose $\nu^* > 0$, Σ_y is an invertible covariance matrix, and $g(\mathbf{x}) = \mathbf{a}^T\mathbf{x} + b$, where $\mathbf{a} \neq \mathbf{0}$. Then*

$$\int_{R^d} \Phi\left(\frac{(-1)^y g(\boldsymbol{\mu}_y)}{\sqrt{\mathbf{a}^T\Sigma_y\mathbf{a}}}\right) f_{\{m^*,\Sigma_y/\nu^*\}}(\boldsymbol{\mu}_y)d\boldsymbol{\mu}_y = \Phi\left(d\right), \tag{8.67}$$

$$\int_{R^d} \left(\Phi\left(\frac{(-1)^y g(\boldsymbol{\mu}_y)}{\sqrt{\mathbf{a}^T\Sigma_y\mathbf{a}}}\right)\right)^2 f_{\{m^*,\Sigma_y/\nu^*\}}(\boldsymbol{\mu}_y)d\boldsymbol{\mu}_y$$
$$= I_{\{d>0\}}\left(2\Phi\left(d\right) - 1\right) + \frac{1}{\pi}\int_0^{\arctan\left(\frac{\sqrt{\nu^*+2}}{\sqrt{\nu^*}}\right)} \exp\left(-\frac{d^2}{2\sin^2\theta}\right)d\theta, \tag{8.68}$$

where

$$d = \frac{(-1)^y g\left(\boldsymbol{m}^*\right)}{\sqrt{\mathbf{a}^T\Sigma_y\mathbf{a}}}\sqrt{\frac{\nu^*}{\nu^*+1}}. \tag{8.69}$$

It follows at once from (8.67) that the Bayesian MMSE error estimator reduces to an integral over the covariance parameter only:

$$E_{\pi^*}\left[\varepsilon_n^y\right] = \int_{\Lambda_y} \Phi\left(\frac{(-1)^y g(\boldsymbol{m}^*)}{\sqrt{\mathbf{a}^T\Sigma_y\mathbf{a}}}\sqrt{\frac{\nu^*}{\nu^*+1}}\right)\pi^*(\Lambda_y)d\Lambda_y. \tag{8.70}$$

We now focus on the case $\Sigma_y = \Lambda_y$. Σ_y is an arbitrary covariance matrix and the parameter space $\boldsymbol{\Lambda}_y$ contains all positive definite matrices. This is called the *general covariance model* in Dalton and Dougherty (2011c, 2012a). Evaluation of the proportionality constant in (8.62) in this case yields an inverse-Wishart distribution (Mardia et al., 1979; O'Hagan and Forster, 2004),

$$\pi^*(\Sigma_y) = \frac{|S^*|^{\kappa^*/2}|\Sigma_y|^{-\frac{\kappa^*+d+1}{2}}}{2^{\kappa^*d/2}\Gamma_d(\kappa^*/2)}\exp\left(-\frac{1}{2}\mathrm{trace}\left(S^*\Sigma_y^{-1}\right)\right), \tag{8.71}$$

where Γ_d is the multivariate gamma function, we require S^* to be positive definite and symmetric (which is true when $\hat{\Sigma}_y$ is invertible), and $\kappa^* > d - 1$.

To evaluate the prior and posterior moments, we employ the integral equalities in the next two lemmas. They involve the *regularized incomplete beta function,*

$$I(x;a,b) = \frac{\Gamma(a+b)}{\Gamma(a)\Gamma(b)} \int_0^x t^{a-1}(1-t)^{b-1}dt\,, \tag{8.72}$$

defined for $0 \leq x \leq 1$, $a > 0$, and $b > 0$.

Lemma 8.3 *(Dalton and Dougherty, 2011c) Let $A \in R$, $\alpha > 0$, and $\beta > 0$. Let $f(x;\alpha,\beta)$ be the inverse-gamma distribution with shape parameter α and scale parameter β. Then*

$$\int_0^{\infty} \Phi\left(\frac{A}{\sqrt{z}}\right) f(z;\alpha,\beta)dz = \frac{1}{2}\left(1 + sgn(A) I\left(\frac{A^2}{A^2+2\beta};\frac{1}{2},\alpha\right)\right), \tag{8.73}$$

where $I(x;a,b)$ is the regularized incomplete beta function.

Lemma 8.4 *(Dalton and Dougherty, 2012a) Let $A \in R$, $\pi/4 < B < \pi/2$, $\mathbf{a} \neq \mathbf{0}$, $\kappa^* > d-1$, S^* be positive definite and symmetric, and $f_W(\Sigma;S^*,\kappa^*)$ be an inverse-Wishart distribution with parameters S^* and κ^*. Then*

$$\begin{aligned} &\int_{\Sigma>0} I_{\{A>0\}}\left(2\Phi\left(\frac{A}{\sqrt{\mathbf{a}^T\Sigma\mathbf{a}}}\right)-1\right) \\ &\quad + \frac{1}{\pi}\int_0^B \exp\left(-\frac{A^2}{\left(2\sin^2\theta\right)\mathbf{a}^T\Sigma\mathbf{a}}\right) d\theta f_W(\Sigma;S^*,\kappa^*)d\Sigma \\ &= I_{\{A>0\}} I\left(\frac{A^2}{A^2+\mathbf{a}^T S^*\mathbf{a}};\frac{1}{2},\frac{\kappa^*-d+1}{2}\right) \\ &\quad + R\left(\sin^2 B,\frac{A^2}{\mathbf{a}^T S^*\mathbf{a}};\frac{\kappa^*-d+1}{2}\right), \end{aligned} \tag{8.74}$$

where the outer integration is over all positive definite matrices, $I(x;a,b)$ is the regularized incomplete beta function, and R is given by an Appell hypergeometric function, F_1, with $R(x,0;a) = \sin^{-1}\left(\sqrt{x}\right)/\pi$ and

$$\begin{aligned} R(x,y;a) = &\frac{\sqrt{y}}{\pi(2a+1)}\left(\frac{x}{x+y}\right)^{a+\frac{1}{2}} \\ &\times F_1\left(a+\frac{1}{2};\frac{1}{2},1;a+\frac{3}{2};\frac{x(y+1)}{x+y},\frac{x}{x+y}\right), \end{aligned} \tag{8.75}$$

for $a > 0$, $0 < x < 1$ and $y > 0$.

Theorem 8.5 *(Dalton and Dougherty, 2011c, 2012a) In the general covariance model, with $\Sigma_y = \Lambda_y$, $\nu^* > 0$, $\kappa^* > d - 1$, and S^* positive definite and symmetric,*

$$E_{\pi^*}\left[\varepsilon_n^y\right] = \frac{1}{2}\left(1 + sgn(A) I\left(\frac{A^2}{A^2 + 2\beta}; \frac{1}{2}, \alpha\right)\right), \tag{8.76}$$

where

$$A = (-1)^y g(m^*)\sqrt{\frac{\nu^*}{\nu^* + 1}}, \tag{8.77}$$

$$\alpha = \frac{\kappa^* - d + 1}{2}, \tag{8.78}$$

$$\beta = \frac{\mathbf{a}^T S^* \mathbf{a}}{2}, \tag{8.79}$$

and

$$E_{\pi^*}\left[\left(\varepsilon_n^y\right)^2\right] = I_{\{A>0\}} I\left(\frac{A^2}{A^2 + 2\beta}; \frac{1}{2}, \alpha\right) + R\left(\frac{\nu^* + 2}{2(\nu^* + 1)}, \frac{A^2}{2\beta}; \alpha\right), \tag{8.80}$$

where I is the regularized incomplete gamma function and R is defined in (8.75).

Proof: Letting $f_W(X; S^*, \kappa^*)$ be the inverse-Wishart distribution (8.71) with parameters S^* and κ^*, it follows from (8.70) that

$$E_{\pi^*}\left[\varepsilon_n^y\right] = \int_{X>0} \Phi\left(\frac{A}{\sqrt{\mathbf{a}^T X \mathbf{a}}}\right) f_W(X; S^*, \kappa^*) dX, \tag{8.81}$$

where the integration is over all positive definite matrices. Let

$$B = \left[\begin{array}{c|c} \multicolumn{2}{c}{\mathbf{a}^T} \\ \hline 0_{d-1\times 1} & I_{d-1\cdot} \end{array}\right]. \tag{8.82}$$

Since $\mathbf{a}$ is nonzero, with a reordering of the dimensions we can guarantee $a_1 \neq 0$. The value of $\mathbf{a}^T S^* \mathbf{a}$ is unchanged by such a redefinition, so without loss of generality assume B is invertible. Next define a change of variables, $Y = BXB^T$. Since B is invertible, Y is positive definite if and only if X is also. Furthermore, the Jacobean determinant of this transformation is $|B|^{d+1}$ (Arnold, 1981; Mathai and Haubold, 2008). Note $\mathbf{a}^T X \mathbf{a} = y_{11}$, where the subscript 11 indexes the upper left element of a matrix. Hence,

$$\begin{aligned} E_{\pi^*}\left[\varepsilon_n^y\right] &= \int_{Y>0} \Phi\left(\frac{A}{\sqrt{y_{11}}}\right) f_W(B^{-1} Y (B^T)^{-1}; S^*, \kappa^*) \frac{dY}{|B|^{d+1}} \\ &= \int_{Y>0} \Phi\left(\frac{A}{\sqrt{y_{11}}}\right) f_W(Y; BS^*B^T, \kappa^*) dY. \end{aligned} \tag{8.83}$$

Since $\Phi\left(\frac{A}{\sqrt{y_{11}}}\right)$ now depends on only one parameter in Y, the other parameters can be integrated out. It can be shown that for any inverse-Wishart random variable X with density $f_W(X; S^*, \kappa^*)$, the marginal distribution of x_{11} is also an inverse-Wishart distribution with density $f_W(x_{11}; s_{11}^*, \kappa^* - d + 1)$ (Muller and Stewart, 2006). In one dimension, this is equivalent to the inverse-gamma distribution $f(x_{11}; (\kappa^* - d + 1)/2, s_{11}^*/2)$. In this case, $(BS^*B^T)_{11} = \mathbf{a}^T S^* \mathbf{a}$, so

$$E_{\pi^*}\left[\varepsilon_n^y\right] = \int_0^\infty \Phi\left(\frac{A}{\sqrt{y_{11}}}\right) f\left(y_{11}; \frac{\kappa^* - d + 1}{2}, \frac{\mathbf{a}^T S^* \mathbf{a}}{2}\right) dy_{11}. \tag{8.84}$$

This integral is solved in Lemma 8.3 and the result for $E_{\pi^*}[\varepsilon_n^y]$ follows.

As for the second moment, from (8.68),

$$\begin{aligned} E_{\pi^*}\left[\left(\varepsilon_n^y\right)^2\right] = \int_{\Sigma>0} & I_{\{A>0\}}\left(2\Phi\left(\frac{A}{\sqrt{\mathbf{a}^T\Sigma\mathbf{a}}}\right) - 1\right) \\ & + \frac{1}{\pi}\int_0^{\arctan\left(\frac{\sqrt{\nu^*+2}}{\sqrt{\nu^*}}\right)} \exp\left(-\frac{A^2}{\left(2\sin^2\theta\right)\mathbf{a}^T\Sigma\mathbf{a}}\right) d\theta f_W(\Sigma; S^*, \kappa^*) d\Sigma. \end{aligned} \tag{8.85}$$

Applying Lemma 8.4 yields the desired second moment. ■

Combining (8.16) and (8.76) gives the Bayesian MMSE error estimator for the general covariance Gaussian model; combining (8.76) and (8.80) with (8.24) gives the conditional MSE of the Bayesian error estimator under the general covariance model. Closed-form expressions for both R and the regularized incomplete beta function I for integer or half-integer values of α are provided in Dalton and Dougherty (2012a).

To illustrate the optimality and unbiasedness of Bayesian MMSE error estimators over all distributions in a model we consider linear discriminant analysis (LDA) classification on the Gaussian general covariance model. For both classes, define the prior parameters $\kappa = \nu = 5d$ and $S = (\nu - 1)\, 0.7413^2 I_d$. For class 0, define $m = [0, 0, \ldots, 0]$ and for class 1, define $m = [1, 0, \ldots, 0]$. In addition, assume a uniform distribution for c. Figure 8.6 shows the RMS and bias results for several error estimates, including the Bayesian MMSE error estimate, as functions of sample size n utilizing Monte-Carlo approximation for RMS and bias over all distributions and samples for 1, 2, and 5 features.

Now consider LDA classification on the Gaussian general covariance model with a fixed sample size, $n = 60$ and fixed $c = 0.5$. For the class-conditional distribution parameters, consider three priors: "low-information," "medium-information," and "high-information" priors, with hyperparameters defined in Table 8.1 (see Dalton and Dougherty (2012b) for a complete description of the Monte-Carlo methodology). All priors are proper probability densities. For each prior model, the parameter m for

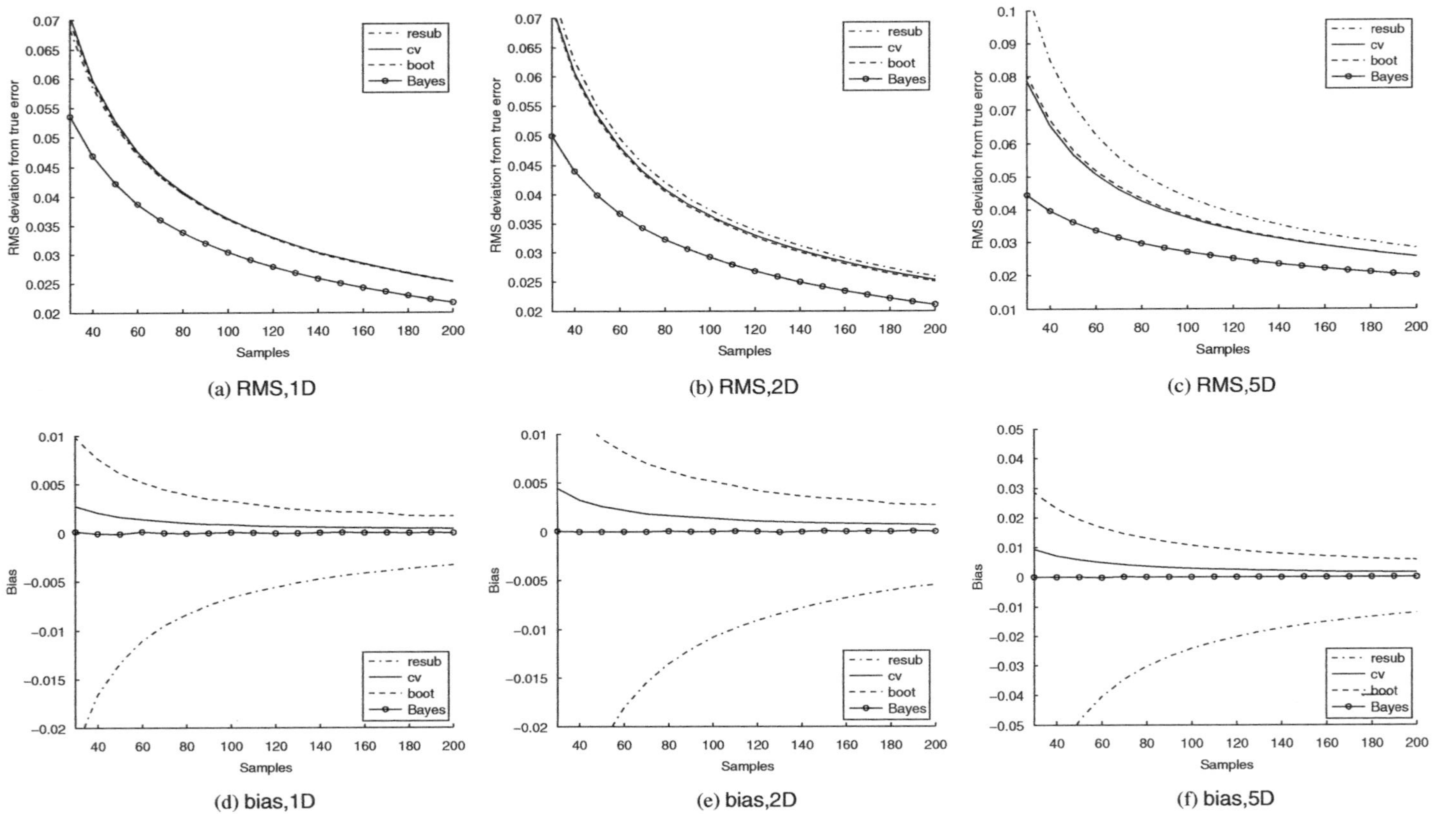

FIGURE 8.6 RMS deviation from true error and bias for linear classification of Gaussian distributions, averaged over all distributions and samples using a proper prior. (a) RMS, $d = 1$; (b) RMS, $d = 2$; (c) RMS, $d = 5$; (d) bias, $d = 1$; (e) bias, $d = 2$; (g) bias, $d = 5$.

TABLE 8.1 The "low-information," "medium-information," and "high-information" priors used in the Gaussian model

Hyperparameter	Low-information	Medium-information	High-information
a priori probability, c	fixed at 0.5	fixed at 0.5	fixed at 0.5
κ, class 0 and 1	$d = 3$	$d = 9$	$d = 54$
S, class 0 and 1	$0.03(\kappa - d - 1)I_d$	$0.03(\kappa - d - 1)I_d$	$0.03(\kappa - d - 1)I_d$
ν, class 0	$d = 6$	$d = 18$	$d = 108$
ν, class 1	$d = 3$	$d = 9$	$d = 54$
m, class 0	$[0, 0, \ldots, 0]$	$[0, 0, \ldots, 0]$	$[0, 0, \ldots, 0]$
m, class 1	$-0.1719[1, 1, \ldots, 1]$	$-0.2281[1, 1, \ldots, 1]$	$-0.2406[1, 1, \ldots, 1]$

class 1 has been calibrated to give an expected true error of 0.25 with one feature. The low-information prior is closer to a flat non-informative prior and models a setting where our knowledge about the distribution parameters is less certain. Conversely, the high-information prior has a relatively tight distribution around the expected parameters and models a situation where we have more certainty. The amount of information in each prior is reflected in the values of κ and ν, which increase as the amount of information in the prior increases. Figure 8.7 shows the estimated densities of the conditional RMS for both the cross-validation and Bayesian error estimates with the low-, medium-, and high-information priors corresponding to each row. The unconditional RMS for each error estimator is printed in each graph for reference. The high variance of these distributions illustrates that different samples condition the RMS to different extents. For example, in part (d), the expected true error is 0.25 and the conditional RMS of the optimal Bayesian error estimator ranges between about 0 and 0.05, depending on the actual observed sample. Meanwhile, the conditional RMS for cross-validation has a much higher variance and is shifted to the right, which is expected since the conditional RMS of the Bayesian error estimator is optimal. Furthermore, the distributions for the conditional RMS of the Bayesian error estimator with high-information priors have very low variance and are shifted to the left relative to the low-information prior, demonstrating that models using more informative, or "tighter," priors have better RMS performance – assuming the probability mass is concentrated close to the true value of $\boldsymbol{\theta}$.

8.5 CONSISTENCY

This section treats the consistency of Bayesian MMSE error estimators, that is, the convergence of the Bayesian MMSE error estimates to the true error. To do so, let $\overline{\boldsymbol{\theta}} \in \Theta$ be the unknown true parameter, S_∞ represent an infinite sample drawn from the true distribution, and S_n denote the first n observations of this sample. The sampling distribution will be specified in the subscript of probabilities and expectations using a notation of the form "$S_\infty | \overline{\boldsymbol{\theta}}$." A sequence of estimates, $\hat{\varepsilon}_n(S_n)$, for $\varepsilon_n(\boldsymbol{\theta}, S_n)$ is *weakly*

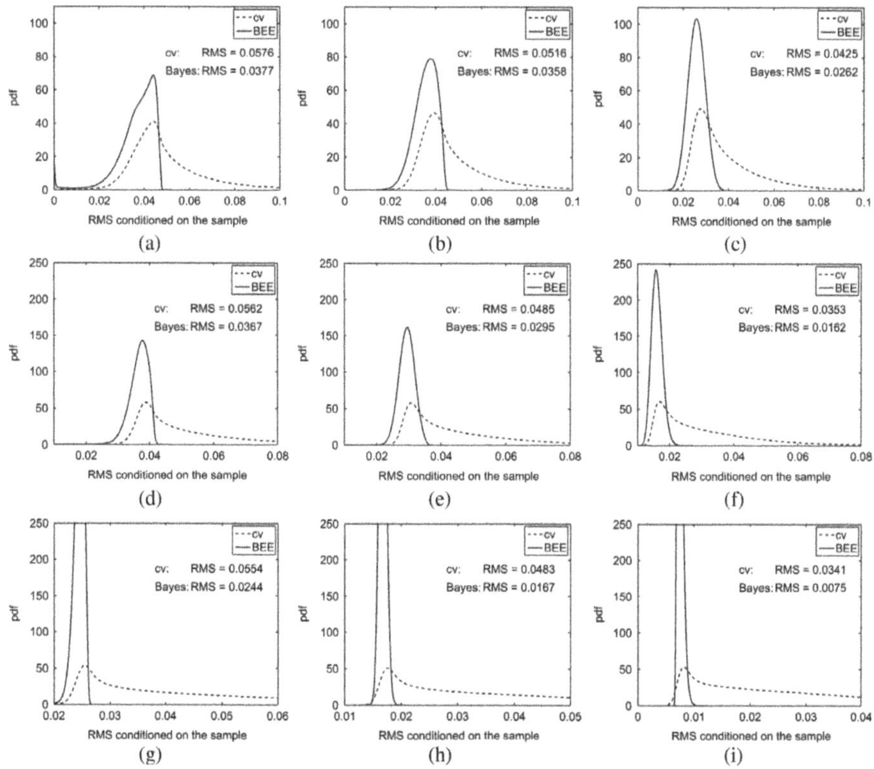

FIGURE 8.7 Probability densities for the conditional RMS of the cross-validation and Bayesian error estimators with sample size $n = 60$. The unconditional RMS for both error estimators is also indicated. (a) Low-information prior, $d = 1$. (b) Low-information prior, $d = 2$. (c) Low-information prior, $d = 5$. (d) Medium-information prior, $d = 1$. (e) Medium-information prior, $d = 2$. (f) Medium-information prior, $d = 5$. (g) High-information prior, $d = 1$. (h) High-information prior, $d = 2$. (i) High-information prior, $d = 5$. Reproduced from Dalton and Dougherty (2012b) with permission from IEEE.

consistent at $\overline{\theta}$ if $\hat{\varepsilon}_n(S_n) - \varepsilon_n(\overline{\theta}, S_n) \to 0$ in probability. If this is true for all $\overline{\theta} \in \Theta$, then $\hat{\varepsilon}_n(S_n)$ is *weakly consistent*. On the other hand, L^2 consistency is defined by convergence in the mean-square:

$$\lim_{n\to\infty} E_{S_n|\overline{\theta}} \left[(\hat{\varepsilon}_n(S_n) - \varepsilon_n(\overline{\theta}, S_n))^2 \right] = 0 . \tag{8.86}$$

It can be shown that L^2 consistency implies weak consistency. *Strong consistency* is defined by almost sure convergence:

$$P_{S_\infty|\overline{\theta}}(\hat{\varepsilon}_n(S_n) - \varepsilon_n(\overline{\theta}, S_n) \to 0) = 1 . \tag{8.87}$$

If $\hat{\varepsilon}_n(S_n) - \varepsilon_n(\overline{\theta}, S_n)$ is bounded, which is always true for classifier error estimation, then strong consistency implies L^2 consistency by the dominated convergence theorem. We are also concerned with convergence of the sample-conditioned MSE. In this sense, a sequence of estimators, $\hat{\varepsilon}_n(S_n)$, possesses the property of *conditional MSE convergence* if for all $\overline{\theta} \in \Theta$, $\mathrm{MSE}(\hat{\varepsilon}_n(S_n)|S_n) \to 0$ a.s., or more precisely,

$$P_{S_\infty|\overline{\theta}}\left(E_{\theta|S_n}\left[\left(\hat{\varepsilon}_n(S_n) - \varepsilon_n(\theta, S_n)\right)^2\right] \to 0\right) = 1. \tag{8.88}$$

We will prove (8.87) and (8.88) assuming fairly weak conditions on the model and classification rule. To do so, it is essential to show that the Bayes posterior of the parameter converges in some sense to a delta function on the true parameter. This concept is formalized via weak* consistency and requires some basic notions from measure theory. Assume the sample space $\mathcal{X}$ and the parameter space $\boldsymbol{\Theta}$ are Borel subsets of complete separable metric spaces, each being endowed with the induced σ-algebra from the Borel σ-algebra on its respective metric space. In the discrete model with bin probabilities p_i and q_i,

$$\begin{aligned}\boldsymbol{\Theta} = \Bigg\{ & [c, p_1, \dots, p_{b-1}, q_1, \dots, q_{b-1}] : \\ & c, p_i, q_i \in [0,1], i = 1, \dots, b-1, \sum_{i=1}^{b-1} p_i \le 1, \sum_{i=1}^{b-1} q_i \le 1 \Bigg\},\end{aligned} \tag{8.89}$$

so $\boldsymbol{\Theta} \subset R \times R^{b-1} \times R^{b-1}$, which is a normed space for which we use the L^1-norm. Letting $\mathcal{B}$ be the Borel σ-algebra on $R \times R^{b-1} \times R^{b-1}$, $\boldsymbol{\Theta} \in \mathcal{B}$ and the σ-algebra on $\boldsymbol{\Theta}$ is the induced σ-algebra $\mathcal{B}_{\boldsymbol{\Theta}} = \{\boldsymbol{\Theta} \cap A : A \in \mathcal{B}\}$. In the Gaussian model,

$$\begin{aligned}\boldsymbol{\Theta} = \{ & [c, \boldsymbol{\mu}_0, \Sigma_0, \boldsymbol{\mu}_1, \Sigma_1] : \\ & c \in [0,1], \boldsymbol{\mu}_0, \boldsymbol{\mu}_1 \in R^d, \Sigma_0 \text{ and } \Sigma_1 \text{ are } d \times d \text{ invertible matrices}\},\end{aligned} \tag{8.90}$$

so $\boldsymbol{\Theta} \subset S = R \times R^d \times R^{d^2} \times R^d \times R^{d^2}$, which is a normed space, for which we use the L^1 -norm. $\boldsymbol{\Theta}$ lies in the Borel σ-algebra on S and the σ-algebra on $\boldsymbol{\Theta}$ is defined in the same manner as in the discrete model.

If λ_n and λ are probability measures on $\boldsymbol{\Theta}$, then $\lambda_n \to \lambda$ weak* (i.e., in the weak* topology on the space of all probability measures over $\boldsymbol{\Theta}$) if and only if $\int f d\lambda_n \to \int f d\lambda$ for all bounded continuous functions f on $\boldsymbol{\Theta}$. Further, if δ_θ is a point mass at $\theta \in \boldsymbol{\Theta}$, then it can be shown that $\lambda_n \to \delta_\theta$ weak* if and only if $\lambda_n(U) \to 1$ for every neighborhood U of θ.

Bayesian modeling parameterizes a family of probability measures, $\{F_\theta : \theta \in \boldsymbol{\Theta}\}$, on $\mathcal{X}$. For a fixed true parameter, $\overline{\theta}$, we denote the sampling distribution by $F_{\overline{\theta}}^{\infty}$, which is an infinite product measure on $\mathcal{X}^\infty$. We say that the Bayes posterior of θ is *weak* consistent* at $\overline{\theta} \in \boldsymbol{\Theta}$ if the posterior probability of the parameter converges weak* to $\delta_{\overline{\theta}}$ for $F^{\infty}_{\overline{\theta}}$-almost all sequences — in other words, if for all bounded continuous functions f on $\boldsymbol{\Theta}$,

$$P_{S_\infty|\overline{\theta}}(E_{\theta|S_n}[f(\theta)] \to f(\overline{\theta})) = 1\,. \tag{8.91}$$

Equivalently, we require the posterior probability (given a fixed sample) of any neighborhood U of the true parameter, $\overline{\theta}$, to converge to 1 almost surely with respect to the sampling distribution, that is,

$$P_{S_\infty|\overline{\theta}}(P_{\theta|S_n}(U) \to 1) = 1\,. \tag{8.92}$$

The posterior is called *weak* consistent* if it is weak* consistent for every $\overline{\theta} \in \boldsymbol{\Theta}$. Following Dalton and Dougherty (2012b), we demonstrate that the Bayes posteriors of c, θ_0, and θ_1 are weak* consistent for both discrete and Gaussian models (in the usual topologies), assuming proper priors.

In the context of Bayesian estimation, if there is only a finite number of possible outcomes and the prior probability does not exclude any neighborhood of the true parameter as impossible, posteriors are known to be weak* consistent (Diaconis and Freedman, 1986; Freedman, 1963). Thus, if the *a priori* class probability c has a beta distribution, which has positive mass in every open interval in $[0, 1]$, then the posterior is weak* consistent. Likewise, since sample points in discrete classification have a finite number of possible outcomes, the posteriors of θ_0 and θ_1 are weak* consistent as $n_0, n_1 \to \infty$, respectively.

In a general Bayesian estimation problem with a proper prior on a finite dimensional parameter space, as long as the true feature–label distribution is included in the parameterized family of distributions and some regularity conditions hold, notably that the likelihood is a bounded continuous function of the parameter that is not flat for a range of values of the parameter and the true parameter is not excluded by the prior as impossible or on the boundary of the parameter space, then the posterior distribution of the parameter approaches a normal distribution centered at the true mean with variance proportional to $1/n$ as $n \to \infty$ (Gelman et al., 2004). These regularity conditions hold in the Gaussian model being employed here for both classes, $y = 0, 1$. Hence the posterior of θ_y is weak* consistent as $n_y \to \infty$.

Owing to the weak* consistency of posteriors for c, θ_0 , and θ_1 in the discrete and Gaussian models, for any bounded continuous function f on $\boldsymbol{\Theta}$, (8.91) holds for all $\overline{\theta} = [\overline{c}, \overline{\theta}_0, \overline{\theta}_1] \in \boldsymbol{\Theta}$. Our aim now is to supply sufficient conditions for the consistency of Bayesian error estimation.

Given a true parameter $\overline{\theta}$ and a fixed infinite sample, for each n suppose that the true error function $\varepsilon_n(\theta, S_n)$ is a real measurable function on the parameter space. Define $f_n(\theta, S_n) = \varepsilon_n(\theta, S_n) - \varepsilon_n(\overline{\theta}, S_n)$. Note that the actual true error $\varepsilon_n(\overline{\theta}, S_n)$ is a constant given S_n for each n and $f_n(\overline{\theta}, S_n) = 0$. Since $\hat{\varepsilon}_n(S_n) = E_{\theta|S_n}\left[\varepsilon_n(\theta, S_n)\right]$ for the Bayesian error estimator, to prove strong consistency we must show

$$P_{S_\infty|\overline{\theta}}(E_{\theta|S_n}[f_n(\theta, S_n)] \to 0) = 1\,. \tag{8.93}$$

For conditional MSE convergence we must show

$$
\begin{aligned}
&P_{S_\infty|\overline{\theta}}\left(E_{\theta|S_n}\left[(E_{\theta|S_n}[\varepsilon_n(\theta,S_n)]-\varepsilon_n(\theta,S_n))^2\right]\to 0\right)\\
&\quad= P_{S_\infty|\overline{\theta}}\Big(E_{\theta|S_n}\Big[(E_{\theta|S_n}[\varepsilon_n(\theta,S_n)-\varepsilon_n(\overline{\theta},S_n)]\\
&\qquad-\varepsilon_n(\theta,S_n)+\varepsilon_n(\overline{\theta},S_n))^2\Big]\to 0\Big)\\
&\quad= P_{S_\infty|\overline{\theta}}\left(E_{\theta|S_n}\left[(E_{\theta|S_n}[f_n(\theta,S_n)]-f_n(\theta,S_n))^2\right]\to 0\right)\\
&\quad= P_{S_\infty|\overline{\theta}}\left(E_{\theta|S_n}\left[f_n^2(\theta,S_n)\right]-\left(E_{\theta|S_n}[f_n(\theta,S_n)]\right)^2\to 0\right)\\
&\quad= 1\,.
\end{aligned}
\tag{8.94}
$$

Hence, both forms of convergence are proved if for any true parameter $\overline{\theta}$ and both $i = 1$ and $i = 2$,

$$
P_{S_\infty|\overline{\theta}}\left(E_{\theta|S_n}\left[f_n^i(\theta,S_n)\right]\to 0\right)=1\,. \tag{8.95}
$$

The next two theorems prove that the Bayesian error estimator is both strongly consistent and conditional MSE convergent if the true error functions $\varepsilon_n(\theta, S_n)$ form equicontinuous sets for fixed samples.

Theorem 8.6 *(Dalton and Dougherty, 2012b) Let $\overline{\theta} \in \boldsymbol{\Theta}$ be an unknown true parameter and let $F(S_\infty) = \{f_n(\bullet, S_n)\}_{n=1}^{\infty}$ be a uniformly bounded collection of measurable functions associated with the sample S_∞, where $f_n(\bullet, S_n) : \boldsymbol{\Theta} \to R$ and $f_n(\bullet, S_n) \leq M(S_\infty)$ for each $n \in \mathbb{N}$. If $F(S_\infty)$ is equicontinuous at $\overline{\theta}$ (almost surely with respect to the sampling distribution for $\overline{\theta}$) and the posterior of θ is weak* consistent at $\overline{\theta}$, then*

$$
P_{S_\infty|\overline{\theta}}(E_{\theta|S_n}[f_n(\theta,S_n)]-f_n(\overline{\theta},S_n)\to 0)=1\,. \tag{8.96}
$$

Proof: The probability in (8.96) has the following lower bound:

$$
\begin{aligned}
&P_{S_\infty|\overline{\theta}}(E_{\theta|S_n}[f_n(\theta,S_n)-f_n(\overline{\theta},S_n)]\to 0)\\
&\quad= P_{S_\infty|\overline{\theta}}(|E_{\theta|S_n}[f_n(\theta,S_n)-f_n(\overline{\theta},S_n)]|\to 0)\\
&\quad\geq P_{S_\infty|\overline{\theta}}(E_{\theta|S_n}[\,|f_n(\theta,S_n)-f_n(\overline{\theta},S_n)|\,]\to 0)\,.
\end{aligned}
\tag{8.97}
$$

Let $d_{\boldsymbol{\Theta}}$ be the metric associated with $\boldsymbol{\Theta}$. For fixed S_∞ and $\epsilon > 0$, if equicontinuity holds for $F(S_\infty)$, there is a $\delta > 0$ such that $|f_n(\theta, S_n) - f_n(\overline{\theta}, S_n)| < \epsilon$ for all $f_n \in F(S_\infty)$

whenever $d_\Theta(\theta, \overline{\theta}) < \delta$. Hence,

$$
\begin{aligned}
&E_{\theta|S_n}[\,|f_n(\theta, S_n) - f_n(\overline{\theta}, S_n)|\,] \\
&\quad = E_{\theta|S_n}[\,|f_n(\theta, S_n) - f_n(\overline{\theta}, S_n)| I_{d_\Theta(\theta,\overline{\theta})<\delta}\,] \\
&\qquad + E_{\theta|S_n}[\,|f_n(\theta, S_n) - f_n(\overline{\theta}, S_n)| I_{d_\Theta(\theta,\overline{\theta})\geq\delta}\,] \\
&\quad \leq E_{\theta|S_n}[\epsilon I_{d_\Theta(\theta,\overline{\theta})<\delta}] + E_{\theta|S_n}[M(S_\infty) I_{d_\Theta(\theta,\overline{\theta})\geq\delta}] \\
&\quad = \epsilon E_{\theta|S_n}[I_{d_\Theta(\theta,\overline{\theta})<\delta}] + M(S_\infty) E_{\theta|S_n}[I_{d_\Theta(\theta,\overline{\theta})\geq\delta}] \\
&\quad = \epsilon P_{\theta|S_n}(d_\Theta(\theta, \overline{\theta}) < \delta) + M(S_\infty) \mathrm{P}_{\theta|S_n}(d_\Theta(\theta, \overline{\theta}) \geq \delta)\,.
\end{aligned}
\tag{8.98}
$$

From the weak* consistency of the posterior of θ at $\overline{\theta}$, (8.92) holds and

$$
\begin{aligned}
&\limsup_{n\to\infty} E_{\theta|S_n}[\,|f_n(\theta, S_n) - f_n(\overline{\theta}, S_n)|\,] \\
&\quad \leq \epsilon \limsup_{n\to\infty} P_{\theta|S_n}(d_\Theta(\theta, \overline{\theta}) < \delta) \\
&\qquad + M(S_\infty) \limsup_{n\to\infty} \mathrm{P}_{\theta|S_n}(d_\Theta(\theta, \overline{\theta}) \geq \delta) \\
&\quad \overset{\text{a.s.}}{=} \epsilon \cdot 1 + M(S_\infty) \cdot 0 = \epsilon\,.
\end{aligned}
\tag{8.99}
$$

Since this is (almost surely) true for all $\epsilon > 0$,

$$
\lim_{n\to\infty} E_{\theta|S_n}[\,|f_n(\theta, S_n) - f_n(\overline{\theta}, S_n)|\,] \overset{\text{a.s.}}{=} 0\,, \tag{8.100}
$$

so that the lower bound for the probability in (8.96) must be 1. ■

Theorem 8.7 *(Dalton and Dougherty, 2012b) Given a Bayesian model and classification rule, if $F^y(S_\infty) = \{\varepsilon_n^y(\bullet, S_n)\}_{n=1}^\infty$ is equicontinuous at $\overline{\theta}_y$ (almost surely with respect to the sampling distribution for $\overline{\theta}_y$) for every $\overline{\theta}_y \in \Theta_y$, for $y = 0, 1$, then the resulting Bayesian error estimator is both strongly consistent and conditional MSE convergent.*

Proof: We may decompose the true error of a classifier by $\varepsilon_n(\theta, S_n) = c\varepsilon_n^0(\theta_0, S_n) + (1 - c)\varepsilon_n^1(\theta_1, S_n)$. It is straightforward to show that $F(S_\infty) = \{\varepsilon_n(\bullet, S_n)\}_{n=1}^\infty$ is (a.s.) equicontinuous at every $\overline{\theta} = \{\overline{c}, \overline{\theta}_0, \overline{\theta}_1\} \in \Theta = [0, 1] \times \Theta_0 \times \Theta_1$. Define $f_n(\theta, S_n) = \varepsilon_n(\theta, S_n) - \varepsilon_n(\overline{\theta}, S_n)$, and note $|f_n(\theta, S_n)| \leq 1$. Since $\{f_n(\bullet, S_n)\}_{n=1}^\infty$ and $\{f_n^2(\bullet, S_n)\}_{n=1}^\infty$ are also (a.s.) equicontinuous at every $\overline{\theta} \in \Theta$, by Theorem 8.6,

$$
P_{S_\infty|\overline{\theta}}\left(E_{\theta|S_n}\left[f_n^i(\theta, S_n)\right] \to 0\right) = 1\,, \tag{8.101}
$$

for both $i = 1$ and $i = 2$. ■

Combining Theorem 8.7 with the following two theorems proves that the Bayesian error estimator is strongly consistent and conditional MSE convergent for both the discrete model with any classification rule and the Gaussian model with any linear classification rule.

Theorem 8.8 *(Dalton and Dougherty, 2012b) In the discrete Bayesian model with any classification rule, $F^y(S_\infty) = \{\varepsilon_n^y(\bullet, S_n)\}_{n=1}^{\infty}$ is equicontinuous at every $\overline{\theta}_y \in \Theta_y$ for both $y = 0$ and $y = 1$.*

Proof: In a b-bin model, suppose we obtain the sequence of classifiers $\psi_n : \{1, \ldots, b\} \to \{0, 1\}$ from a given sample. The error of classifier ψ_n contributed by class 0 at parameter $\theta_0 = [p_1, \ldots p_{b-1}] \in \Theta_0$ is

$$\varepsilon_n^0(\theta_0, S_n) = \sum_{i=1}^{b} p_i I_{\psi_n(i)=1} \,. \tag{8.102}$$

For any fixed sample S_∞, fixed true parameter $\overline{\theta}_0 = [\overline{p}_1, \ldots, \overline{p}_{b-1}]$, and any $\theta_0 = [p_1, \ldots, p_{b-1}]$,

$$\begin{aligned} |\varepsilon_n^0(\theta_0, S_n) - \varepsilon_n^0(\overline{\theta}_0, S_n)| &= \left| \sum_{i=1}^{b} (p_i - \overline{p}_i)\, I_{\psi_n(i)=1} \right| \\ &= \left| \sum_{i=1}^{b-1} (p_i - \overline{p}_i)\, I_{\psi_n(i)=1} - \sum_{i=1}^{b-1} (p_i - \overline{p}_i)\, I_{\psi_n(b)=1} \right| \\ &\leq 2 \sum_{i=1}^{b-1} |p_i - \overline{p}_i| = 2\|\theta_0 - \overline{\theta}_0\| \,. \end{aligned} \tag{8.103}$$

Since $\overline{\theta}_0$ was arbitrary, $F^0(S_\infty)$ is equicontinuous. Similarly, $F^1(S_\infty) = \{\sum_{i=1}^{b} q_i I_{\psi_n(i)=0}\}_{n=1}^{\infty}$ is equicontinuous, which completes the proof. ■

Theorem 8.9 *(Dalton and Dougherty, 2012b) In the Gaussian Bayesian model with d features and any linear classification rule, $F^y(S_\infty) = \{\varepsilon_n^y(\bullet, S_n)\}_{n=1}^{\infty}$ is equicontinuous at every $\overline{\theta}_y \in \Theta_y$, for $y = 0, 1$.*

Proof: Given S_∞, suppose we obtain a sequence of linear classifiers $\psi_n : R^d \to \{0, 1\}$ with discriminant functions $g_n(\mathbf{x}) = \mathbf{a}_n^T \mathbf{x} + b_n$ defined by vectors $\mathbf{a}_n = (a_{n1}, \ldots, a_{nd})$ and constants b_n. If $\mathbf{a}_n = \mathbf{0}$ for some n, then the classifier and classifier errors are constant. In this case, $|\varepsilon_n^y(\theta_y, S_n) - \varepsilon_n^y(\overline{\theta}_y, S_n)| = 0$ for all $\theta_y, \overline{\theta}_y \in \Theta_y$, so this classifier does not effect the equicontinuity of $F^y(S_\infty)$. Hence, without loss of generality assume

$\mathbf{a}_n \neq \mathbf{0}$, so that the error of classifier ψ_n contributed by class y at parameter $\boldsymbol{\theta}_y = [\boldsymbol{\mu}_y, \Sigma_y]$ is given by

$$\varepsilon_n^y(\boldsymbol{\theta}_y, S_n) = \Phi\left(\frac{(-1)^y g_n(\boldsymbol{\mu}_y)}{\sqrt{\mathbf{a}_n^T \Sigma_y \mathbf{a}_n}}\right). \tag{8.104}$$

Since scaling g_n does not effect the decision of classifier ψ_n and $\mathbf{a}_n \neq \mathbf{0}$, without loss of generality we also assume g_n is normalized so that $\max_i |a_{ni}| = 1$ for all n.

Treating both classes at the same time, it is enough to show that $\{g_n(\boldsymbol{\mu})\}_{n=1}^{\infty}$ is equicontinuous at every $\overline{\boldsymbol{\mu}} \in R^d$ and $\{\mathbf{a}_n^T \Sigma \mathbf{a}_n\}_{n=1}^{\infty}$ is equicontinuous at every positive definite $\overline{\Sigma}$ (considering one fixed $\overline{\Sigma}$ at a time, by positive definiteness $\mathbf{a}_n^T \overline{\Sigma} \mathbf{a}_n > 0$). For any fixed but arbitrary $\overline{\boldsymbol{\mu}} = [\overline{\mu}_1, \dots, \overline{\mu}_d]$ and any $\boldsymbol{\mu} \in R^d$,

$$\begin{aligned} |g_n(\boldsymbol{\mu}) - g_n(\overline{\boldsymbol{\mu}})| &= \left| \sum_{i=1}^{d} a_{ni} \left(\mu_i - \overline{\mu}_i\right) \right| \\ &\leq \max_i |a_{ni}| \sum_{i=1}^{d} |\mu_i - \overline{\mu}_i| \\ &= \|\boldsymbol{\mu} - \overline{\boldsymbol{\mu}}\|. \end{aligned} \tag{8.105}$$

Hence, $\{g_n(\boldsymbol{\mu})\}_{n=1}^{\infty}$ is equicontinuous. For any fixed $\overline{\Sigma}$, denote $\overline{\sigma}_{ij}$ as its ith row, jth column element and use similar notation for an arbitrary matrix, Σ . Then,

$$\begin{aligned} |\mathbf{a}_n^T \Sigma \mathbf{a}_n - \mathbf{a}_n^T \overline{\Sigma} \mathbf{a}_n| &= |\mathbf{a}_n^T \left(\Sigma - \overline{\Sigma}\right) \mathbf{a}_n| \\ &= \left| \sum_{i=1}^{d} \sum_{j=1}^{d} a_{ni} a_{nj} \left(\sigma_{ij} - \overline{\sigma}_{ij}\right) \right| \\ &\leq \max_i |a_{ni}|^2 \sum_{i=1}^{d} \sum_{j=1}^{d} |\sigma_{ij} - \overline{\sigma}_{ij}| \\ &= \|\Sigma - \overline{\Sigma}\|. \end{aligned} \tag{8.106}$$

Hence, $\{\mathbf{a}_n^T \Sigma \mathbf{a}_n\}_{n=1}^{\infty}$ is equicontinuous. ■

8.6 CALIBRATION

When one can employ a Bayesian framework but an analytic solution for the Bayesian MMSE error estimator is not available, one can use a classical *ad hoc* error estimator; however, rather than using it directly, it can be calibrated based on the Bayesian framework by computing a calibration function mapping error estimates (from the specified error estimation rule) to their calibrated values (Dalton and Dougherty, 2012c). Once a calibration function has been established, it may be applied by post-processing a final error estimate with a simple lookup table.

An optimal calibration function is associated with four assumptions: a fixed sample size n, a Bayesian model with a proper prior $\pi(\boldsymbol{\theta}) = \pi(c)\pi(\boldsymbol{\theta}_0)\pi(\boldsymbol{\theta}_1)$, a fixed classification rule (including possibly a feature-selection scheme), and a fixed (uncalibrated) error estimation rule with estimates denoted by $\hat{\varepsilon}_{\bullet}$. Given these assumptions, the *optimal MMSE calibration function* is the expected true error conditioned on the observed error estimate:

$$E[\varepsilon_n|\hat{\varepsilon}_{\bullet}] = \int_0^1 \varepsilon_n f(\varepsilon_n|\hat{\varepsilon}_{\bullet})d\varepsilon_n = \frac{\int_0^1 \varepsilon_n f(\varepsilon_n, \hat{\varepsilon}_{\bullet})d\varepsilon_n}{f(\hat{\varepsilon}_{\bullet})}, \tag{8.107}$$

where $f(\varepsilon_n, \hat{\varepsilon}_{\bullet})$ is the unconditional joint density between the true and estimated errors and $f(\hat{\varepsilon}_{\bullet})$ is the unconditional marginal density of the estimated error. The MMSE calibration function may be used to calibrate any error estimator to have optimal MSE performance for the assumed model. Evaluated at a particular value of $\hat{\varepsilon}_{\bullet}$, it is called the *MMSE calibrated error estimate* and will be denoted by $\hat{\varepsilon}_{\bullet}^{\text{cal}}$. Based on the property, $E[E[X|Y]] = E[X]$, of conditional expectation,

$$E[\hat{\varepsilon}_{\bullet}^{\text{cal}}] = E[E[\varepsilon_n|\hat{\varepsilon}_{\bullet}]] = E[\varepsilon_n], \tag{8.108}$$

so that calibrated error estimators are unbiased.

They possess another interesting property. We say that an arbitrary estimator, $\hat{\varepsilon}_{\bullet}$, has *ideal regression* if $E[\varepsilon_n|\hat{\varepsilon}_{\bullet}] = \hat{\varepsilon}_{\bullet}$ almost surely. This means that, given the estimate, the mean of the true error is equal to the estimate. Hence, the regression of the true error on the estimate is the 45-degree line through the origin. Based upon results from measure theory, it is shown in Dalton and Dougherty (2012c) that both Bayesian and calibrated error estimators possess ideal regression: $E[\varepsilon_n|\hat{\varepsilon}] = \hat{\varepsilon}$ and $E[\varepsilon_n|\hat{\varepsilon}_{\bullet}^{\text{cal}}] = \hat{\varepsilon}_{\bullet}^{\text{cal}}$.

If an analytic representation for the joint density $f(\varepsilon_n, \hat{\varepsilon}_{\bullet}|\boldsymbol{\theta})$ between the true and estimated errors for fixed distributions is available, then

$$f(\varepsilon_n, \hat{\varepsilon}_{\bullet}) = \int_{\boldsymbol{\Theta}} f(\varepsilon_n, \hat{\varepsilon}_{\bullet}|\boldsymbol{\theta})\pi(\boldsymbol{\theta})d\boldsymbol{\theta}. \tag{8.109}$$

Now, $f(\hat{\varepsilon}_{\bullet})$ may either be found directly from $f(\varepsilon_n, \hat{\varepsilon}_{\bullet})$ or from analytic representation of $f(\hat{\varepsilon}_{\bullet}|\boldsymbol{\theta})$ via

$$f(\hat{\varepsilon}_{\bullet}) = \int_{\boldsymbol{\Theta}} f(\hat{\varepsilon}_{\bullet}|\boldsymbol{\theta})\pi(\boldsymbol{\theta})d\boldsymbol{\theta}. \tag{8.110}$$

If analytical results for $f(\varepsilon_n, \hat{\varepsilon}_\bullet|\theta)$ and $f(\hat{\varepsilon}_\bullet|\theta)$ are not available, then $E[\varepsilon_n|\hat{\varepsilon}_\bullet]$ may be found via Monte-Carlo approximation by simulating the model and classification procedure to generate a large collection of true and estimated error pairs.

We illustrate the performance of MMSE calibrated error estimation using synthetic data from a general covariance Gaussian model assuming normal-inverse-Wishart priors on the distribution parameters and using LDA classification. We consider three priors: low-information, medium-information, or high-information priors (see Dalton and Dougherty (2012c) for a complete description of the Monte-Carlo methodology). Fix $c = 0.5$, set $m = 0$ for class 0, and choose m for class 1 to give an expected true error of about 0.28 with one feature. After the simulation is complete, the synthetically generated true and estimated error pairs are used to estimate the joint densities $f(\varepsilon_n, \hat{\varepsilon})$ and $f(\varepsilon_n, \hat{\varepsilon}_\bullet)$, where $\hat{\varepsilon}_\bullet$ can be cross-validation, bootstrap, or bolstering. Figure 8.8 shows performance results for $d = 2$ and $n = 30$, where the left, middle, and right columns contain plots for low-, medium-, and high-information priors, respectively. The first row shows the RMS conditioned on the true error,

$$\mathrm{RMS}[\hat{\varepsilon}_\circ\,|\varepsilon_n] = \sqrt{E\left[\left(\varepsilon_n - \hat{\varepsilon}\right)^2|\varepsilon_n\right]}, \tag{8.111}$$

where $\hat{\varepsilon}_\circ$ equals $\hat{\varepsilon}_\bullet$, $\hat{\varepsilon}_\bullet^{\mathrm{cal}}$, or $\hat{\varepsilon}$. The second row has probability densities for the RMS conditioned on the sample, $\mathrm{RMS}(\hat{\varepsilon}|S_n)$, over all samples. For comparison, legends in the bottom row also show the unconditional RMS for all error estimators (averaged over both the prior and sampling distributions). The behavior of the RMS conditioned on the true error in the third row is typical: uncalibrated error estimators tend to be best for low true errors and Bayesian error estimators are usually best for moderate true errors. The latter is also true for calibrated error estimators, which have true-error-conditioned RMS plots usually tracking just above the Bayesian error estimator. The distribution of $\mathrm{RMS}(\hat{\varepsilon}|S_n)$ for calibrated error estimators tends to have more mass toward lower values of RMS than uncalibrated error estimators, with the Bayesian error estimator being even more shifted to the left. Finally, the performances of all calibrated error estimators tend to be close to each other, with perhaps the calibrated bolstered error estimator performing slightly better than the others.

8.7 CONCLUDING REMARKS

We conclude this chapter with some remarks on application and on extension of the Bayesian theory. Since performance depends on the prior distribution, one would like to construct a prior distribution on existing knowledge relating to the feature–label distribution, balancing the desire for a tight prior with the danger of concentrating the mass away from the true value of the parameter. This problem has been addressed in the context of genomic classification via optimization paradigms that utilize the

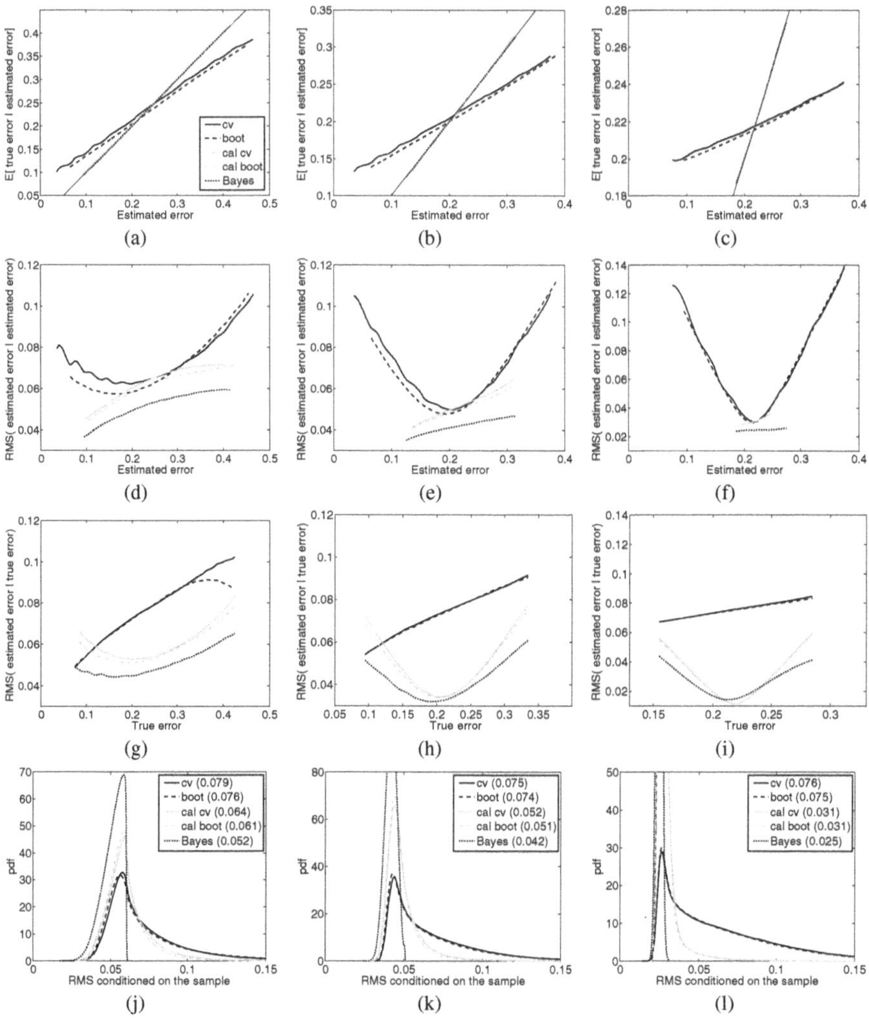

FIGURE 8.8 Expected true error given estimated error, conditional RMS given estimated error, conditional RMS given true error, and probability densities for the sample-conditioned RMS for the Gaussian model with $d = 2$, $n = 30$, and LDA. Low-, medium, and high-information priors are shown left to right. Cross-validation, bootstrap, bolstering, the corresponding calibrated error estimators for each of these error estimators, as well as the Bayesian error estimator are shown in each plot. (a) Low-information, $E[\varepsilon_n \mid \hat{\varepsilon}]$; (b) medium-information, $E[\varepsilon_n \mid \hat{\varepsilon}]$; (c) high-information, $E[\varepsilon_n \mid \hat{\varepsilon}]$; (d) low-information, $\text{RMS}(\hat{\varepsilon} \mid \hat{\varepsilon})$; (e) medium-information, $\text{RMS}(\hat{\varepsilon} \mid \hat{\varepsilon})$; (f) high-information, $\text{RMS}(\hat{\varepsilon} \mid \hat{\varepsilon})$; (g) low-information, $\text{RMS}(\hat{\varepsilon} \mid \varepsilon_n)$; (h) medium-information, $\text{RMS}(\hat{\varepsilon} \mid \varepsilon_n)$; (i) high-information, $\text{RMS}\ (\hat{\varepsilon} \mid \varepsilon_n)$; (j) low-information, pdf of conditional RMS; (k) medium-information, pdf of conditional RMS; (l) high-information, pdf of conditional RMS.

incomplete prior information contained in gene regulatory pathways (Esfahani and Dougherty, 2014b).

A second issue concerns application when we are not in possession of an analytic representation of the estimator, as we are for both discrete classification and linear classification in the Gaussian model. In this case, the relevant integrals can be evaluated using Monte-Carlo integration to obtain an approximation of the Bayesian MMSE error estimate (Dalton and Dougherty, 2011a).

There is the issue of Gaussianity itself. Since the true feature–label distribution is extremely unlikely to be truly Gaussian, for practical application one must consider the robustness of the estimation procedure to different degrees of lack of Gaussianity (Dalton and Dougherty, 2011a). Moreover, if one has a large set of potential features, one can apply a Gaussianity test and only use features passing the test. While this might mean discarding potentially useful features, the downside is countered by the primary requirement of having an accurate error estimate. Going further, in many applications a Gaussian assumption cannot be utilized. In this case a fully Monte-Carlo approach can be taken (Knight et al., 2014).

In Section 7.3 we presented the double-asymptotic theory for error-estimator second-order moments. There is also an asymptotic theory for Bayesian MMSE error estimators, but here the issue is more complicated than just assuming limiting conditions on p/n_0 and p/n_1. The Bayesian setting also requires limiting conditions on the hyperparameters that, in combination with the basic limiting conditions on the dimension and sample sizes, are called *Bayesian-Kolmogorov asymptotic conditions* in Zollanvari and Dougherty (2014).

Finally, and perhaps most significantly, if one is going to make modeling assumptions regarding an uncertainty class of feature–label distributions in order to estimate the error, then, rather than applying some heuristically chosen classification rule, one can find a classifier that possesses minimal expected error across the uncertainty class relative to the posterior distribution (Dalton and Dougherty, 2013a,b). The consequence of this approach is optimal classifier design along with optimal MMSE error estimation.

EXERCISES

8.1 Redo Figure 8.3 for sample size 40.

8.2 Redo Figure 8.4 for expected true error 0.1 and 0.4.

8.3 Redo Figure 8.2 with bin size 12.

8.4 Redo Figure 8.6 for $d = 3$.

8.5 Redo Figure 8.6 with $d = 5$ but with c fixed at $c = 0.5, 0.8$.

8.6 Prove Lemma 8.1.

8.7 For the uncertainty class $\boldsymbol{\Theta}$ and the posterior distribution $\pi^*(\theta_y)$, $y = 0, 1$, the *effective class-conditional density* is the average of the state-specific conditional densities relative to the posterior distribution:

$$f_{\Theta}(\mathbf{x}|y) = \int_{\boldsymbol{\Theta}_y} f_{\theta_y}(\mathbf{x}|y)\,\pi^*(\theta_y)d\theta_y.$$

Let ψ be a fixed classifier given by $\psi(\mathbf{x}) = 0$ if $\mathbf{x} \in R_0$ and $\psi(\mathbf{x}) = 1$ if $\mathbf{x} \in R_1$, where R_0 and R_1 are measurable sets partitioning the sample space. Show that the MMSE Bayesian error estimator is given by (Dalton and Dougherty, 2013a)

$$\hat{\varepsilon}(S_n) = \mathrm{E}_{\pi^*}[c]\int_{R_1} f_{\Theta}(\mathbf{x}|0)d\mathbf{x} + (1 - \mathrm{E}_{\pi^*}[c])\int_{R_0} f_{\Theta}(\mathbf{x}|1)d\mathbf{x}.$$

8.8 Define an *optimal Bayesian classifier* ψ_{OBC} to be any classifier satisfying

$$\mathrm{E}_{\pi^*}[\varepsilon(\theta, \psi_{\mathrm{OBC}})] \leq \mathrm{E}_{\pi^*}[\varepsilon(\theta, \psi)]$$

for all $\psi \in C$, where C is the set of all classifiers with measurable decision regions. Use the result of Exercise 8.7 to prove that an optimal Bayesian classifier ψ_{OBC} satisfying the preceding inequality for all $\psi \in C$ exists and is given pointwise by (Dalton and Dougherty, 2013a)

$$\psi_{\mathrm{OBC}}(\mathbf{x}) = \begin{cases} 0 & \text{if } \mathrm{E}_{\pi^*}[c]f(\mathbf{x}|0) \geq (1 - \mathrm{E}_{\pi^*}[c])f(\mathbf{x}|1), \\ 1 & \text{otherwise.} \end{cases}$$

8.9 Use Theorem 8.1 to prove that, for the discrete model with Dirichlet priors, the effective class-conditional densities are given by

$$f_{\Theta}(j|y) = \frac{U_j^y + \alpha_j^y}{n_y + \sum_{i=1}^b \alpha_i^y}.$$

for $y = 0, 1$.

8.10 Put Exercises 8.8 and 8.9 together to obtain an optimal Bayesian classifier for the discrete model with Dirichlet priors.

8.11 Put Exercises 8.7 and 8.9 together to obtain the expected error of an optimal Bayesian classifier for the discrete model with Dirichlet priors.

APPENDIX A

BASIC PROBABILITY REVIEW

A.1 SAMPLE SPACES AND EVENTS

A *sample space* S is the collection of all outcomes of an experiment. An *event* E is a subset $E \subseteq S$. Event E is said to *occur* if it contains the outcome of the experiment.

Special Events:

- Inclusion: $E \subseteq F$ iff the occurrence of E implies the occurrence of F.
- Union: $E \cup F$ occurs iff E, F, or both E and F occur.
- Intersection: $E \cap F$ occurs iff both E and F occur. If $E \cap F = \emptyset$, then E and F are mutually exclusive.
- Complement: E^c occurs iff E does not occur.
- If $E_1, E_2, \ldots$ is an *increasing* sequence of events, that is, $E_1 \subseteq E_2 \subseteq \ldots$, then

$$\lim_{n\to\infty} E_n = \bigcup_{n=1}^{\infty} E_n. \tag{A.1}$$

- If $E_1, E_2, \ldots$ is a *decreasing* sequence of events, that is, $E_1 \supseteq E_2 \supseteq \ldots$, then

$$\lim_{n\to\infty} E_n = \bigcap_{n=1}^{\infty} E_n. \tag{A.2}$$

Error Estimation for Pattern Recognition, First Edition. Ulisses M. Braga-Neto and Edward R. Dougherty.

- If $E_1, E_2, \ldots$ is *any* sequence of events, then

$$[E_n \ i.o.] = \limsup_{n\to\infty} E_n = \bigcap_{n=1}^{\infty} \bigcup_{i=n}^{\infty} E_i. \tag{A.3}$$

  We can see that $\limsup_{n\to\infty} E_n$ occurs iff E_n occurs for an infinite number of n, that is, E_n *occurs infinitely often.*
- If $E_1, E_2, \ldots$ is *any* sequence of events, then

$$\liminf_{n\to\infty} E_n = \bigcup_{n=1}^{\infty} \bigcap_{i=n}^{\infty} E_i. \tag{A.4}$$

  We can see that $\liminf_{n\to\infty} E_n$ occurs iff E_n occurs for all but a finite number of n, that is, E_n *eventually occurs for all n.*

A.2 DEFINITION OF PROBABILITY

A collection $\mathcal{F} \subset \mathcal{P}(S)$ of events is a *σ-algebra* if it is closed under complementation and countable intersection and union. If S is a topological space, the *Borel algebra* is the σ-algebra generated by complement and countable union and intersections of the open sets in S. A *measurable* function between two sample spaces S and S', furnished with σ-algebras $\mathcal{F}$ and $\mathcal{F}'$, respectively, is a mapping $f : S \to S'$ such that for every $E \in \mathcal{F}'$, the pre-image

$$f^{-1}(E) = \{x \in S \mid f(x) \in E\} \tag{A.5}$$

belongs to $\mathcal{F}$. A real-valued function is Borel measurable if it is measurable with respect to the Borel algebra of the usual open sets in R.

A *probability space* is a triple $(S, \mathcal{F}, P)$ consisting of a sample space S, a σ-algebra $\mathcal{F}$ containing all the events of interest, and a *probability measure* P, i.e., a real-valued function defined on each event $E \in \mathcal{F}$ such that

1. $0 \leq P(E) \leq 1$,
2. $P(S) = 1$,
3. For a sequence of mutually exclusive events $E_1, E_2, \ldots$

$$P\left(\bigcup_{i=1}^{\infty} E_i\right) = \sum_{i=1}^{\infty} P(E_i). \tag{A.6}$$

Some direct consequences of the probability axioms are given next.

- $P(E^c) = 1 - P(E)$.
- If $E \subseteq F$ then $P(E) \leq P(F)$.

- $P(E \cup F) = P(E) + P(F) - P(E \cap F)$.
- (*Boole's Inequality*) For *any* sequence of events $E_1, E_2, \ldots,$

$$P\left(\bigcup_{i=1}^{\infty} E_i\right) \leq \sum_{i=1}^{\infty} P(E_i). \tag{A.7}$$

- (Continuity of Probability Measure) If $E_1, E_2, \ldots$ is an increasing or decreasing sequence of events, then

$$P\left(\lim_{n\to\infty} E_n\right) = \lim_{n\to\infty} P(E_n). \tag{A.8}$$

A.3 BOREL-CANTELLI LEMMAS

Lemma A.1 *(First Borel-Cantelli lemma) For* any *sequence of events* $E_1, E_2, \ldots,$

$$\sum_{n=1}^{\infty} P(E_n) < \infty \quad \Rightarrow \quad P([E_n \ i.o.]) = 0. \tag{A.9}$$

Proof:

$$\begin{aligned} P\left(\bigcap_{n=1}^{\infty}\bigcup_{i=n}^{\infty} E_i\right) &= P\left(\lim_{n\to\infty}\bigcup_{i=n}^{\infty} E_i\right) \\ &= \lim_{n\to\infty} P\left(\bigcup_{i=n}^{\infty} E_i\right) \\ &\leq \lim_{n\to\infty}\sum_{i=n}^{\infty} P(E_i) \\ &= 0. \end{aligned} \tag{A.10}$$

■

Lemma A.2 *(Second Borel-Cantelli lemma) For an* independent *sequence of events* $E_1, E_2, \ldots,$

$$\sum_{n=1}^{\infty} P(E_n) = \infty \quad \Rightarrow \quad P([E_n \ i.o.]) = 1. \tag{A.11}$$

Proof: Given an independent sequence of events $E_1, E_2, \ldots,$ note that

$$[E_n \ i.o.] = \bigcap_{n=1}^{\infty}\bigcup_{i=n}^{\infty} E_i = \lim_{n\to\infty}\bigcup_{i=n}^{\infty} E_i. \tag{A.12}$$

Therefore,

$$P([E_n\ i.o.]) = P\left(\lim_{n\to\infty}\bigcup_{i=n}^{\infty}E_i\right) = \lim_{n\to\infty}P\left(\bigcup_{i=n}^{\infty}E_i\right) = 1 - \lim_{n\to\infty}P\left(\bigcap_{i=n}^{\infty}E_i^c\right), \quad (A.13)$$

where the last equality follows from DeMorgan's Law. Now, note that, by independence,

$$P\left(\bigcap_{i=n}^{\infty}E_i^c\right) = \prod_{i=n}^{\infty}P\left(E_i^c\right) = \prod_{i=n}^{\infty}\left(1 - P(E_i)\right). \quad (A.14)$$

From the inequality $1 - x \leq e^{-x}$ we obtain

$$P\left(\bigcap_{i=n}^{\infty}E_i^c\right) \leq \prod_{i=n}^{\infty}\exp(-P(E_i)) = \exp\left(-\sum_{i=n}^{\infty}P(E_i)\right) = 0 \quad (A.15)$$

since, by assumption, $\sum_{i=n}^{\infty}P(E_i) = \infty$, for all n. From (A.13) and (A.15) it follows that $P([E_n\ i.o.]) = 1$, as required. ■

A.4 CONDITIONAL PROBABILITY

Given that an event F has occurred, for E to occur, $E \cap F$ has to occur. In addition, the sample space gets *restricted* to those outcomes in F, so a normalization factor $P(F)$ has to be introduced. Therefore,

$$P(E \mid F) = \frac{P(E \cap F)}{P(F)}. \quad (A.16)$$

By the same token, one can condition on any number of events:

$$P(E \mid F_1, F_2, \ldots, F_n) = \frac{P(E \cap F_1 \cap F_2 \cap \ldots \cap F_n)}{P(F_1 \cap F_2 \cap \ldots \cap F_n)}. \quad (A.17)$$

Note: For simplicity, it is usual to write $P(E \cap F) = P(E, F)$ to indicate the *joint probability* of E and F.

A few useful formulas:

- $P(E, F) = P(E \mid F)P(F)$
- $P(E_1, E_2, \ldots, E_n) = P(E_n \mid E_1, \ldots, E_{n-1})P(E_{n-1} \mid E_1, \ldots, E_{n-2}) \cdots P(E_2 \mid E_1)$ $P(E_1)$
- $P(E) = P(E, F) + P(E, F^c) = P(E \mid F)P(F) + P(E \mid F^c)(1 - P(F))$
- $P(E) = \sum_{i=1}^{n} P(E, F_i) = \sum_{i=1}^{n} P(E \mid F_i)P(F_i)$, whenever the events F_i are mutually exclusive and $\bigcup F_i \supseteq E$

- (*Bayes Theorem*):

$$P(E \mid F) = \frac{P(F \mid E)P(E)}{P(F)} = \frac{P(F \mid E)P(E)}{P(F \mid E)P(E) + P(F \mid E^c)(1 - P(E)))} \quad \text{(A.18)}$$

Events E and F are independent if the occurrence of one does not carry information as to the occurrence of the other. That is,

$$P(E \mid F) = P(E) \quad \text{and} \quad P(F \mid E) = P(F). \quad \text{(A.19)}$$

It is easy to see that this is equivalent to the condition

$$P(E, F) = P(E)P(F). \quad \text{(A.20)}$$

If E and F are independent, so are the pairs (E,F^c), (E^c,F), and (E^c,F^c). However, E being independent of F and G does not imply that E is independent of $F \cap G$. Furthermore, three events E, F, G are independent if $P(E, F, G) = P(E)P(F)P(G)$ and each pair of events is independent.

A.5 RANDOM VARIABLES

A *random variable* X defined on a probability space $(S, \mathcal{F}, P)$ is a Borel-measurable function $X : S \to R$. Thus, a random variable assigns to each outcome $\omega \in S$ a real number.

Given a set of real numbers A, we define an event

$$\{X \in A\} = X^{-1}(A) \subseteq S. \quad \text{(A.21)}$$

It can be shown that all probability questions about a random variable X can be phrased in terms of the probabilities of a simple set of events:

$$\{X \in (-\infty, a]\}, \quad a \in R. \quad \text{(A.22)}$$

These events can be written more simply as $\{X \leq a\}$, for $a \in R$.

The *probability distribution function* (PDF) of a random variable X is the function $F_X : R \to [0, 1]$ given by

$$F_X(a) = P(\{X \leq a\}), \quad a \in R. \quad \text{(A.23)}$$

For simplicity, henceforth we will often remove the braces around statements involving random variables; e.g., we will write $F_X(a) = P(X \leq a)$, for $a \in R$.

Properties of a PDF:

1. F_X is non-decreasing: $a \leq b \Rightarrow F_X(a) \leq F_X(b)$.
2. $\lim_{a\to-\infty} F_X(a) = 0$ and $\lim_{a\to+\infty} F_X(a) = 1$.
3. F_X is right-continuous: $\lim_{b\to a_+} F_X(b) = F_X(a)$.

The *joint PDF* of two random variables X and Y is the joint probability of the events $\{X \leq a\}$ and $\{Y \leq b\}$, for $a, b \in R$. Formally, we define a function $F_{XY} : R \times R \rightarrow [0, 1]$ given by

$$F_{XY}(a, b) = P(\{X \leq a\} \cap \{Y \leq b\}) = P(X \leq a, Y \leq B), \quad a, b \in R. \tag{A.24}$$

This is the probability of the "lower-left quadrant" with corner at (a, b). Note that $F_{XY}(a, \infty) = F_X(a)$ and $F_{XY}(\infty, b) = F_Y(b)$. These are called the *marginal PDFs*.

The *conditional PDF* of X given Y is just the conditional probability of the corresponding special events. Formally, it is a function $F_{X|Y} : R \times R \rightarrow [0, 1]$ given by

$$\begin{aligned} F_{X|Y}(a, b) &= P(X \leq a \mid Y \leq b) \\ &= \frac{P(X \leq a, Y \leq B)}{P(Y \leq b)} = \frac{F_{XY}(a, b)}{F_Y(b)}, \quad a, b \in R. \end{aligned} \tag{A.25}$$

The random variables X and Y are *independent random variables* if $F_{X|Y}(a, b) = F_X(a)$ and $F_{Y|X}(b, a) = F_Y(b)$, that is, if $F_{XY}(a, b) = F_X(a)F_Y(b)$, for $a, b \in R$.

Bayes' Theorem for random variables is

$$F_{X|Y}(a, b) = \frac{F_{Y|X}(b, a)F_X(a)}{F_Y(b)}, \quad a, b \in R. \tag{A.26}$$

The notion of a *probability density function* (pdf) is fundamental in probability theory. However, it is a secondary notion to that of a PDF. In fact, all random variables must have a PDF, but not all random variables have a pdf.

If F_X is everywhere continuous and differentiable, then X is said to be a *continuous* random variable and the pdf f_X of X is given by

$$f_X(a) = \frac{dF_X}{dx}(a), \quad a \in R. \tag{A.27}$$

Probability statements about X can then be made in terms of integration of f_X. For example,

$$\begin{aligned} F_X(a) &= \int_{-\infty}^{a} f_X(x)dx, \quad a \in R \\ P(a \leq X \leq b) &= \int_{a}^{b} f_X(x)dx, \quad a, b \in R \end{aligned} \tag{A.28}$$

A few useful continuous random variables:

- Uniform(a, b)

$$f_X(x) = \frac{1}{b - a}, \quad a < x < b. \tag{A.29}$$

- The univariate Gaussian $N(\mu, \sigma)$, $\sigma > 0$:

$$f_X(x) = \frac{1}{\sqrt{2\pi\sigma^2}} \exp\left(-\frac{(x-\mu)^2}{2\sigma^2}\right), \quad x \in R. \tag{A.30}$$

- Exponential(λ), $\lambda > 0$:

$$f_X(x) = \lambda e^{-\lambda x}, \quad x \geq 0. \tag{A.31}$$

- Gamma(λ), $\lambda, t > 0$:

$$f_X(x) = \frac{\lambda e^{-\lambda x}(\lambda x)^{t-1}}{\Gamma(t)}, \quad x \geq 0. \tag{A.32}$$

- Beta(a, b):

$$f_X(x) = \frac{1}{B(a,b)} x^{a-1}(1-x)^{b-1}, \quad 0 < x < 1. \tag{A.33}$$

A.6 DISCRETE RANDOM VARIABLES

If F_X is not everywhere differentiable, then X does not have a pdf. A useful particular case is that where X assumes countably many values and F_X is a *staircase function*. In this case, X is said to be a *discrete* random variable.

One defines the *probability mass function* (PMF) of a discrete random variable X to be:

$$p_X(a) = P(X = a) = F_X(a) - F_X(a^-), \quad a \in R. \tag{A.34}$$

(Note that for any continuous random variable X, $P(X = a) = 0$ so there is no PMF to speak of.)

A few useful discrete random variables:

- Bernoulli(p):

$$\begin{aligned} p_X(0) &= P(X = 0) = 1 - p \\ p_X(1) &= P(X = 0) = p \end{aligned} \tag{A.35}$$

- Binomial(n,p):

$$p_X(i) = P(X = i) = \binom{n}{i} p^i (1-p)^{n-i}, \quad i = 0, 1, \ldots, n. \tag{A.36}$$

- Poisson (λ):

$$p_X(i) = P(X = i) = e^{-\lambda}\frac{\lambda^i}{i!}, \quad i = 0, 1, \ldots \tag{A.37}$$

A.7 EXPECTATION

The mean value of a random variable X is an average of its values weighted by their probabilities. If X is a continuous random variable, this corresponds to

$$E[X] = \int_{-\infty}^{\infty} x f_X(x)\, dx. \tag{A.38}$$

If X is discrete, then this can be written as

$$E[X] = \sum_{i:p_X(x_i)>0} x_i p_X(x_i). \tag{A.39}$$

Given a random variable X, and a Borel measurable function $g : R \to R$, then $g(X)$ is also a random variable. If there is a pdf, it can be shown that

$$E[g(X)] = \int_{-\infty}^{\infty} g(x) f_X(x)\, dx. \tag{A.40}$$

If X is discrete, then this becomes

$$E[g(X)] = \sum_{i:p_X(x_i)>0} g(x_i) p_X(x_i). \tag{A.41}$$

An an immediate corollary, one gets $E[aX + c] = aE[X] + c$.

For jointly distributed random variables (X, Y), and a measurable $g : R \times R \to R$, then, in the continuous case,

$$E[g(X, Y)] = \int_{-\infty}^{\infty} \int_{-\infty}^{\infty} g(x, y) f_{XY}(x, y)\, dxdy, \tag{A.42}$$

with a similar expression in the discrete case.

Some properties of expectation:

- Linearity: $E[a_1 X_1 + \ldots + a_n X_n] = a_1 E[X_1] + \ldots + a_n E[X_n]$.
- If X and Y are *uncorrelated*, then $E[XY] = E[X]E[Y]$ (independence always implies uncorrelatedness, but the converse is true only in special cases; for example, jointly Gaussian or multinomial random variables).
- If $P(X \geq Y) = 1$, then $E[X] > E[Y]$.

- *Holder's Inequality*: for $1 < r < \infty$ and $1/r + 1/s = 1$, we have

$$E[|XY|] \leq E[|X|^r]^{1/r} E[|Y|^s]^{1/s}. \tag{A.43}$$

The special case $r = s = 2$ results in the *Cauchy–Schwartz Inequality*: $E[|XY|] \leq \sqrt{E[X^2]E[Y^2]}$.

The expectation of a random variable X is affected by its *probability tails*, given by $F_X(a) = P(X \leq a)$ and $1 - F_X(a) = P(X \geq a)$. If the probability tails fail to vanish sufficiently fast (X has "fat tails"), then $E[X]$ will not be finite, and the expectation is undefined. For a nonnegative random variable X (i.e., one for which $P(X \geq 0) = 1$), there is only one probability tail, the upper tail $P(X > a)$, and there is a simple formula relating $E[X]$ to it:

$$E[X] = \int_0^\infty P(X > x)\, dx. \tag{A.44}$$

A small $E[X]$ constrains the upper tail to be thin. This is guaranteed by Markov's inequality: if X is a nonnegative random variable,

$$P(X \geq a) \leq \frac{E[X]}{a}, \quad \text{for all } a > 0. \tag{A.45}$$

Finally, a particular result that is of interest to our purposes relates an exponentially vanishing upper tail of a nonnegative random variable to a bound on its expectation.

Lemma A.3 *If X is a nonnegative random variable such that $P(X > t) \leq ce^{-at^2}$, for all $t > 0$ and given $a, c > 0$, we have*

$$E[X] \leq \sqrt{\frac{1 + \log c}{a}}. \tag{A.46}$$

Proof: Note that $P(X^2 > t) = P(X > \sqrt{t}) \leq ce^{-at}$. From (A.44) we get:

$$E[X^2] = \int_0^\infty P(X^2 > t)\, dt = \int_0^u P(X^2 > t)\, dt + \int_u^\infty P(X^2 > t)\, dt$$
$$\leq u + \int_u^\infty ce^{-at}\, dt = u + \frac{c}{a}e^{-au}. \tag{A.47}$$

By direct differentiation, it is easy to verify that the upper bound on the right hand side is minimized at $u = (\log c)/a$. Substituting this value back into the bound leads to $E[X^2] \leq (1 + \log c)/a$. The result then follows from the fact that $E[X] \leq \sqrt{E[X^2]}$. ■

A.8 CONDITIONAL EXPECTATION

If X and Y are jointly continuous random variables and $f_Y(y) > 0$, we define:

$$E[X \mid Y = y] = \int_{-\infty}^{\infty} x f_{X|Y}(x, y)\, dx = \int_{-\infty}^{\infty} x \frac{f_{XY}(x, y)}{f_Y(y)}\, dx. \tag{A.48}$$

If X, Y are jointly discrete and $p_Y(y_j) > 0$, this can be written as:

$$E[X \mid Y = y_j] = \sum_{i:p_X(x_i)>0} x_i p_{X|Y}(x_i, y_j) = \sum_{i:p_X(x_i)>0} x_i \frac{p_{XY}(x_i, y_j)}{p_Y(y_j)}. \tag{A.49}$$

Conditional expectations have all the properties of usual expectations. For example,

$$E\left[\sum_{i=1}^{n} X_i \,\middle|\, Y = y\right] = \sum_{i=1}^{n} E[X_i \mid Y = y]. \tag{A.50}$$

Given a random variable X, the mean $E[X]$ is a deterministic parameter. However, $E[X \mid Y]$ is a function of random variable Y, and therefore a random variable itself. One can show that its mean is precisely $E[X]$:

$$E[E[X \mid Y]] = E[X]. \tag{A.51}$$

What this says in the continuous case is

$$E[X] = \int_{-\infty}^{\infty} E[X \mid Y = y] f_Y(y)\, dy \tag{A.52}$$

and, in the discrete case,

$$E[X] = \sum_{i:p_Y(y_i)>0} E[X \mid Y = y_i] P(Y = y_i). \tag{A.53}$$

Suppose one is interested in predicting the value of an unknown random variable Y. One wants the predictor $\hat{Y}$ to be optimal according to some criterion. The criterion most widely used is the *Mean Square Error* (MSE), defined by

$$MSE(\hat{Y}) = E[(\hat{Y} - Y)^2]. \tag{A.54}$$

It can be shown that the MMSE (minimum MSE) constant (no-information) estimator of Y is just the mean $E[Y]$, while, given partial information through an observed random variable $X = x$, the overall MMSE estimator is the conditional mean $E[Y \mid X = x]$. The function $\eta(x) = E[Y \mid X = x]$ is called the *regression* of Y on X.

A.9 VARIANCE

The *variance* Var(X) says how the values of X are spread around the mean $E[X]$:

$$\mathrm{Var}(X) = E[(X - E[X])^2] = E[X^2] - (E[X])^2. \tag{A.55}$$

The MSE of the best constant estimator $\hat{Y} = E[Y]$ of Y is just the variance of Y.

Some properties of variance:

- $\mathrm{Var}(aX + c) = a^2\mathrm{Var}(X)$.
- *Chebyshev's Inequality*: With $X = |Z - E[Z]|^2$ and $a = \tau^2$ in Markov's Inequality (A.45), one obtains

$$P(|Z - E[Z]| \geq \tau) \leq \frac{\mathrm{Var}(Z)}{\tau^2}, \quad \text{for all } \tau > 0. \tag{A.56}$$

If X and Y are jointly distributed random variables, we define

$$\mathrm{Var}(X \mid Y) = E[(X - E[X \mid Y])^2 \mid Y] = E[X^2 \mid Y] - (E[X \mid Y])^2. \tag{A.57}$$

The *Conditional Variance Formula* states that

$$\mathrm{Var}(X) = E[\mathrm{Var}(X \mid Y)] + \mathrm{Var}(E[X \mid Y]). \tag{A.58}$$

This breaks down the *total variance* in a "within-rows" component and an "across-rows" component.

The covariance of two random variables X and Y is given by:

$$\mathrm{Cov}(X, Y) = E[(X - E[X])(Y - E[Y])] = E[XY] - E[X]E[Y]. \tag{A.59}$$

Two variables X and Y are uncorrelated if and only if $\mathrm{Cov}(X, Y) = 0$. Jointly Gaussian X and Y are independent if and only if they are uncorrelated (in general, independence implies uncorrelatedness but not vice versa). It can be shown that

$$\mathrm{Var}(X_1 + X_2) = \mathrm{Var}(X_1) + \mathrm{Var}(X_2) + 2\mathrm{Cov}(X_1, X_2). \tag{A.60}$$

So the variance is distributive over sums if all variables are *pairwise uncorrelated*.

The correlation coefficient ρ between two random variables X and Y is given by:

$$\rho(X, Y) = \frac{\mathrm{Cov}(X, Y)}{\sqrt{\mathrm{Var}(X)\mathrm{Var}(Y)}}. \tag{A.61}$$

Properties of the correlation coefficient:

1. $-1 \leq \rho(X, Y) \leq 1$
2. X and Y are uncorrelated iff $\rho(X, Y) = 0$.

3. (Perfect linear correlation):

$$\rho(X, Y) = \pm 1 \Leftrightarrow Y = a \pm bX, \text{ where } b = \frac{\sigma_y}{\sigma_x}. \tag{A.62}$$

A.10 VECTOR RANDOM VARIABLES

A vector random variable or *random vector* $\mathbf{X} = (X_1, \ldots, X_d)$ is a collection of random variables. Its distribution is the joint distribution of the component random variables. The mean of $\mathbf{X}$ is the vector $\boldsymbol{\mu} = (\mu_1, \ldots, \mu_d)$, where $\mu_i = E[X_i]$. The *covariance matrix* Σ is a $d \times d$ matrix given by

$$\Sigma = E[(\mathbf{X} - \boldsymbol{\mu})(\mathbf{X} - \boldsymbol{\mu})^T], \tag{A.63}$$

where $\Sigma_{ii} = \text{Var}(X_i)$ and $\Sigma_{ij} = \text{Cov}(X_i, X_j)$. Matrix Σ is real symmetric and thus diagonalizable:

$$\Sigma = UDU^T, \tag{A.64}$$

where U is the matrix of eigenvectors and D is the diagonal matrix of eigenvalues. All eigenvalues are nonnegative (Σ is *positive semi-definite*). In fact, except for "degenerate" cases, all eigenvalues are positive, and so Σ is invertible (Σ is said to be *positive definite* in this case).

It is easy to check that the random vector

$$\mathbf{Y} = \Sigma^{-\frac{1}{2}}(\mathbf{X} - \boldsymbol{\mu}) = D^{-\frac{1}{2}} U^T (\mathbf{X} - \boldsymbol{\mu}) \tag{A.65}$$

has zero mean and covariance matrix I_d (so that all components of $\mathbf{Y}$ are zero-mean, unit-variance, and uncorrelated). This is called the *whitening or Mahalanobis transformation.*

Given n *independent and identically distributed* (i.i.d.) sample observations $\mathbf{X}_1, \ldots, \mathbf{X}_n$ of the random vector $\mathbf{X}$, then the *sample mean* estimator is given by

$$\hat{\boldsymbol{\mu}} = \frac{1}{n} \sum_{i=1}^{n} \mathbf{X}_i. \tag{A.66}$$

It can be shown that this estimator is *unbiased* (that is, $E[\hat{\boldsymbol{\mu}}] = \boldsymbol{\mu}$) and *consistent* (that is, $\hat{\boldsymbol{\mu}}$ converges in probability to $\boldsymbol{\mu}$ as $n \to \infty$, c.f. Section A.12). The *sample covariance* estimator is given by

$$\hat{\Sigma} = \frac{1}{n-1} \sum_{i=1}^{n} (\mathbf{X}_i - \hat{\boldsymbol{\mu}})(\mathbf{X}_i - \hat{\boldsymbol{\mu}})^T. \tag{A.67}$$

This is an unbiased and consistent estimator of Σ.

A.11 THE MULTIVARIATE GAUSSIAN

The random vector $\mathbf{X}$ has a multivariate Gaussian distribution with mean $\boldsymbol{\mu}$ and covariance matrix Σ (assuming Σ invertible, so that also $\det(\Sigma) > 0$) if its density is given by

$$f_{\mathbf{X}}(\mathbf{x}) = \frac{1}{\sqrt{(2\pi)^d \det(\Sigma)}} \exp\left(-\frac{1}{2}(\mathbf{x}-\boldsymbol{\mu})^T \Sigma^{-1} (\mathbf{x}-\boldsymbol{\mu})\right). \tag{A.68}$$

We write $\mathbf{X} \sim N_d(\boldsymbol{\mu}, \Sigma)$.

The multivariate Gaussian has elliptical contours of the form

$$(\mathbf{x}-\boldsymbol{\mu})^T \Sigma^{-1} (\mathbf{x}-\boldsymbol{\mu}) = c^2, \quad c > 0. \tag{A.69}$$

The axes of the ellipsoids are given by the eigenvectors of Σ and the length of the axes are proportional to its eigenvalues.

If $d = 1$, one gets the univariate Gaussian distribution $X \sim N(\mu, \sigma^2)$. With $\mu = 0$ and $\sigma = 1$, the PDF of X is given by

$$P(X \le x) = \Phi(x) = \int_{-\infty}^{x} \frac{1}{2\pi} e^{-\frac{u^2}{2}}\, du. \tag{A.70}$$

It is clear that the function $\Phi(\cdot)$ satisfies the property $\Phi(-x) = 1 - \Phi(x)$.

If $d = 2$, one gets the bivariate Gaussian distribution, a case of special interest to our purposes. Let $\mathbf{X} = (X, Y)$ be a bivariate Gaussian vector, in which case one says that X and Y are *jointly Gaussian*. It is easy to show that (A.68) reduces in this case to

$$\begin{aligned} f_{X,Y}(x,y) = {} & \frac{1}{2\pi\sigma_x\sigma_y\sqrt{1-\rho^2}} \\ & \times \exp\left\{-\frac{1}{2(1-\rho^2)}\left[\left(\frac{x-\mu_x}{\sigma_x}\right)^2 + \left(\frac{y-\mu_y}{\sigma_y}\right)^2 - 2\rho\frac{(x-\mu_x)(y-\mu_y)}{\sigma_x\sigma_y}\right]\right\}, \end{aligned} \tag{A.71}$$

where $E[X] = \mu_x$, $\mathrm{Var}(X) = \sigma_x^2$, $E[Y] = \mu_y$, $\mathrm{Var}(Y) = \sigma_y^2$, and ρ is the correlation coefficient between X and Y. With $\mu_x = \mu_y = 0$ and $\sigma_x = \sigma_y = 1$, the PDF of X is given by

$$\begin{aligned} P(X \le x, Y \le y) &= \Phi(x, y; \rho) \\ &= \int_{-\infty}^{x}\int_{-\infty}^{y} \frac{1}{2\pi\sqrt{1-\rho^2}} \exp\left\{-\frac{u^2+v^2-2\rho uv}{2(1-\rho^2)}\right\} du dv. \end{aligned} \tag{A.72}$$

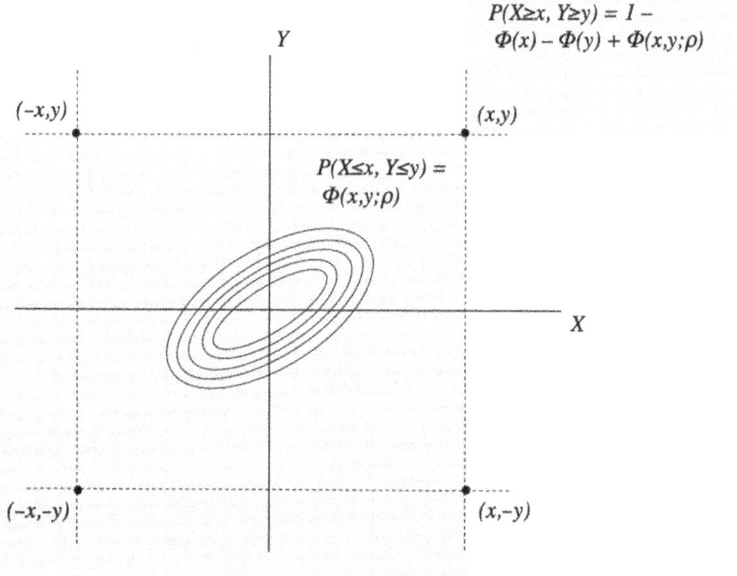

FIGURE A.1 Bivariate Gaussian density and some probabilities of interest.

By considering Figure A.1, it becomes apparent that the bivariate function $\Phi(\cdot, \cdot; \rho)$ satisfies the following relationships:

$$\begin{aligned}
\Phi(x, y; \rho) &= P(X \geq -x, Y \geq -y), \\
\Phi(x, y; -\rho) &= P(X \geq -x, Y \leq y) = P(X \leq x, Y \geq -y), \\
\Phi(x, -y; \rho) &= P(X \leq x) - P(X \leq x, Y \geq -y), \\
\Phi(-x, y; \rho) &= P(Y \leq y) - P(X \geq -x, Y \leq y), \\
\Phi(-x, -y; \rho) &= P(X \geq x, Y \geq y) = 1 - P(X \leq x) - P(Y \leq y) + P(X \leq x, Y \leq y),
\end{aligned} \tag{A.73}$$

where $\Phi(\cdot)$ is the univariate standard Gaussian PDF, defined previously. It follows that

$$\begin{aligned}
\Phi(x, -y; \rho) &= \Phi(x) - \Phi(x, y; -\rho), \\
\Phi(-x, y; \rho) &= \Phi(y) - \Phi(x, y; -\rho), \\
\Phi(-x, -y; \rho) &= 1 - \Phi(x) - \Phi(y) + \Phi(x, y; \rho).
\end{aligned} \tag{A.74}$$

These are the bivariate counterparts to the univariate relationship $\Phi(-x) = 1 - \Phi(x)$.

The following are additional useful properties of a multivariate Gaussian random vector $\mathbf{X} \sim N(\boldsymbol{\mu}, \Sigma)$:

- The density of each component X_i is univariate Gaussian $N(\mu_i, \Sigma_{ii})$.
- The components of $\mathbf{X}$ are independent *if and only if* they are uncorrelated, that is, Σ is a diagonal matrix.
- The *whitening transformation* $\mathbf{Y} = \Sigma^{-\frac{1}{2}}(\mathbf{X} - \boldsymbol{\mu})$ produces a multivariate Gaussian $\mathbf{Y} \sim N(0, I_p)$ (so that all components of $\mathbf{Y}$ are zero-mean, unit-variance, and uncorrelated Gaussian random variables).
- In general, if A is a nonsingular $p \times p$ matrix and $\mathbf{c}$ is a p-vector, then $\mathbf{Y} = A\mathbf{X} + \mathbf{c} \sim N_p(A\boldsymbol{\mu} + \mathbf{c}, A\Sigma A^T)$.
- The random vectors $A\mathbf{X}$ and $B\mathbf{X}$ are independent iff $A\Sigma B^T = 0$.
- If $\mathbf{Y}$ and $\mathbf{X}$ are jointly multivariate Gaussian, then the distribution of $\mathbf{Y}$ given $\mathbf{X}$ is again multivariate Gaussian.
- The best MMSE predictor $E[\mathbf{Y} \mid \mathbf{X}]$ is a linear function of $\mathbf{X}$.

A.12 CONVERGENCE OF RANDOM SEQUENCES

A *random sequence* $\{X_n; n = 1, 2, \dots\}$ is a sequence of random variables. One would like to study the behavior of such random sequences as n grows without bound. The standard modes of convergence for random sequences are

- *"Sure"* convergence: $X_n \to X$ surely if for all outcomes $\xi \in S$ in the sample space one has $\lim_{n\to\infty} X_n(\xi) = X(\xi)$.
- *Almost-sure* convergence or convergence *with probability one*: $X_n \to X$ *(a.s.)* if pointwise convergence fails only for an event of probability zero, that is,

$$P\left(\left\{\xi \in S \mid \lim_{n\to\infty} X_n(\xi) = X(\xi)\right\}\right) = 1. \tag{A.75}$$

- L^p-convergence: $X_n \to X$ in L^p, for $p > 0$, also denoted by $X_n \xrightarrow{L^p} X$, if $E[|X_n|^p] < \infty$ for $n = 1, 2, \dots$, $E[|X|^p] < \infty$, and the *p-norm* of the difference between X_n and X converges to zero:

$$\lim_{n\to\infty} E[|X_n - X|^p] = 0. \tag{A.76}$$

  The special case of L^2 convergence is also called *mean-square* (m.s.) convergence.

- Convergence in *Probability*: $X_n \to X$ in probability, also denoted by $X_n \xrightarrow{P} X$, if the "probability of error" converges to zero:

$$\lim_{n\to\infty} P(|X_n - X| > \tau) = 0, \quad \text{for all } \tau > 0. \tag{A.77}$$

- Convergence *in Distribution*: $X_n \to X$ in distribution, also denoted by $X_n \xrightarrow{D} X$, if the corresponding PDFs converge:

$$\lim_{n\to\infty} F_{X_n}(a) = F_X(a), \tag{A.78}$$

at all points $a \in R$ where F_X is continuous.

The following relationships between modes of convergence hold (Chung, 1974):

$$\left.\begin{array}{r}\text{Sure} \Rightarrow \text{Almost-sure}\\ \text{Mean-square}\end{array}\right\} \Rightarrow \text{Probability} \Rightarrow \text{Distribution.} \tag{A.79}$$

Therefore, sure convergence is the strongest mode of convergence and convergence in distribution is the weakest. In practice, sure convergence is rarely used, and almost-sure convergence is the strongest mode of convergence employed. On the other hand, convergence is distribution is really convergence of PDFs, and does not have the same properties one expects from convergence, which the other modes have. (e.g., convergence of X_n to X and Y_n to Y in distribution does not imply in general that $X_n + Y_n$ converges to $X + Y$ in distribution, unless X_n and Y_n are independent for all $n = 1, 2, \ldots$)

Stronger relations between the modes of convergence can be proved for special cases. In particular, mean-square convergence and convergence in probability can be shown to be equivalent for *uniformly bounded* sequences. A random sequence $\{X_n; n = 1, 2, \ldots\}$ is uniformly bounded if there exists a finite $K > 0$, *which does not depend on n*, such that

$$|X_n| \le K, \text{with probability } 1, \text{ for all } n = 1, 2, \ldots \tag{A.80}$$

meaning that $P(|X_n| < K) = 1$, for all $n = 1, 2, \ldots$ The classification error rate sequence $\{\varepsilon_n; n = 1, 2, \ldots\}$ is an example of a uniformly bounded random sequence, with $K = 1$. We have the following theorem.

Theorem A.1 *Let $\{X_n; n = 1, 2, \ldots\}$ be a uniformly bounded random sequence. The following statements are equivalent.*

(1) $X_n \to X$ in L^p, for some $p > 0$.
(2) $X_n \to X$ in L^q, for all $q > 0$.
(3) $X_n \to X$ in probability.

Proof: First note that we can assume without loss of generality that $X = 0$, since $X_n \to X$ if and only if $X_n - X \to 0$, and $X_n - X$ is also uniformly bounded, with $E[|X_n - X|^p] < \infty$. Showing that (1) $\Leftrightarrow$ (2) requires showing that $X_n \to 0$ in L^p, for some $p > 0$ implies that $X_n \to 0$ in L^q, for all $q > 0$. First observe that $E[|X_n|^q] \le E[K^q] = K^q < \infty$, for all $q > 0$. If $q > p$, the result is immediate. Let

$0 < q < p$. With $X = X_n^q$, $Y = 1$ and $r = p/q$ in Holder's Inequality (A.43), we can write

$$E[|X_n|^q] \le E[|X_n|^p]^{q/p}. \tag{A.81}$$

Hence, if $E[|X_n|^p] \to 0$, then $E[|X_n|^q] \to 0$, proving the assertion. To show that (2) $\Leftrightarrow$ (3), first we show the direct implication by writing Markov's Inequality (A.45) with $X = |X_n|^p$ and $a = \tau^p$:

$$P(|X_n| \ge \tau) \le \frac{E[|X_n|^p]}{\tau^p}, \quad \text{for all } \tau > 0. \tag{A.82}$$

The right-hand side goes to 0 by hypothesis, and thus so does the left-hand side, which is equivalent to (A.77) with $X = 0$. To show the reverse implication, write

$$E[|X_n|^p] = E[|X_n|^p I_{|X_n|<\tau}] + E[|X_n|^p I_{|X_n|\ge\tau}] \le \tau^p + K^p P(|X_n| \ge \tau). \tag{A.83}$$

By assumption, $P(|X_n| \ge \tau) \to 0$, for all $\tau > 0$, so that $\lim E[|X_n|^p] \le \tau^p$. Letting $\tau \to 0$ then yields the desired result. ■

As a simple corollary, we have the following result.

Corollary A.1 *If $\{X_n; n = 1, 2, \ldots\}$ is a uniformly bounded random sequence and $X_n \to X$ in probability, then $E[X_n] \to E[X]$.*

Proof: From the previous theorem, $X_n \to X$ in L^1, i.e., $E[|X_n - X|] \to 0$. But $|E[X_n - X]| \le E[|X_n - X|]$, so $|E[X_n - X]| \to 0 \Rightarrow E[X_n - X] \to 0$. ■

A.13 LIMITING THEOREMS

The following two theorems are the classical limiting theorems for random sequences, the proofs of which can be found in any advanced text in Probability Theory, for example, Chung (1974).

Theorem A.2 *(Law of Large Numbers) Given an i.i.d. random sequence $\{X_n; n = 1, 2, \ldots\}$ with common finite mean μ,*

$$\frac{1}{n}\sum_{i=1}^{n} X_i \to \mu, \quad \textit{with probability 1.} \tag{A.84}$$

Mainly for historical reasons, the previous theorem is sometimes called the *Strong* Law of Large Numbers, with the weaker result involving only convergence in probability being called the *Weak* Law of Large Numbers.

Theorem A.3 *(Central Limit Theorem) Given an i.i.d. random sequence* $\{X_n; n = 1, 2, \ldots\}$ *with common finite mean* μ *and common finite variance* σ^2,

$$\frac{1}{\sigma\sqrt{n}}\left(\sum_{i=1}^{n} X_i - n\mu\right) \xrightarrow{D} N(0, 1). \tag{A.85}$$

The previous limiting theorems concern behavior of a sum of n random variables as n approach infinity. It is also useful to have an idea of how partial sums differ from expected values for finite n. This problem is addressed by the so-called *concentration inequalities*, the most famous of which is Hoeffding's inequality.

Theorem A.4 *(Hoeffding's inequality) Given an independent (not necessarily identically distributed) random sequence* $\{X_n; n = 1, 2, \ldots\}$ *such that* $P(a \leq X_n \leq b) = 1$, *for all* $n = 1, 2, \ldots$, *the sequence of partial sums* $S_n = \sum_{i=1}^{n} X_i$ *satisfies*

$$P(|S_n - E[S_n]| \geq \tau) \leq 2e^{-\frac{2\tau^2}{n(a-b)^2}}, \quad \textit{for all } \tau > 0. \tag{A.86}$$

APPENDIX B

VAPNIK–CHERVONENKIS THEORY

Vapnik–Chervonenkis (VC) Theory (Devroye et al., 1996; Vapnik, 1998) introduces intuitively satisfying metrics for classification complexity, via quantities known as the shatter coefficients and the VC dimension, and uses them to uniformly bound the difference between apparent and true errors over a family of classifiers, in a distribution-free manner. The VC theorem also leads to a simple bound on the expected classifier design error for certain classification rules, which obey the empirical risk minimization principle.

The main result of VC theory is the VC theorem, which is related to the Glivenko–Cantelli theorem and the theory of empirical processes. All bounds in VC theory are worst-case, as there are no distributional assumptions, and thus they can be very loose for a particular feature–label distribution and small sample sizes. Nevertheless, the VC theorem remains a powerful tool for the analysis of the large sample behavior of both the true and apparent (i.e., resubstitution) classification errors.

B.1 SHATTER COEFFICIENTS

Intuitively, the complexity of a classification rule must have to do with its ability to "pick out" subsets of a given set of points. For a given n, consider a set of points $\mathbf{x}_1, \ldots, \mathbf{x}_n$ in R^d. Given a set $A \subseteq R^d$, then

$$A \cap \{\mathbf{x}_1, \ldots, \mathbf{x}_n\} \subseteq \{\mathbf{x}_1, \ldots, \mathbf{x}_n\} \tag{B.1}$$

Error Estimation for Pattern Recognition, First Edition. Ulisses M. Braga-Neto and Edward R. Dougherty.

is the subset of $\{\mathbf{x}_1, \ldots, \mathbf{x}_n\}$ "picked out" by A. Now, consider a family $\mathcal{A}$ of measurable subsets of R^d, and let

$$N_{\mathcal{A}}(\mathbf{x}_1, \ldots, \mathbf{x}_n) = \left|\{A \cap \{\mathbf{x}_1, \ldots, \mathbf{x}_n\} \mid A \in \mathcal{A}\}\right|, \tag{B.2}$$

that is, the total number of subsets of $\{\mathbf{x}_1, \ldots, \mathbf{x}_n\}$ that can be picked out by sets in $\mathcal{A}$. The nth *shatter coefficient* of the family $\mathcal{A}$ is defined as

$$s(\mathcal{A}, n) = \max_{\{\mathbf{x}_1, \ldots, \mathbf{x}_n\}} N_{\mathcal{A}}(\mathbf{x}_1, \ldots, \mathbf{x}_n). \tag{B.3}$$

The shatter coefficients $s(\mathcal{A}, n)$ measure the richness (the size, the complexity) of $\mathcal{A}$. Note also that $s(\mathcal{A}, n) \leq 2^n$ for all n.

B.2 THE VC DIMENSION

The Vapnik–Chervonenkis (VC) dimension is a measure of the size, that is, the complexity, of a class of classifiers C. It agrees very naturally with our intuition of complexity as the ability of a classifier to cut up the space finely.

From the previous definition of shatter coefficient, we have that if $s(\mathcal{A}, n) = 2^n$, then there is a set of points $\{\mathbf{x}_1, \ldots, \mathbf{x}_n\}$ such that $N_{\mathcal{A}}(\mathbf{x}_1, \ldots, \mathbf{x}_n) = 2^n$, and we say that $\mathcal{A}$ *shatters* $\{\mathbf{x}_1, \ldots, \mathbf{x}_n\}$. On the other hand, if $s(\mathcal{A}, n) < 2^n$, then any set of points $\{\mathbf{x}_1, \ldots, \mathbf{x}_n\}$ contains at least one subset that cannot be picked out by any member of $\mathcal{A}$. In addition, we must have $s(\mathcal{A}, m) < 2^m$, for all $m > n$.

The VC dimension $V_{\mathcal{A}}$ of $\mathcal{A}$ (assuming $|\mathcal{A}| \geq 2$) is the largest integer $k \geq 1$ such that $s(\mathcal{A}, k) = 2^k$. If $s(\mathcal{A}, n) = 2^n$ for all n, then $V_{\mathcal{A}} = \infty$. The VC dimension of $\mathcal{A}$ is thus the *maximal number of points in* R^d *that can be shattered by* $\mathcal{A}$. Clearly, $V_{\mathcal{A}}$ measures the complexity of $\mathcal{A}$.

Some simple examples:

- Let $\mathcal{A}$ be the class of half-lines: $\mathcal{A} = \{(-\infty, a] \mid a \in R\}$, then

$$s(\mathcal{A}, n) = n + 1 \quad \text{and} \quad V_{\mathcal{A}} = 1. \tag{B.4}$$

- Let $\mathcal{A}$ be the class of intervals: $\mathcal{A} = \{[a, b] \mid a, b \in R\}$, then

$$s(\mathcal{A}, n) = \frac{n(n+1)}{2} + 1 \quad \text{and} \quad V_{\mathcal{A}} = 2. \tag{B.5}$$

- Let $\mathcal{A}_d$ be the class of "half-rectangles" in R^d: $\mathcal{A}_d = \{(-\infty, a_1] \times \cdots \times (-\infty, a_d] \mid (a_1, \ldots, a_d) \in R^d\}$, then $V_{\mathcal{A}_d} = d$.
- Let $\mathcal{A}_d$ be the class of rectangles in R^d: $\mathcal{A}_d = \{[a_1, b_1] \times \cdots \times [a_d, b_d] \mid (a_1, \ldots, a_d, b_1, \ldots, b_d) \in R^{2d}\}$, then $V_{\mathcal{A}_d} = 2d$.

Note that in the examples above, the VC dimension is equal to the number of parameters. While this is intuitive, it is not true in general. In fact, one can find a one-parameter family $\mathcal{A}$ for which $V_{\mathcal{A}} = \infty$. So one should be careful about naively attributing complexity to the number of parameters.

Note also that the last two examples generalize the first two, by means of cartesian products. It can be shown that in such cases

$$s(\mathcal{A}_d, n) \leq s(\mathcal{A}, n)^d, \quad \text{for all } n. \tag{B.6}$$

A general bound for shatter coefficients is

$$s(\mathcal{A}, n) \leq \sum_{i=0}^{V_{\mathcal{A}}} \binom{n}{i}, \quad \text{for all } n. \tag{B.7}$$

The first two examples achieve this bound, so it is tight. From this bound it also follows, in the case that $V_{\mathcal{A}} < \infty$, that

$$s(\mathcal{A}, n) \leq (n+1)^{V_{\mathcal{A}}}. \tag{B.8}$$

B.3 VC THEORY OF CLASSIFICATION

The preceding concepts can be applied to define the complexity of a class $\mathcal{C}$ of classifiers (i.e., a classification rule). Given a classifier $\psi \in \mathcal{C}$, let us define the set $A_\psi = \{\mathbf{x} \in R^d \mid \psi(\mathbf{x}) = 1\}$, that is, the 1-decision region (this specifies the classifier completely, since the 0-decision region is simply A_ψ^c). Let $\mathcal{A}_{\mathcal{C}} = \{A_\psi \mid \psi \in \mathcal{C}\}$, that is, the family of all 1-decision regions produced by $\mathcal{C}$. We define the shatter coefficients $\mathcal{S}(\mathcal{C}, n)$ and VC dimension $V_{\mathcal{C}}$ for $\mathcal{C}$ as

$$\begin{aligned} \mathcal{S}(\mathcal{C}, n) &= s(\mathcal{A}_{\mathcal{C}}, n), \\ V_{\mathcal{C}} &= V_{\mathcal{A}_{\mathcal{C}}}. \end{aligned} \tag{B.9}$$

All the results discussed previously apply in the new setting; for example, following (B.8), if $V_{\mathcal{C}} < \infty$, then

$$\mathcal{S}(\mathcal{C}, n) \leq (n+1)^{V_{\mathcal{C}}}. \tag{B.10}$$

Next, results on the VC dimension and shatter coefficients for commonly used classification rules are given.

B.3.1 Linear Classification Rules

Linear classification rules are those that produce hyperplane decision-boundary classifiers. This includes NMC, LDA, Perceptrons, and Linear SVMs. Let $\mathcal{C}$ be the class of hyperplane decision-boundary classifiers in R^d. Then, as proved in Devroye et al. (1996, Cor. 13.1),

$$\mathcal{S}(C,n) = 2\sum_{i=0}^{d}\binom{n-1}{i},$$

$$V_C = d+1. \tag{B.11}$$

The fact that $V_C = d+1$ means that there is a set of $d+1$ points that can be shattered by oriented hyperplanes in R^d, but no set of $d+2$ points (in general position) can be shattered—a familiar fact in the case $d = 2$.

The VC dimension of linear classification rules thus increases linearly with the number of variables. Note however that the fact that all linear classification rules have the same VC dimension does not mean they will perform the same in small-sample cases.

B.3.2 *k*NN Classification Rule

For $k = 1$, clearly any set of points can be shattered (just use the points as training data). This is also true for any $k > 1$. Therefore, $V_C = \infty$.

Classes C with finite VC dimension are called *VC classes*. Thus, the class C_k of kNN classifiers is not a VC class, for each $k > 1$. Classification rules that have infinite VC dimension are not necessarily useless. For example, there is empirical evidence that 3NN is a good rule in small-sample cases. In addition, the Cover-Hart Theorem says that the asymptotic kNN error rate is near the Bayes error. However, the worst-case scenario if $V_C = \infty$ is indeed very bad, as shown in Section B.4.

B.3.3 Classification Trees

A binary tree with a depth of k levels of splitting nodes has at most $2^k - 1$ splitting nodes and at most 2^k leaves. Therefore, for a classification tree with data-independent splits (that is, fixed-partition tree classifiers), we have that

$$\mathcal{S}(C,n) = \begin{cases} 2^n, & n \le 2^k \\ 2^{2^k}, & n > 2^k \end{cases} \tag{B.12}$$

and it follows that $V_C = 2^k$. The shatter coefficients and VC dimension thus grow very fast (exponentially) with the number of levels. The case is different for data-dependent decision trees (e.g., CART and BSP). If stopping or pruning criteria are not strict enough, one may have $V_C = \infty$ in those cases.

B.3.4 Nonlinear SVMs

It is easy to see that the shatter coefficients and VC dimension correspond to those of linear classification in the transformed high-dimensional space. More precisely, if the *minimal* space where the kernel can be written as a dot product is m, then $V_C = m + 1$.

For example, for the polynomial kernel

$$K(\mathbf{x}, \mathbf{y}) = (\mathbf{x}^T \mathbf{y})^p = (x_1 y_1 + \cdots + x_d y_d)^p, \tag{B.13}$$

we have $m = \binom{d+p-1}{p}$, that is, the number of distinct powers of $x_i y_i$ in the expansion of $K(\mathbf{x}, \mathbf{y})$, so $V_C = \binom{d+p-1}{p} + 1$.

For certain kernels, such as the Gaussian kernel,

$$K(\mathbf{x}, \mathbf{y}) = \exp\left(\frac{-||\mathbf{x} - \mathbf{y}||^2}{\sigma^2} \right), \tag{B.14}$$

the minimal space is infinite dimensional, so $V_C = \infty$.

B.3.5 Neural Networks

A basic result for neural networks is proved in (Devroye et al., 1996, Theorem 30.5). For the class C_k of neural networks with k neurons in one hidden layer, and arbitrary sigmoids,

$$V_{C_k} \geq 2 \left\lfloor \frac{k}{2} \right\rfloor d \tag{B.15}$$

where $\lfloor x \rfloor$ is the largest integer $\leq x$. If k is even, this simplifies to $V_C \geq kd$.

For threshold sigmoids, $V_{C_k} < \infty$. In fact,

$$S(C_k, n) \leq (ne)^\gamma \quad \text{and} \quad V_{C_k} \leq 2\gamma \log_2(e\gamma), \tag{B.16}$$

where $\gamma = kd + 2k + 1$ is the number of weights. The threshold sigmoid achieves the smallest possible VC dimension among all sigmoids. In fact, there are sigmoids for which $V_{C_k} = \infty$ for $k \geq 2$.

B.3.6 Histogram Rules

For a histogram rule with a finite number of partitions b, it is easy to see that the the shatter coefficients are given by

$$S(C, n) = \begin{cases} 2^n, & n < b, \\ 2^b, & n \geq b. \end{cases} \tag{B.17}$$

Therefore, the VC dimension is $V_C = b$.

B.4 VAPNIK–CHERVONENKIS THEOREM

The celebrated Vapnik–Chervonenkis theorem uses the shatter coefficients $\mathcal{S}(C,n)$ and the VC dimension V_C to bound

$$P(|\hat{\varepsilon}[\psi] - \varepsilon[\psi]| > \tau), \quad \text{for all } \tau > 0, \tag{B.18}$$

in a distribution-free manner, where $\varepsilon[\psi]$ is the true classification error of a classifier $\psi \in C$ and $\hat{\varepsilon}_n[\psi]$ is the *empirical error* of ψ given data S_n:

$$\hat{\varepsilon}_n[\psi] = \frac{1}{n}\sum_{i=1}^{n} |y_i - \psi(x_i)|. \tag{B.19}$$

Note that if S_n could be assumed independent of every $\psi \in C$, then $\hat{\varepsilon}_n[\psi]$ would be an independent test-set error, and we could use Hoeffding's Inequality (c.f. Section A.13) to get

$$P(|\hat{\varepsilon}[\psi] - \varepsilon[\psi]| > \tau) \leq 2e^{-2n\tau^2}, \quad \text{for all } \tau > 0. \tag{B.20}$$

However, that is not sufficient. If one wants to study $|\hat{\varepsilon}[\psi] - \varepsilon[\psi]|$ for any distribution, and any classifier $\psi \in C$, in particular a designed classifier ψ_n, one cannot assume independence from S_n.

The solution is to bound $P(|\hat{\varepsilon}[\psi] - \varepsilon[\psi]| > \tau)$ uniformly for all possible $\psi \in C$, that is, to find a (distribution-free) bound for the probability of the worst-case scenario

$$P\left(\sup_{\psi\in C} |\hat{\varepsilon}[\psi] - \varepsilon[\psi]| > \tau\right), \quad \text{for all } \tau > 0. \tag{B.21}$$

This is the purpose of the Vapnik–Chervonenkis theorem, given next. This is a deep theorem, a proof of which can be found in Devroye et al. (1996).

Theorem B.1 *(Vapnik–Chervonenkis theorem.) Let $\mathcal{S}(C,n)$ be the nth shatter coefficient for class C, as defined previously. Regardless of the distribution of (X, Y),*

$$P\left(\sup_{\psi\in C} |\hat{\varepsilon}[\psi] - \varepsilon[\psi]| > \tau\right) \leq 8\mathcal{S}(C,n)e^{-n\tau^2/32}, \quad \textit{for all } \tau > 0. \tag{B.22}$$

If V_C is finite, we can use the inequality (B.10) to write the bound in terms of V_C:

$$P\left(\sup_{\psi\in C} |\hat{\varepsilon}[\psi] - \varepsilon[\psi]| > \tau\right) \leq 8(n+1)^{V_C}e^{-n\tau^2/32}, \quad \text{for all } \tau > 0. \tag{B.23}$$

Therefore, if V_C is finite, the term $e^{-n\tau^2/32}$ dominates, and the bound decreases exponentially fast as $n \to \infty$.

If, on the other hand, $V_C = \infty$, we cannot make $n \gg V_C$, and a worst-case bound can be found that is independent of n (this means that there exists a situation where the design error cannot be reduced no matter how large n may be). More specifically, for every $\delta > 0$, and every classification rule associated with C, it can be shown (Devroye et al., 1996, Theorem 14.3) that there is a feature–label distribution for $(\mathbf{X}, Y)$ with $\varepsilon_C = 0$ but

$$E[\varepsilon_{n,C} - \varepsilon_C] = E[\varepsilon_{n,C}] > \frac{1}{2e} - \delta, \quad \text{for all } n > 1. \tag{B.24}$$

APPENDIX C

DOUBLE ASYMPTOTICS

Traditional large-sample theory in statistics deals with asymptotic behavior as the sample size $n \to \infty$ with dimensionality p fixed. Double asymptotics require both the sample size $n \to \infty$ and the dimensionality $p \to \infty$ at a fixed rate $n/p \to \lambda$. For simplicity, we assume this scenario in this section; all of the definitions and results are readily extendable to the case where the sample sizes n_0 and n_1 are treated independently, with $n_0, n_1 \to \infty$, $p \to \infty$, and $n_0/p \to \lambda_0$ and $n_1/p \to \lambda_1$. We start by recalling that the *double limit* of a double sequence $\{x_{n,p}\}$ of real or complex numbers, if it exists, is a finite number

$$L = \lim_{\substack{n \to \infty \\ p \to \infty}} x_{n,p}, \tag{C.1}$$

such that for every $\tau > 0$, there exists N with the property that $|x_{n,p} - L| < \tau$ for all $n \geq N$ and $p \geq N$. In this case, $\{x_{n,p}\}$ is said to be *convergent* (to L). Double limits are not equal, in general, to iterated limits,

$$\lim_{\substack{n \to \infty \\ p \to \infty}} x_{n,p} \neq \lim_{n \to \infty} \lim_{p \to \infty} x_{n,p} \neq \lim_{p \to \infty} \lim_{n \to \infty} x_{n,p}, \tag{C.2}$$

unless certain conditions involving the uniform convergence of one of the iterated limits are satisfied (Moore and Smith, 1922). A *linear subsequence* $\{x_{n_i,p_i}\}$ of a given double sequence $\{x_{n,p}\}$ is defined by subsequences of increasing indices $\{n_i\}$ and

Error Estimation for Pattern Recognition, First Edition. Ulisses M. Braga-Neto and Edward R. Dougherty.

$\{p_i\}$ out of the sequence of positive integers. It is well known that if an ordinary sequence converges to a limit L then all of its subsequences also converge to L. This property is preserved in the case of double sequences and their linear subsequences, as stated next.

Theorem C.1 *If a double sequence $\{x_{n,p}\}$ converges to a limit L then all of its linear subsequences $\{x_{n_i,p_i}\}$ also converge to L.*

The following definition introduces the notion of limit of interest here.

Definition C.1 *Let $\{x_{n,p}\}$ be a double sequence. Given a number L and a number $\lambda > 0$, if*

$$\lim_{i\to\infty} x_{n_i,p_i} = L \tag{C.3}$$

for all linear subsequences $\{x_{n_i,p_i}\}$ such that $\lim_{i\to\infty} n_i/p_i = \lambda$, then we write

$$\lim_{\substack{n \to \infty \\ p \to \infty \\ n/p \to \lambda}} x_{n,p} = L, \tag{C.4}$$

and say that $\{x_{n,p}\}$ is convergent (to L) at a rate $n/p \to \lambda$.

From Theorem C.1 and Definition C.1, one obtains the following corollary.

Corollary C.1 *If a double sequence $\{x_{n,p}\}$ converges to a limit L, then*

$$\lim_{\substack{n \to \infty \\ p \to \infty \\ n/p \to \lambda}} x_{n,p} = L, \quad \textit{for all } \lambda > 0. \tag{C.5}$$

In other words, if $\{x_{n,p}\}$ is convergent, then it is convergent at any rate of increase between n and p, with the same limit in each case. The interesting situations occur when $\{x_{n,p}\}$ is not convergent. For an example, consider the double sequence

$$x_{n,p} = \frac{n}{n+p}. \tag{C.6}$$

In this case, $\lim_{\substack{n \to \infty \\ p \to \infty}} x_{n,p}$ does not exist. In fact,

$$\lim_{\substack{n \to \infty \\ p \to \infty \\ n/p \to \lambda}} \frac{n}{n+p} = \frac{\lambda}{\lambda+1}, \tag{C.7}$$

which depends on λ, that is, the ratio of increase between n and p.

The limit introduced in Definition C.1 has many of the properties of ordinary limits. It in fact reduces to the ordinary limit if the double sequence is constant in one of the variables. If $\{x_{n,p}\}$ is constant in p, then one can define an ordinary sequence $\{x_n\}$ via $x_n = x_{n,p}$, in which case it is easy to check that $\{x_{n,p}\}$ converges at rate $n/p \to \lambda$ if and only $\{x_n\}$ converges, with

$$\lim_{\substack{n \to \infty \\ p \to \infty \\ n/p \to \lambda}} x_{n,p} = \lim_{n \to \infty} x_n. \tag{C.8}$$

A similar statement holds if $\{x_{n,p}\}$ is constant in n.

Most of the definitions and results on convergence of random sequences discussed in Appendix A apply, with the obvious modifications, to double sequences $\{X_{n,p}\}$ of random variables. In particular, we define the following notions of convergence:

- Convergence in Probability:

$$X_{n,p} \underset{\substack{n \to \infty \\ p \to \infty \\ n/p \to \lambda}}{\overset{P}{\longrightarrow}} X \quad \text{if} \quad \lim_{\substack{n \to \infty \\ p \to \infty \\ n/p \to \lambda}} P(|X_{n,p} - X| > \tau) = 0, \quad \text{for all } \tau > 0. \tag{C.9}$$

- Convergence in Distribution:

$$X_{n,p} \underset{\substack{n \to \infty \\ p \to \infty \\ n/p \to \lambda}}{\overset{D}{\longrightarrow}} X \quad \text{if} \quad \lim_{\substack{n \to \infty \\ p \to \infty \\ n/p \to \lambda}} F_{X_{n,p}}(a) = F_X(a), \tag{C.10}$$

  at all points $a \in R$ where F_X is continuous.

Notice the following important facts about convergence in probability and in distribution; the proofs are straightforward extensions of the corresponding well-known classical results.

Theorem C.2 *If* $\{X_{n,p}\}$ *and* $\{Y_{n,p}\}$ *are* independent, *and*

$$X_{n,p} \underset{\substack{n \to \infty \\ p \to \infty \\ n/p \to \lambda}}{\overset{D}{\longrightarrow}} X \quad \textit{and} \quad Y_{n,p} \underset{\substack{n \to \infty \\ p \to \infty \\ n/p \to \lambda}}{\overset{D}{\longrightarrow}} Y, \tag{C.11}$$

then

$$X_{n,p} + Y_{n,p} \underset{\substack{n \to \infty \\ p \to \infty \\ n/p \to \lambda}}{\overset{D}{\longrightarrow}} X + Y. \tag{C.12}$$

Theorem C.3 *For any* $\{X_{n,p}\}$ *and* $\{Y_{n,p}\}$, *if*

$$X_{n,p} \underset{\substack{n\to\infty \\ p\to\infty \\ n/p\to\lambda}}{\overset{P}{\longrightarrow}} X \quad \text{and} \quad Y_{n,p} \underset{\substack{n\to\infty \\ p\to\infty \\ n/p\to\lambda}}{\overset{P}{\longrightarrow}} Y, \tag{C.13}$$

then

$$X_{n,p} + Y_{n,p} \underset{\substack{n\to\infty \\ p\to\infty \\ n/p\to\lambda}}{\overset{P}{\longrightarrow}} X + Y,$$

$$X_{n,p}Y_{n,p} \underset{\substack{n\to\infty \\ p\to\infty \\ n/p\to\lambda}}{\overset{P}{\longrightarrow}} XY. \tag{C.14}$$

Theorem C.4 (Double-Asymptotic Slutsky Theorem) *If*

$$X_{n,p} \underset{\substack{n\to\infty \\ p\to\infty \\ n/p\to\lambda}}{\overset{D}{\longrightarrow}} X \quad \text{and} \quad Y_{n,p} \underset{\substack{n\to\infty \\ p\to\infty \\ n/p\to\lambda}}{\overset{D}{\longrightarrow}} c, \tag{C.15}$$

where c *is a constant, then*

$$X_{n,p}Y_{n,p} \underset{\substack{n\to\infty \\ p\to\infty \\ n/p\to\lambda}}{\overset{D}{\longrightarrow}} cX. \tag{C.16}$$

The previous result also holds for convergence in probability, as a direct application of Theorem C.3.

The following is a double asymptotic version of the Central Limit Theorem (CLT) theorem.

Theorem C.5 (Double-Asymptotic Central Limit Theorem) *Let* $\{X_{n,p}\}$ *be a double sequence of independent random variables, with* $E[X_{n,p}] = \mu_n$ *and* $Var(X_{n,p}) = \sigma_n^2 < \infty$, *where* μ_n *and* σ_n *do not depend on* p. *In addition, assume that*

$$\lim_{\substack{n\to\infty \\ p\to\infty \\ n/p\to\lambda}} \mu_n p = m < \infty \quad \text{and} \quad \lim_{\substack{n\to\infty \\ p\to\infty \\ n/p\to\lambda}} \sigma_n^2 p = s^2 < \infty. \tag{C.17}$$

If $(X_{n,p} - \mu_n)/\sigma_n$ *is identically distributed for all* n, p, *and* $S_{n,p} = X_{n,1} + \cdots + X_{n,p}$, *then*

$$S_{n,p} \underset{\substack{n\to\infty \\ p\to\infty \\ n/p\to\lambda}}{\overset{D}{\longrightarrow}} N(m, s^2), \tag{C.18}$$

where $N(m, s^2)$ *denotes the Gaussian distribution with mean* m *and variance* s^2.

Proof: Define $Y_{n,m} = (X_{n,m} - \mu_n)/\sigma_n$. By hypothesis, the random variables $Y_{n,m}$ are identically distributed, with $E[Y_{n,m}] = 0$ and $\mathrm{Var}[Y_{n,m}] = 1$, for all n and $m = 1, \ldots, p$. We have

$$\frac{S_{n,p} - \mu_n p}{\sigma_n \sqrt{p}} = \frac{1}{\sqrt{p}} \sum_{m=1}^{p} Y_{n,m} \underset{\substack{n \to \infty \\ p \to \infty \\ n/p \to \lambda}}{\overset{D}{\longrightarrow}} N(0,1), \tag{C.19}$$

by a straightforward extension of the ordinary CLT. The desired result now follows by application of Theorems C.2 and C.4. ■

A double sequence $\{X_{n,p}\}$ of random variable is uniformly bounded if there exists a finite $K > 0$, which does not depend on n and p, such that $P(|X_{n,p}| \leq K) = 1$ for all n and p. The following is the double-asymptotic version of Corollary A.1.

Theorem C.6 *If $\{X_{n,p}\}$ is a uniformly bounded random double sequence, then*

$$X_{n,p} \underset{\substack{n \to \infty \\ p \to \infty \\ n/p \to \lambda}}{\overset{P}{\longrightarrow}} X \Rightarrow E[X_{n,p}] \underset{\substack{n \to \infty \\ p \to \infty \\ n/p \to \lambda}}{\longrightarrow} E[X]. \tag{C.20}$$

Finally, the following result states that continuous functions preserve convergence in probability and in distribution.

Theorem C.7 (Double-Asymptotic Continuous Mapping Theorem) *Assume that $f : R \to R$ is continuous almost surely with respect to random variable X, in the sense that $P(f(X)$ is continuous$) = 1$. Then*

$$X_{n,p} \underset{\substack{n \to \infty \\ p \to \infty \\ n/p \to \lambda}}{\overset{D,P}{\longrightarrow}} X \Rightarrow f(X_{n,p}) \underset{\substack{n \to \infty \\ p \to \infty \\ n/p \to \lambda}}{\overset{D,P}{\longrightarrow}} f(X). \tag{C.21}$$

The previous result also holds for the double-asymptotic version of almost-sure convergence, just as the classical Continuous Mapping Theorem.

BIBLIOGRAPHY

Afra, S. and Braga-Neto, U. (2011). Peaking phenomenon and error estimation for support vector machines. In: Proceedings of the IEEE International Workshop on Genomic Signal Processing and Statistics (GENSIPS'2011), San Antonio, TX.

Afsari, B., Braga-Neto, U., and Geman, D. (2014). Rank discriminants for predicting phenotypes from RNA expression. *Annals of Applied Statistics*, 8(3):1469–1491.

Agresti, A. (1992). A survey of exact inference for contingency tables. *Statistical Science*, 7(1):131–153.

Agresti, A. (2002). *Categorical Data Analysis*, 2nd edition. John Wiley & Sons, New York.

Ambroise, C. and McLachlan, G. (2002). Selection bias in gene extraction on the basis of microarray gene expression data. *Proceedings of the National Academy of Sciences*, 99(10):6562–6566.

Anderson, T. (1951). Classification by multivariate analysis. *Psychometrika*, 16:31–50.

Anderson, T. (1973). An asymptotic expansion of the distribution of the studentized classification statistic W. *The Annals of Statistics*, 1:964–972.

Anderson, T. (1984). *An Introduction to Multivariate Statistical Analysis*, 2nd edition. John Wiley & Sons, New York.

Arnold, S. F. (1981). *The Theory of Linear Models and Multivariate Analysis*. John Wiley & Sons, New York.

Bartlett, P., Boucheron, S., and Lugosi, G. (2002). Model selection and error estimation. *Machine Learning*, 48:85–113.

Bishop, C. (1995). *Neural Networks for Pattern Recognition*. Oxford University Press, New York.

Error Estimation for Pattern Recognition, First Edition. Ulisses M. Braga-Neto and Edward R. Dougherty.

Boulesteix, A. (2010). Over-optimism in bioinformatics research. *Bioinformatics*, 26(3):437–439.

Bowker, A. (1961). A representation of hotelling's t^2 and Anderson's classification statistic w in terms of simple statistics. In: *Studies in Item Analysis and Prediction*, Solomon, H. (editor). Stanford University Press, pp. 285–292.

Bowker, A. and Sitgreaves, R. (1961). An asymptotic expansion for the distribution function of the w-classification statistic. In: *Studies in Item Analysis and Prediction*, Solomon, H. (editor). Stanford University Press, pp. 292–310.

Braga-Neto, U. (2007). Fads and Fallacies in the Name of Small-Sample Microarray Classification. *IEEE Signal Processing Magazine*, Special Issue on Signal Processing Methods in Genomics and Proteomics, 24(1):91–99.

Braga-Neto, U. and Dougherty, E. (2004a). Bolstered error estimation. *Pattern Recognition*, 37(6):1267–1281.

Braga-Neto, U. and Dougherty, E. (2004b). Is cross-validation valid for microarray classification? *Bioinformatics*, 20(3):374–380.

Braga-Neto, U. and Dougherty, E. (2005). Exact performance of error estimators for discrete classifiers. *Pattern Recognition*, 38(11):1799–1814.

Braga-Neto, U. and Dougherty, E. (2010). Exact correlation between actual and estimated errors in discrete classification. *Pattern Recognition Letters*, 31(5):407–412.

Braga-Neto, U., Zollanvari, A., and Dougherty, E. (2014). Cross-validation under separate sampling: strong bias and how to correct it. *Bioinformatics,* 30(23):3349–3355.

Butler, K. and Stephens, M. (1993). The distribution of a sum of binomial random variables. Technical Report ADA266969, Stanford University Department of Statistics, Stanford, CA. Available at: http://handle.dtic.mil/100.2/ADA266969.

Chen, T. and Braga-Neto, U. (2010). Exact performance of cod estimators in discrete prediction. *EURASIP Journal on on Advances in Signal Processing*, 2010:487893.

Chernick, M. (1999). *Bootstrap Methods: A Practitioner's Guide*. John Wiley & Sons, New York.

Chung, K. L. (1974). *A Course in Probability Theory*, 2nd edition. Academic Press, New York.

Cochran, W. and Hopkins, C. (1961). Some classification problems with multivariate qualitative data. *Biometrics*, 17(1):10–32.

Cover, T. (1969). Learning in pattern recognition. In: *Methodologies of Pattern Recognition*, Watanabe, S. (editor). Academic Press, New York, pp. 111–132.

Cover, T. and Hart, P. (1967). Nearest-neighbor pattern classification. *IEEE Transactions on Information Theory*, 13:21–27.

Cover, T. and van Campenhout, J. (1977). On the possible orderings in the measurement selection problem. *IEEE Transactions on Systems, Man, and Cybernetics*, 7:657–661.

Cramér, H. (1946). *Mathematical Methods of Statistics*. Princeton University Press, Princeton, NJ.

Dalton, L. and Dougherty, E. (2011a). Application of the bayesian MMSE error estimator for classification error to gene-expression microarray data. *IEEE Transactions on Signal Processing*, 27(13):1822–1831.

Dalton, L. and Dougherty, E. (2011b). Bayesian minimum mean-square error estimation for classification error part I: definition and the bayesian MMSE error estimator for discrete classification. *IEEE Transactions on Signal Processing*, 59(1):115–129.

Dalton, L. and Dougherty, E. (2011c). Bayesian minimum mean-square error estimation for classification error part II: linear classification of gaussian models. *IEEE Transactions on Signal Processing*, 59(1):130–144.

Dalton, L. and Dougherty, E. (2012a). Exact sample conditioned MSE performance of the bayesian MMSE estimator for classification error part I: representation. *IEEE Transactions on Signal Processing*, 60(5):2575–2587.

Dalton, L. and Dougherty, E. (2012b). Exact sample conditioned MSE performance of the bayesian MMSE estimator for classification error part II: performance analysis and applications. *IEEE Transactions on Signal Processing*, 60(5):2588–2603.

Dalton, L. and Dougherty, E. (2012c). Optimal MSE calibration of error estimators under bayesian models. *Pattern Recognition*, 45(6):2308–2320.

Dalton, L. and Dougherty, E. (2013a). Optimal classifiers with minimum expected error within a bayesian framework – part I: discrete and gaussian models. *Pattern Recognition*, 46(5):1301–1314.

Dalton, L. and Dougherty, E. (2013b). Optimal classifiers with minimum expected error within a bayesian framework – part II: properties and performance analysis. *Pattern Recognition*, 46(5):1288–1300.

Deev, A. (1970). Representation of statistics of discriminant analysis and asymptotic expansion when space dimensions are comparable with sample size. *Dokl. Akad. Nauk SSSR*, 195:759–762. [in Russian].

Deev, A. (1972). Asymptotic expansions for distributions of statistics w, m, and $w^\star$ in discriminant analysis. *Statistical Methods of Classification*, 31:6–57. [in Russian].

DeGroot, M. (1970). *Optimal Statistical Decisions*. McGraw-Hill.

Devroye, L., Gyorfi, L., and Lugosi, G. (1996). *A Probabilistic Theory of Pattern Recognition*. Springer, New York.

Devroye, L. and Wagner, T. (1979). Distribution-free performance bounds for potential function rules. *IEEE Transactions on Information Theory*, 25:601–604.

Diaconis, P. and Freedman, D. (1986). On the consistency of Bayes estimates. *The Annals of Statistics*, 14(1):1–26.

Dougherty, E. (2012). Prudence, risk, and reproducibility in biomarker discovery. *BioEssays*, 34(4):277–279.

Dougherty, E. and Bittner, M. (2011). *Epistemology of the Cell: A Systems Perspective on Biological Knowledge*. John Wiley & Sons, New York.

Dougherty, E., Hua, J., and Sima, C. (2009). Performance of feature selection methods. *Current Genomics*, 10(6):365–374.

Duda, R. and Hart, P. (2001). *Pattern Classification*, 2nd edition. John Wiley & Sons, New York.

Dudoit, S., Fridlyand, J., and Speed, T. (2002). Comparison of discrimination methods for the classification of tumors using gene expression data. *Journal of the American Statistical Association*, 97(457):77–87.

Efron, B. (1979). Bootstrap methods: another look at the jacknife. *Annals of Statistics*, 7: 1–26.

Efron, B. (1983). Estimating the error rate of a prediction rule: improvement on cross-validation. *Journal of the American Statistical Association*, 78(382):316–331.

Efron, B. and Tibshirani, R. (1997). Improvements on cross-validation: the .632+ bootstrap method. *Journal of the American Statistical Association*, 92(438):548–560.

Esfahani, S. and Dougherty, E. (2014a). Effect of separate sampling on classification accuracy. *Bioinformatics*, 30(2):242–250.

Esfahani, S. and Dougherty, E. (2014b). Incorporation of biological pathway knowledge in the construction of priors for optimal bayesian classification. *IEEE/ACM Transactions on Computational Biology and Bioinformatics*, 11(1):202–218.

Evans, M., Hastings, N., and Peacock, B. (2000). *Statistical Distributions*, 3rd edition. John Wiley & Sons, New York.

Farago, A. and Lugosi, G. (1993). Strong universal consistency of neural network classifiers. *IEEE Transactions on Information Theory*, 39:1146–1151.

Feller, W. (1968). *An Introduction to Probability Theory and Its Applications*, Vol. 1. John Wiley & Sons, New York.

Feynman, R. (1985). *QED: The Strange Theory of Light and Matter*. Princeton University Press, Princeton, NJ.

Fisher, R. (1915). Frequency distribution of the values of the correlation coefficient in samples from an indefinitely large population. *Biometrika*, 10:507–521.

Fisher, R. (1921). On the "probable error" of a coefficient of correlation deduced from a small sample. *Metron*, 1:3–32.

Fisher, R. (1928). The general sampling distribution of the multiple correlation coefficient. *Proceedings of the Royal Society Series A*, 121:654–673.

Fisher, R. (1936). The use of multiple measurements in taxonomic problems. *Annals of Eugenics*, 7:179–188.

Foley, D. (1972). Considerations of sample and feature size. *IEEE Transactions on Information Theory*, IT-18(5):618–626.

Freedman, D. A. (1963). On the asymptotic behavior of Bayes' estimates in the discrete case. *The Annals of Mathematical Statistics*, 34(4):1386–1403.

Fujikoshi, Y. (2000). Error bounds for asymptotic approximations of the linear discriminant function when the sample sizes and dimensionality are large. *Journal of Multivariate Analysis*, 73:1–17.

Gelman, A., Carlin, J. B., Stern, H. S., and Rubin, D. B. (2004). *Bayesian Data Analysis*, 2nd edition. Chapman & Hall/CRC, Boca Raton, FL.

Geman, D., d'Avignon, C., Naiman, D., Winslow, R., and Zeboulon, A. (2004). Gene expression comparisons for class prediction in cancer studies. In: Proceedings of the 36th Symposium on the Interface: Computing Science and Statistics, Baltimore, MD.

Genz, A. (1992). Numerical computation of multivariate normal probabilities. *Journal of Computational and Graphical Statistics*, 1:141–149.

Genz, A. and Bretz, F. (2002). Methods for the computation of multivariate t-probabilities. *Journal of Statistical Computation and Simulation*, 11:950–971.

Glick, N. (1972). Sample-based classification procedures derived from density estimators. *Journal of the American Statistical Association*, 67(337):116–122.

Glick, N. (1973). Sample-based multinomial classification. *Biometrics*, 29(2):241–256.

Glick, N. (1978). Additive estimators for probabilities of correct classification. *Pattern Recognition*, 10:211–222.

Gordon, L. and Olshen, R. (1978). Asymptotically efficient solutions to the classification problem. *Annals of Statistics*, 6:525–533.

Hanczar, B., Hua, J., and Dougherty, E. (2007). Decorrelation of the true and estimated classifier errors in high-dimensional settings. *EURASIP Journal on Bioinformatics and Systems Biology*, 2007: 38473.

Hand, D. (1986). Recent advances in error rate estimation. *Pattern Recognition Letters*, 4:335–346.

Harter, H. (1951). On the distribution of Wald's classification statistics. *Annals of Mathematical Statistics*, 22:58–67.

Haussler, H., Kearns, M., and Tishby, N. (1994). Rigorous learning curves from statistical mechanics. In: Proceedings of the 7th Annual ACM Conference on Computer Learning Theory, San Mateo, CA.

Hills, M. (1966). Allocation rules and their error rates. *Journal of the Royal Statistical Society. Series B (Methodological)*, 28(1):1–31.

Hirji, K., Mehta, C., and Patel, N. (1987). Computing distributions for exact logistic regression. *Journal of the American Statistical Association*, 82(400):1110–1117.

Hirst, D. (1996). Error-rate estimation in multiple-group linear discriminant analysis. *Technometrics*, 38(4):389–399.

Hua, J., Tembe, W., and Dougherty, E. (2009). Performance of feature-selection methods in the classification of high-dimension data. *Pattern Recognition*, 42:409–424.

Hua, J., Xiong, Z., and Dougherty, E. (2005a). Determination of the optimal number of features for quadratic discriminant analysis via the normal approximation to the discriminant distribution. *Pattern Recognition*, 38:403–421.

Hua, J., Xiong, Z., Lowey, J., Suh, E., and Dougherty, E. (2005b). Optimal number of features as a function of sample size for various classification rules. *Bioinformatics*, 21(8):1509–1515.

Hughes, G. (1968). On the mean accuracy of statistical pattern recognizers. *IEEE Transactions on Information Theory*, IT-14(1):55–63.

Imhof, J. (1961). Computing the distribution of quadratic forms in normal variables. *Biometrika*, 48(3/4):419–426.

Jain, A. and Chandrasekaran, B. (1982). Dimensionality and sample size considerations in pattern recognition practice. In: *Classification, Pattern Recognition and Reduction of Dimensionality*, Krishnaiah, P. and Kanal, L. (editors). Volume 2 of *Handbook of Statistics*, Chapter 39. North-Holland, Amsterdam, pp. 835–856.

Jain, A. and Waller, W. (1978). On the optimal number of features in the classification of multivariate gaussian data. *Pattern Recognition*, 10:365–374.

Jain, A. and Zongker, D. (1997). Feature selection: evaluation, application, and small sample performance. *IEEE Transactions on Pattern Analysis and Machine Intelligence*, 19(2):153–158.

Jiang, X. and Braga-Neto, U. (2014). A Naive-Bayes approach to bolstered error estimation in high-dimensional spaces. In: Proceedings of the IEEE International Workshop on Genomic Signal Processing and Statistics (GENSIPS'2014), Atlanta, GA.

John, S. (1961). Errors in discrimination. *Annals of Mathematical Statistics*, 32:1125–1144.

Johnson, N., Kotz, S., and Balakrishnan, N. (1994). *Continuous Univariate Distributions*, Vol. 1. John Wiley & Sons, New York.

Johnson, N., Kotz, S., and Balakrishnan, N. (1995). *Continuous Univariate Distributions*, Vol. 2. John Wiley & Sons, New York.

Kaariainen, M. (2005). Generalization error bounds using unlabeled data. In: Proceedings of COLT'05.

Kaariainen, M. and Langford, J. (2005). A comparison of tight generalization bounds. In: Proceedings of the 22nd International Conference on Machine Learning, Bonn, Germany.

Kabe, D. (1963). Some results on the distribution of two random matrices used in classification procedures. *Annals of Mathematical Statistics*, 34:181–185.

Kan, R. (2008). From moments of sum to moments of product. *Journal of Multivariate Analysis*, 99:542–554.

Kanal, L. and Chandrasekaran, B. (1971). On dimensionality and sample size in statistical pattern classification. *Pattern Recognition*, 3(3):225–234.

Mardia K. V., Kent J. T., and Bibby, J. M. (1979). *Multivariate Analysis*. Academic Press, London.

Kim, S., Dougherty, E., Barrera, J., Chen, Y., Bittner, M., and Trent, J. (2002). Strong feature sets from small samples. *Computational Biology*, 9:127–146.

Klotz, J. (1966). The wilcoxon, ties, and the computer. *Journal of the American Statistical Association*, 61(315):772–787.

Knight, J., Ivanov, I., and Dougherty, E. (2008). Bayesian multivariate poisson model for RNA-seq classification. In: Proceedings of the IEEE International Workshop on Genomic Signal Processing and Statistics (GENSIPS'2013), Houston, TX.

Kohavi, R. (1995). A study of cross-validation and bootstrap for accuracy estimation and model selection. In: Proceedings of 14th International Joint Conference on Artificial Intelligence (IJCAI), Montreal, CA, pp. 1137–1143.

Kudo, M. and Sklansky, J. (2000). Comparison of algorithms that select features for pattern classifiers. *Pattern Recognition*, 33:25–41.

Lachenbruch, P. (1965). Estimation of error rates in discriminant analysis. PhD thesis, University of California at Los Angeles, Los Angeles, CA.

Lachenbruch, P. and Mickey, M. (1968). Estimation of error rates in discriminant analysis. *Technometrics*, 10:1–11.

Lugosi, G. and Pawlak, M. (1994). On the posterior-probability estimate of the error rate of nonparametric classification rules. *IEEE Transactions on Information Theory*, 40(2):475–481.

Mathai, A. M. and Haubold, H. J. (2008). *Special Functions for Applied Scientists*. Springer, New York.

McFarland, H. and Richards, D. (2001). Exact misclassification probabilities for plug-in normal quadratic discriminant functions. I. The equal-means case. *Journal of Multivariate Analysis*, 77:21–53.

McFarland, H. and Richards, D. (2002). Exact misclassification probabilities for plug-in normal quadratic discriminant functions. II. The heterogeneous case. *Journal of Multivariate Analysis*, 82:299–330.

McLachlan, G. (1976). The bias of the apparent error in discriminant analysis. *Biometrika*, 63(2):239–244.

McLachlan, G. (1987). Error rate estimation in discriminant analysis: recent advances. In: *Advances in Multivariate Analysis*, Gupta, A. (editor). Dordrecht: Reidel, pp. 233–252.

McLachlan, G. (1992). *Discriminant Analysis and Statistical Pattern Recognition*. John Wiley & Sons, New York.

Meshalkin, L. and Serdobolskii, V. (1978). Errors in the classification of multi-variate observations. *Theory of Probability and its Applications*, 23:741–750.

Molinaro, A., Simon, R., and Pfeiffer, M. (2005). Prediction error estimation: a comparison of resampling methods. *Bioinformatics*, 21(15):3301–3307.

Moore, E. and Smith, H. (1922). A general theory of limits. *Bell System Technical Journal*, 26(2):147–160.

Moran, M. (1975). On the expectation of errors of allocation associated with a linear discriminant function. *Biometrika*, 62(1):141–148.

Muller, K. E. and Stewart, P. W. (2006). *Linear Model Theory: Univariate, Multivariate, and Mixed Models*. John Wiley & Sons, Hoboken, NJ.

Nijenhuis, A. and Wilf, H. (1978). *Combinatorial Algorithms*, 2nd edition. Academic Press, New York.

O'Hagan, A. and Forster, J. (2004). *Kendalls Advanced Theory of Statistics, Volume 2B: Bayesian Inference*, 2nd edition. Hodder Arnold, London.

Okamoto, M. (1963). An asymptotic expansion for the distribution of the linear discriminant function. *Annals of Mathematical Statistics*, 34:1286–1301. Correction: *Annals of Mathematical Statistics*, 39:1358–1359, 1968.

Pikelis, V. (1976). Comparison of methods of computing the expected classification errors. *Automat. Remote Control*, 5(7):59–63. [in Russian].

Price, R. (1964). Some non-central f-distributions expressed in closed form. *Biometrika*, 51:107–122.

Pudil, P., Novovicova, J., and Kittler, J. (1994). Floating search methods in feature selection. *Pattern Recognition Letters*, 15:1119–1125.

Raiffa, H. and Schlaifer, R. (1961). *Applied Statistical Decision Theory*. MIT Press.

Raudys, S. (1972). On the amount of a priori information in designing the classification algorithm. *Technical Cybernetics*, 4:168–174. [in Russian].

Raudys, S. (1978). Comparison of the estimates of the probability of misclassification. In: Proceedings of the 4th International Conference on Pattern Recognition, Kyoto, Japan, pp. 280–282.

Raudys, S. (1997). On dimensionality, sample size and classification error of nonparametric linear classification algorithms. *IEEE Transactions on Pattern Analysis and Machine Intelligence*, 19:667–671.

Raudys, S. (1998). Evolution and generalization of a single neurone ii. complexity of statistical classifiers and sample size considerations. *Neural Networks*, 11:297–313.

Raudys, S. and Jain, A. (1991). Small sample size effects in statistical pattern recognition: recommendations for practitioners. *IEEE Transactions on Pattern Analysis and Machine Intelligence*, 13(3):4–37.

Raudys, S. and Pikelis, V. (1980). On dimensionality, sample size, classification error, and complexity of classification algorithm in pattern recognition. *IEEE Transactions on Pattern Analysis and Machine Intelligence*, 2:242–252.

Raudys, S. and Young, D. (2004). Results in statistical discriminant analysis: a review of the former soviet union literature. *Journal of Multivariate Analysis*, 89:1–35.

Sayre, J. (1980). The distributions of the actual error rates in linear discriminant analysis. *Journal of the American Statistical Association*, 75(369):201–205.

Schiavo, R. and Hand, D. (2000). Ten more years of error rate research. *International Statistical Review*, 68(3):295–310.

Serdobolskii, V. (2000). *Multivariate Statistical Analysis: A High-Dimensional Approach*. Springer.

Shao, J. (1993). Linear model selection by cross-validation. *Journal of the American Statistical Association*, 88:486–494.

Sima, C., Attoor, S., Braga-Neto, U., Lowey, J., Suh, E., and Dougherty, E. (2005a). Impact of error estimation on feature-selection algorithms. *Pattern Recognition*, 38(12): 2472–2482.

Sima, C., Braga-Neto, U., and Dougherty, E. (2005b). Bolstered error estimation provides superior feature-set ranking for small samples. *Bioinformatics*, 21(7):1046–1054.

Sima, C., Braga-Neto, U., and Dougherty, E. (2011). High-dimensional bolstered error estimation. *Bioinformatics*, 27(21):3056–3064.

Sima, C. and Dougherty, E. (2006a). Optimal convex error estimators for classification. *Pattern Recognition*, 39(6):1763–1780.

Sima, C. and Dougherty, E. (2006b). What should be expected from feature selection in small-sample settings. *Bioinformatics*, 22(19):2430–2436.

Sima, C. and Dougherty, E. (2008). The peaking phenomenon in the presence of feature selection. *Pattern Recognition*, 29:1667–1674.

Sitgreaves, R. (1951). On the distribution of two random matrices used in classification procedures. *Annals of Mathematical Statistics*, 23:263–270.

Sitgreaves, R. (1961). Some results on the distribution of the W-classification. In: *Studies in Item Analysis and Prediction*, Solomon, H. (editor). Stanford University Press, pp. 241–251.

Smith, C. (1947). Some examples of discrimination. *Annals of Eugenics*, 18:272–282.

Snapinn, S. and Knoke, J. (1985). An evaluation of smoothed classification error-rate estimators. *Technometrics*, 27(2):199–206.

Snapinn, S. and Knoke, J. (1989). Estimation of error rates in discriminant analysis with selection of variables. *Biometrics*, 45:289–299.

Stone, C. (1977). Consistent nonparametric regression. *Annals of Statistics*, 5:595–645.

Stone, M. (1974). Cross-validatory choice and assessment of statistical predictions. *Journal of the Royal Statistical Society. Series B (Methodological)*, 36:111–147.

Teichroew, D. and Sitgreaves, R. (1961). Computation of an empirical sampling distribution for the *w*-classification statistic. In: *Studies in Item Analysis and Prediction*, Solomon, H. (editor). Stanford University Press, pp. 285–292.

Toussaint, G. (1974). Bibliography on estimation of misclassification. *IEEE Transactions on Information Theory*, IT-20(4):472–479.

Toussaint, G. and Donaldson, R. (1970). Algorithms for recognizing contour-traced handprinted characters. *IEEE Transactions on Computers*, 19:541–546.

Toussaint, G. and Sharpe, P. (1974). An efficient method for estimating the probability of misclassification applied to a problem in medical diagnosis. *IEEE Transactions on Information Theory*, IT-20(4):472–479.

Tutz, G. (1985). Smoothed additive estimators for non-error rates in multiple discriminant analysis. *Pattern Recognition*, 18(2):151–159.

Vapnik, V. (1998). *Statistical Learning Theory*. John Wiley & Sons, New York.

Verbeek, A. (1985). A survey of algorithms for exact distributions of test statistics in rxc contingency tables with fixed margins. *Computational Statistics and Data Analysis*, 3:159–185.

Vu, T. (2011). The bootstrap in supervised learning and its applications in genomics/proteomics. PhD thesis, Texas A&M University, College Station, TX.

Vu, T. and Braga-Neto, U. (2008). Preliminary study on bolstered error estimation in high-dimensional spaces. In: GENSIPS'2008 – IEEE International Workshop on Genomic Signal Processing and Statistics, Phoenix, AZ.

Vu, T., Sima, C., Braga-Neto, U., and Dougherty, E. (2014). Unbiased bootstrap error estimation for linear discrimination analysis. *EURASIP Journal on Bioinformatics and Systems Biology*, 2014:15.

Wald, A. (1944). On a statistical problem arising in the classification of an individual into one of two groups. *Annals of Mathematical Statistics*, 15:145–162.

Wald, P. and Kronmal, R. (1977). Discriminant functions when covariances are unequal and sample sizes are moderate. *Biometrics*, 33:479–484.

Webb, A. (2002). *Statistical Pattern Recognition*, 2nd edition. John Wiley & Sons, New York.

Wyman, F., Young, D., and Turner, D. (1990). A comparison of asymptotic error rate expansions for the sample linear discriminant function. *Pattern Recognition*, 23(7):775–783.

Xiao, Y., Hua, J., and Dougherty, E. (2007). Quantification of the impact of feature selection on cross-validation error estimation precision. *EURASIP Journal on Bioinformatics and Systems Biology, 2007:16354.*

Xu, Q., Hua, J., Braga-Neto, U., Xiong, Z., Suh, E., and Dougherty, E. (2006). Confidence intervals for the true classification error conditioned on the estimated error. *Technology in Cancer Research and Treatment*, 5(6):579–590.

Yousefi, M. and Dougherty, E. (2012). Performance reproducibility index for classification. *Bioinformatics*, 28(21):2824–2833.

Yousefi, M., Hua, J., and Dougherty, E. (2011). Multiple-rule bias in the comparison of classification rules. *Bioinformatics*, 27(12):1675–1683.

Yousefi, M., Hua, J., Sima, C., and Dougherty, E. (2010). Reporting bias when using real data sets to analyze classification performance. *Bioinformatics*, 26(1):68–76.

Zhou, X. and Mao, K. (2006). The ties problem resulting from counting-based error estimators and its impact on gene selection algorithms. *Bioinformatics*, 22:2507–2515.

Zollanvari, A. (2010). Analytical study of performance of error estimators for linear discriminant analysis with applications in genomics. PhD thesis, Texas A&M University, College Station, TX.

Zollanvari, A., Braga-Neto, U., and Dougherty, E. (2009). On the sampling distribution of resubstitution and leave-one-out error estimators for linear classifiers. *Pattern Recognition*, 42(11):2705–2723.

Zollanvari, A., Braga-Neto, U., and Dougherty, E. (2010). Joint sampling distribution between actual and estimated classification errors for linear discriminant analysis. *IEEE Transactions on Information Theory*, 56(2):784–804.

Zollanvari, A., Braga-Neto, U., and Dougherty, E. (2011). Analytic study of performance of error estimators for linear discriminant analysis. *IEEE Transactions on Signal Processing*, 59(9):1–18.

Zollanvari, A., Braga-Neto, U., and Dougherty, E. (2012). Exact representation of the second-order moments for resubstitution and leave-one-out error estimation for linear discriminant analysis in the univariate heteroskedastic gaussian model. *Pattern Recognition*, 45(2):908–917.

Zollanvari, A. and Dougherty, E. (2014). Moments and root-mean-square error of the bayesian MMSE estimator of classification error in the gaussian model. *Pattern Recognition*, 47(6):2178–2192.

Zolman, J. (1993). *Biostatistics: Experimental Design and Statistical Inference*. Oxford University Press, New York.

AUTHOR INDEX

SUBJECT INDEX

Error Estimation for Pattern Recognition, First Edition. Ulisses M. Braga-Neto and Edward R. Dougherty.